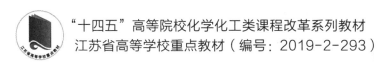

"十四五"高等院校化学化工类课程改革系列教材

江苏省高等学校重点教材（编号：2019-2-293）

化工工艺学（双语版）

CHEMICAL TECHNOLOGY

主　编　　王金权

副主编　　端木传嵩

　　　　　李　进

特配电子资源

微信扫码

◎ 配套资料

◎ 拓展阅读

◎ 互动交流

南京大学出版社

图书在版编目(CIP)数据

化工工艺学 / 王金权主编. — 南京：南京大学出版社，2023.11

ISBN 978-7-305-23594-8

Ⅰ. ①化… Ⅱ. ①王… Ⅲ. ①化工过程—工艺学 Ⅳ. ①TQ02

中国版本图书馆 CIP 数据核字(2020)第 126785 号

出版发行　南京大学出版社
社　　址　南京市汉口路 22 号　　　　邮　编　210093
书　　名　**化工工艺学**
　　　　　HUAGONG GONGYIXUE
主　　编　王金权
责任编辑　刘　飞　　　　　　　编辑热线　025-83592146

照　　排　南京南琳图文制作有限公司
印　　刷　南京鸿图印务有限公司
开　　本　787 mm×1092 mm　1/16　印张 19　字数 458 千
版　　次　2023 年 11 月第 1 版　2023 年 11 月第 1 次印刷
ISBN 978-7-305-23594-8
定　　价　58.00 元

网址：http://www.njupco.com
官方微博：http://weibo.com/njupco
官方微信号：njuyuexue
销售咨询热线：(025) 83594756

前　言

在经济全球化的背景下,为了推动我国高等教育的国际化,培养具有国际化视野的创新型人才,教育部于2001年明确指出要在国内高校积极推动双语教学。目前,国内高校几乎都开设了双语教学课程,并对双语教学进行了广泛的研究。随着我国对外交流的拓展和深入,国外留学生的汉语教学,尤其是专业汉语教学也越来越受到关注。

化工工艺学是化工与制药类专业的基础课,对于培养化学化工类及相近各专业学生的化工专业知识起着至关重要的作用。通过调研发现,目前有不少化工类专业英语书籍出版,但是这些书籍不能同时满足国内学生及国外学生的教学要求。因此,汇编一本既能满足国内学生,又能满足国外学生学习的双语教材,就非常有必要了。

通过对国内外《化工工艺学》《化学工艺学》教材及相关资料调研后,结合我校教学实际,我们发现米镇涛主编的《化学工艺学》教材的主体内容非常适合作为双语教学的中文内容。因此,我们通过对米镇涛主编的《化学工艺学》教材的部分内容进行节选、改编,并参考了宁春花等编写的《聚合物合成工艺学》内容,进行了翻译,从而形成了本书。

本书编译力求做到简明扼要、内容充实,能够适用于国内高校化工及相关专业的化工工艺学课程双语教学。本教材可作为化工、制药、高分子等工程类高年级本科生及研究生的授课书籍,也可供来华留学生、化学化工专业的科技人员及中等英语水平的其他人员自修参考。

教材的编写分为中文和英文两大部分,中文与英文部分内容严格对照但又各自独立。全书中英文各分10个章节,主要内容涉及化工工艺学基础、烃类热裂解、芳烃转化过程、合成气的生产过程、加氢与脱氢过程、烃类选择性氧化、羰基化过程、聚合物生产方法与工艺、绿色化工。第1、2、3、5、9章由王金权编译;第4章由李进编译,第 6 章由徐纬川编译,第 7 章、第 8 章由端木传嵩编译,第 10 章由季晶玲编译,最后由王金权统稿。另外,有部分研究生、本科生在试用本教材手稿过程中,提出了宝贵意见,在此一并感谢。

限于编者水平和本领域知识发展、更新迅速,书中不当之处敬请读者批评指正。

编 者

2023 年 10 月

目　录

第1章 绪 论

1.1 化工工艺学研究内容

化学工业又称化学加工工业,泛指生产过程中化学方法占主要地位的过程工业。由原料到化学品的转化要通过化学工艺来实现。化工工艺即化学生产工艺,是指原料主要经过化学反应转变为产品的方法和过程,包括实现这种转变的全部化学和物理的措施。

化学工艺是在化学、物理和其他科学成就的基础上,研究综合利用各种原料生产化工产品的原理、方法、流程和设备的一门学科,目的是创立技术先进、经济合理、生产安全、环境无害的生产过程。

化工工艺具有过程工业的特点,即生产不同的化学产品要采用不同的化工工艺。即使生产相同产品,但原料路线不相同时,也要采用不同的化工工艺。尽管如此,化工工艺学所涉及的内容是相同的,一般包括原料的选择和预处理、生产方法的选择及方法原理、设备(反应器和其他)的选择及其结构和操作、催化剂的选择和使用、操作条件的影响和选定、流程组织、生产控制、产品规格和副产物的分离与利用、能量的回收和利用、对不同工艺路线和流程的技术经济评价等问题。

化工工艺学是化学工业的基础科学。化工工艺是以过程为研究目的,重点解决整个生产过程(流程)的组织、优化;将各单项化学工程技术在以产品为目标的前提下集成,解决各单元间的匹配、链接;在确保产品质量条件下,实现全系统的能量、物料及安全污染诸因素的最优化。化工工艺学是将化学工程学的先进技术运用到具体生产过程中,以化工产品为目标的过程技术。化学工程学主要研究化学工业和其他过程工业生产中所进行的化学过程和物理过程的共同规律,它的一个重要任务就是研究有关工程因素对过程和装置的效应,特别是放大中的效应。化工工艺与化学工程相配合,可以解决化工过程开发、装置设计、流程组织、操作原理及方法等方面的问题。此外,解决化工生产实际中的问题也需要这两门学科的理论指导。化学工业的发展促进了化工工艺学和化学工程学两门学科不断发展和完善,它们反过来也能促进化学工业迅速发展和提高。

1.2 化学工业的发展、地位与作用

化学工业是在人类生活和生产需要的基础上发展起来的,反过来,化工生产的发展也推动了社会的发展。

18 世纪以前,化工生产均为作坊式手工工艺,像早期的制陶、酿造、冶炼等。18 世

纪初叶建成了第一个典型的化工厂,即以含硫矿石和硝石为原料的铅室法硫酸厂。1791 年路布兰法制碱工艺出现,满足了纺织、玻璃、肥皂等工业对碱的大量需求,有力地推动了当时在英国开始的产业革命。从 18 世纪到 20 世纪初期,接触法制硫酸取代了铅室法,索尔维法(氨碱法)制碱取代了路布兰法,以酸、碱为基础的无机化工初具规模。同期,随着钢铁工业的发展,炼焦过程产生的大量焦炉气、粗苯和煤焦油得到重视和应用。在德国,首创了肥料工业和煤化学工业,人类进入了化学合成的时代,染料、农药、香料、医药等有机化工迅速发展,化肥和农药在农作物增产中起了重要作用。20 世纪初,化学家 F. 哈伯发明了合成氨技术,并于 1913 年在化学工程师 C. 博施的协助下建成世界上第一个合成氨厂,促使氮肥工业迅速发展。

自 20 世纪初期,石油和天然气得到大量开采和利用,向人类提供了各种燃料和丰富的化工原料。

1931 年氯丁橡胶实现工业化和 1937 年聚己二酰己二胺(尼龙 66)合成以后,高分子化工蓬勃发展起来,到 20 世纪 50 年代初期形成了大规模生产塑料、合成橡胶和合成纤维的工业,人类进入了合成材料的时代,更进一步地推动了工、农业生产水平和科学技术的发展,人类生活水平得到了显著的提高。

石油化工和高分子化工发展的同时,为满足人们生活更高的需求,高附加值、功能性化学品的合成成为现代化工发展方向之一,其产品批量小、品种多、技术含量高、更新快。提高化工生产的精细化率已成为世界化学工业发展的重要指标。

随着生物技术的发展,化学工业与生物技术的相互渗透与结合,也是当今化学工业的发展方向,现已初步形成具有广阔发展前景的生物化工产业,给传统的化学工业增添了新的活力。

综上所述,化学工业为工农业、现代交通运输业、国防军事、尖端科技等领域提供了必不可少的化学品和能源,保证并促进了这些部门的发展和技术进步。化学工业与人类生活更是息息相关,在现代人类生活中,从衣、食、住、行、战胜疾病等物质生活到文化艺术、娱乐消遣等精神生活都离不开化工产品。有些化工产品的开发、生产和应用对工业革命、农业发展和人类生活水平起到划时代的促进作用。

1.3 现代化学工业的特点和发展方向

1.3.1 现代化学工业的特点

(1) 原料、生产方法和产品的多样性与复杂性

用同一种原料可以制造多种不同的化工产品;同一种产品可采用不同原料,或不同方法和工艺路线来生产;一个产品可以有不同用途,而不同产品可能会有相同用途。由于这些多样性,化学工业能够为人类提供越来越多的新物质、新材料和新能源。同时,多数化工产品的生产过程是多步骤的,有的步骤很复杂,其影响因素也是复杂的。

(2) 向大型化、综合化、精细化发展

装置规模增大,其单位容积、单位时间的产出率随之显著增大。生产的综合化可以

使资源和能源得到充分、合理的利用,可以就地利用副产物和"废料",将它们转化成有用产品,做到没有废物排放或排放最少。综合化不应局限于不同化工厂的联合体,也应该是化工厂与其他工厂联合的综合性企业。

精细化不仅指生产小批量的化工产品,更主要的是指生产技术含量高、附加产值高的具有优异性能或功能的产品,并且要能适应变化快的市场需求,不断改变产品品种和型号。化学工艺更精细化,深入到分子内部的原子水平上进行化学品的合成,使产品的生产更加高效、节能、省资源。

(3) 多学科合作、技术密集型生产

现代化学工业是高度自动化和机械化的生产,并进一步朝着智能化发展。当今化学工业的持续发展越来越多地依靠高新技术迅速将科研成果转化为生产力,如生物与化学工程、微电子与化学、材料与化工等不同学科的相互结合,可创造出更多优良的新物质和新材料;计算机技术的高水平发展,已经使化工生产实现了远程自动化控制,也将给化学品的合成提供强有力的智能化工具;将组合化学、计算化学与计算机相结合,可以准确地进行新分子、新材料的设计与合成,节省大量实验时间和人力。因此,现代化学工业需要高水平、有创造性和开拓能力的多种学科、不同专业的技术专家,以及受过良好教育及训练的、熟悉生产技术的操作和管理人员。

(4) 重视能量合理利用,积极采用节能工艺和方法

化工生产是原料物质主要经化学变化转化为产品物质的过程,同时伴随能量的传递和转换,即必须消耗能量。化工生产部门是耗能大户,合理用能和节能显得极为重要。许多生产过程的先进性体现在其低能耗工艺或节能工艺上。

(5) 资金密集,投资回收速度快,利润高

现代化学工业的装备复杂,技术程度高,产品更新迅速,需要大量的投资。然而,化工产品产值较高,成本低、利润高,一旦工厂建成投产,可很快收回投资并获利。化学工业的产值成为各国国民经济总产值指标的重要组成部分。

(6) 安全与环境保护问题日益突出

化工生产中易燃、易爆、有毒仍然是现代化工企业首要解决的问题,要采用安全的生产工艺,就要有可靠的安全技术保障、严格的规章制度及其监督机构。创建清洁生产环境,大力发展绿色化工,采用无毒无害的方法和过程,生产环境友好的产品,这是化学工业赖以持续发展的关键之一。

1.3.2 化学工业发展的方向

人类生活和生产的不断发展,也带来了市场竞争激烈、自然资源和能源减少、环境污染加剧等问题,化学工业同样面临着这些问题的挑战,要走可持续发展的道路,必须做好以下几方面工作。

(1) 面向市场竞争激烈的形势,积极开发高新技术,缩短新技术、新工艺工业化的周期,加快产品更新和升级的速度。

(2) 最充分、最彻底地利用原料。除了发展大型的综合性生产企业,使原料、产品和副产品得到综合利用外,提倡设计和开发原子经济性反应。

（3）大力发展绿色化工。包括采用无毒、无害的原料、溶剂和催化剂；应用反应选择性高的工艺和催化剂；将副产物或废物转化为有用的物质；采用原子经济性反应，提高原料中原子的利用率，实现零排放；淘汰污染环境和破坏生态平衡的产品，开发和生产环境友好产品等。

（4）化工过程要高效、节能和智能化。

（5）实施废物再生利用工程。

欲将以上几方面付诸实现，需要所有化学家和化学工程师的艰苦努力，也需要多学科、多部门的精诚合作，更需依赖于科学的不断进步和高新技术的发展。

1.4　化工原料资源及其加工

1.4.1　无机化学矿及其加工

1. 主要无机化学矿

无机化学矿主要用于生产无机化合物和冶炼金属，其矿物资源的开采和选矿称为化学矿山行业，在我国属于化工行业之一。化学矿山的产品非常繁多，仅列举主要矿物产品如下。

（1）盐矿

岩盐、海盐或湖盐等，用于制造纯碱、烧碱、盐酸和氯乙烯等。

（2）硫矿

硫黄（S）、硫铁矿（FeS_2）等，用于生产硫酸和硫黄。

（3）磷矿

氟磷灰石[$Ca_5F(PO_4)_3$]、氯磷灰石[$Ca_5Cl(PO_4)_3$]等，用于生产磷肥、磷酸及磷酸盐等。

（4）钾盐矿

钾石盐（KCl 和 NaCl 混合物）、光卤石（$KCl \cdot MgCl_2 \cdot 6H_2O$）、钾盐镁矾（$KCl \cdot MgSO_4 \cdot 3H_2O$）。

（5）铝土矿

水硬铝石（$\alpha\text{-}Al_2O_3 \cdot H_2O$）和三水铝石（$Al_2O_3 \cdot 3H_2O$）的混合物。

稀有金属矿和贵金属储量少，但实用价值高，是极为宝贵的资源。如铍、锂、铕、钽等是高新技术需要的材料；钛的耐热及耐腐蚀性强，是钢的竞争对手，也是烯烃聚合催化剂的主要成分之一，有些多相催化剂中也含有 TiO_2；铂、钯、铑、铼等贵金属是重要的催化剂材料，如果没有它们，催化化学难于发展，许多重要化学品和新材料将不能合成。

除了少数品位和质量高的矿物开采出来不需经初步加工即可利用外，大多数矿物需要在开采地进行选矿和初步加工，以除去其中无用的杂质，并加工成一定规格的形状。矿物初步加工的主要方法有分级、粉碎、团固和烧结、精选、脱水和除尘等，应根据使用部门对原料的要求来选用其中部分或全部方法。

2. 磷矿和硫铁矿的加工

磷矿和硫铁矿是无机化学矿产量最大的两个产品。多数磷矿为氟磷灰石 $[Ca_5F(PO_4)_3]$，经过分级、水洗脱泥、浮选等方法选矿除去杂质，成为商品磷矿。硫铁矿包括黄铁矿（立方晶系 FeS_2）、白铁矿（斜方晶系 FeS_2）和磁硫铁矿（Fe_nS_{n+1}），其中主要是黄铁矿。

磷矿是生产磷肥、磷酸、单质磷、磷化物和磷酸盐的原料。85%以上的磷矿用于制造磷肥，过去生产的普通过磷酸钙含磷量低，已被淘汰，现在产量最大的磷肥品种为磷酸铵类，属于氮磷复合肥料，其他磷肥有重过磷酸钙、硝酸磷肥和钙镁磷肥等。生产磷肥的方法有两大类。

（1）酸法（又称湿法）

酸法是指用硫酸或硝酸等无机酸来处理磷矿石，最常用的是硫酸。硫酸与磷矿反应生成磷酸和硫酸钙结晶，主反应式为

$$Ca_5F(PO_4)_3 + 5H_2SO_4 + 5nH_2O \longrightarrow 3H_3PO_4 + 5CaSO_4 \cdot nH_2O + HF \qquad (1-1)$$

（2）热法

热法是指利用高温分解磷矿石，并进一步制成可被农作物吸收的磷酸盐。热法还可以生产元素磷、五氧化二磷和磷酸。热法生产要消耗较多的电能和热能。

硫铁矿可用于制硫酸，世界上硫酸总产量的一半以上用于生产磷肥和氮肥。硫铁矿生产硫酸的主要反应式为

$$4FeS_2 + 11O_2 \longrightarrow 2Fe_2O_3 + 8SO_2 \qquad (1-2)$$

$$SO_2 + \frac{1}{2}O_2 \longrightarrow SO_3 \qquad (1-3)$$

$$SO_3 + H_2O \longrightarrow H_2SO_4 \qquad (1-4)$$

1.4.2　石油及其加工

石油化工自 20 世纪 50 年代开始蓬勃发展，至今，有机化工、高分子化工、精细化工及氮肥工业等产品大约有 90%来源于石油和天然气。90%左右的有机化工产品的上游原料可归结为三烯（乙烯、丙烯、丁二烯）、三苯（苯、甲苯、二甲苯）、乙炔、萘和甲醇。其中的三烯主要由石油制取，三苯、萘和甲醇可由石油、天然气和煤制取。

1. 石油的组成

石油是一种有气味的棕黑色或黄褐色黏稠状液体，密度与组成有关，相对密度大约在 0.75～1.0。有些油田常伴生油田气。石油是由相对分子质量不同、组成和结构不同、数量众多的化合物构成的混合物，其中化合物的沸点从常温到 500℃以上均有。石油中含量最大的两种元素是 C 和 H，其质量含量分别为碳 83%～87%、氢 11%～14%，两者主要以碳氢化合物形式存在。其他元素的含量因产地不同而有较大的波动，硫含量为 0.02%～5.5%，氮含量为 0.02%～1.7%，氧含量为 0.08%～1.82%。而 Ni、V、Fe、Cu 等金属元素只含微量，由十亿分之几到百万分之几。地下与石油共存的水相中溶有 K、Na、Ca、Mg 等的氯化物，易于脱除。石油中的化合物可以分为烃类、非

烃类以及胶质和沥青三大类。

（1）烃类化合物

烃类即碳氢化合物，在石油中占绝大部分，约几万种。主要包括链式饱和烃、环烷烃、芳香烃等。这些烃类化合物都是有机化工的基本原料，许多烃类还是汽油、航空煤油、柴油的组分。

石油中几乎没有烯烃和炔烃这两类化合物，然而它们却是石油化工的重要原料，尤其是烯烃更为重要，只有通过对石油的化学加工才能获得这些化合物。

（2）非烃化合物

非烃化合物指含有碳、氢及其他杂原子的有机化合物，如硫化物、氮化物、含氧化合物、金属有机化合物等。

非烃化合物的含量虽然很低，但对石油加工过程以及石油产品的性质有很大影响，有的会使催化剂中毒，有的会腐蚀管道和设备，有的使用时污染环境等，所以在石油加工时均应该预先将其脱除和回收利用，脱硫、脱氮、脱金属是石油化学加工重要的过程之一。

（3）胶质和沥青质

原油经蒸馏加工后，沸点高于500℃的馏分是渣油，在渣油中含有相当数量的胶质和沥青质，它们是由各种结构不同、相对分子质量很大的化合物组成的混合物，多为稠环环烷烃、稠环芳香烃和含 S、N 等杂原子的环状化合物。

2. 石油的常压蒸馏和减压蒸馏

为了充分利用宝贵的石油资源，要对石油进行一次加工和二次加工，在生产汽油、航空煤油、柴油、锅炉燃油和液化气的同时，制取各类化工原料。

石油开采出来尚未加工时称为原油，一次加工方法为常压蒸馏和减压蒸馏。蒸馏是一种利用液体混合物中各组分挥发度的差别（沸点不同）进行分离的方法，是一种没有化学反应的传质、传热物理过程，主要设备是蒸馏塔。常压、减压蒸馏流程有以下三类。

（1）燃料型

以生产汽油、煤油、柴油等为主，没有充分利用石油资源，现已很少采用。

（2）燃料-润滑油型

除生产轻质和重质燃料油外，还生产各种品种的润滑油和石蜡。

（3）燃料-化工型

除生产汽油、煤油、柴油等燃料油外，还从石脑油馏分中抽提芳烃，利用石脑油或柴油热裂解制取烯烃和芳烃等重要有机化工基本原料，炼油的气体副产物也是化工原料。有的工厂还采用燃料-润滑油-化工型流程，主要产品是燃料和化工产品。

原油常压、减压蒸馏工艺流程示意图见图 1-1。

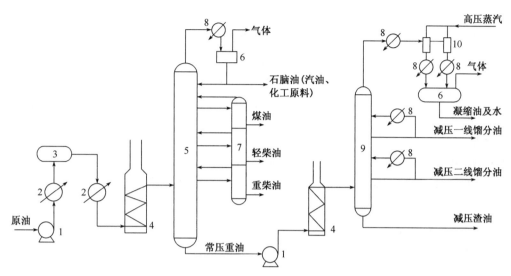

1—输油泵；2—换热器；3—脱盐罐；4—加热炉；5—常压蒸馏塔；6—贮液罐；
7—汽提塔；8—冷凝冷却器；9—减压蒸馏塔；10—蒸汽喷射泵。

图1-1 原油常压、减压蒸馏工艺流程示意图

3. 馏分油的化学加工

常压、减压蒸馏只能将原油切割成几个馏分，生产的燃料量有限，不能满足需求，直接能用作化工原料的也仅是塔顶出来的气体。为了生产更多的燃料和化工原料，需要对各个馏分油进行二次加工。加工的方法很多，主要是化学加工方法，下面简要介绍主要的几种加工过程。

（1）催化重整

催化重整是在含铂的催化剂作用下加热汽油馏分（石脑油），使其中的烃类分子重新排列形成新分子的工艺过程。催化重整装置能提供高辛烷值汽油，还为化纤、橡胶、塑料和精细化工提供苯、甲苯、二甲苯等芳烃原料以及液化气和溶剂油，并副产氢气。

铂重整工艺流程示意图见图1-2。

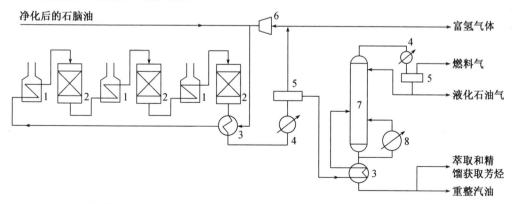

1—加热炉；2—重整反应器；3—热交换器；4—冷却冷凝器；5—油气分离器；
6—循环氢压缩机；7—分馏塔；8—再沸器。

图1-2 铂重整工艺流程示意图

（2）催化裂化

催化裂化是在催化剂作用下加热重质馏分油，使大分子烃类化合物裂化而转化成高质量的汽油，并副产柴油、锅炉燃油、液化气等产品的加工过程。

原料可以是直馏柴油、重柴油、减压柴油或润滑油馏分，甚至可以是渣油焦化制石油焦后的焦化馏分油。它们所含烃类分子中的碳数大多在 18 个以上。

（3）催化加氢裂化

催化加氢裂化是指在催化剂存在及高氢压下，加热重质油使其发生各类加氢和裂化反应，转变成航空煤油、柴油、汽油（或重整原料）等产品的加工过程。

加氢裂化过程见图 1-3。

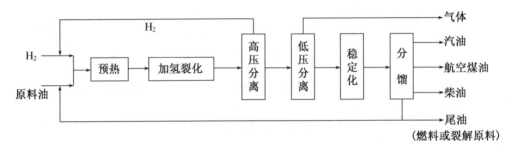

图 1-3　加氢裂化过程

（4）烃类热裂解

烃类热裂解不用催化剂，是将烃类加热到 750℃～900℃ 使其发生热裂解，反应相当复杂，主要是高碳烷烃裂解生成低碳烯烃和二烯烃，同时伴有脱氢、芳构化和结焦等许多反应。裂解后对产物进行冷却、冷凝，得到裂解气和裂解汽油两大类混合物。

烃类热裂解的主要目的是制取乙烯和丙烯，同时副产丁烯、丁二烯、苯、甲苯、二甲苯、乙苯等芳烃及其他化工原料。它是每个石油化工厂必不可少的首要过程。

石油的二次加工还有烷基化、异构化、焦化等，可获得高辛烷值汽油和化工原料。

由石油制取燃料和化工原料的主要途径见图 1-4。

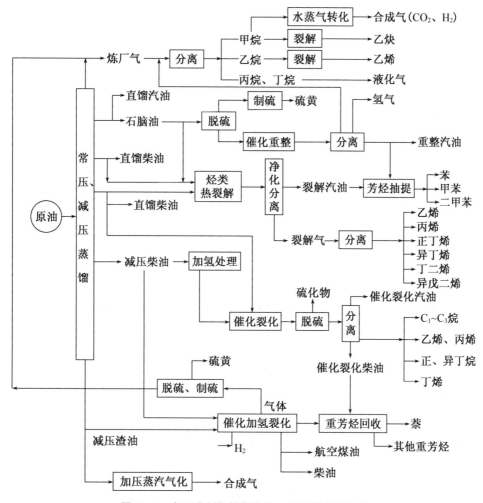

图1-4 由石油制取燃料和化工原料的主要途径

1.4.3 天然气及其加工

天然气的主要成分是甲烷。甲烷含量高于90%的天然气称为干气；$C_2 \sim C_4$烷烃含量在15%～20%或以上的天然气称为湿气。有的天然气与石油共生（油田伴生气）。

天然气的热值高、污染少，是一种清洁能源，在能源结构中的比例逐年提高。它是石油化工的重要原料资源。天然气的加工利用主要有以下几方面。

（1）天然气制氢气和合成氨。

（2）天然气经合成气路线的催化转化制燃料和化工产品。

（3）天然气直接催化转化成化工产品。

（4）天然气热裂解制化工产品。

（5）甲烷的氯化、硝化、氨氧化和硫化制化工产品。

（6）湿性天然气中$C_2 \sim C_4$烷烃的利用。

1.4.4　煤及其加工

煤是由含碳、氢的多种结构的大分子有机物和少量硅、铝、铁、钙、镁的无机矿物质组成。由于成煤过程的时间不同,有泥煤、褐煤、烟煤和无烟煤之分。按质量分数计,泥煤含碳 60%～70%,褐煤含碳 70%～80%,烟煤含碳 80%～90%,无烟煤含碳高达 90%～93%。

已知的煤储量要比石油储量大十几倍,煤的综合利用可同时为能源、化工和冶金提供有价值的原料。煤化工范畴内加工路线主要有以下几种。

（1）煤干馏

在隔绝空气条件下加热煤,使其分解生成焦炭、煤焦油、粗苯和焦炉气的过程。

（2）煤气化

在高温（900℃～1 300℃）下使煤、焦炭或半焦等固体燃料与气化剂反应,转化成主要含有氢、一氧化碳等气体的过程。

（3）煤液化

煤经化学加工转化为液体燃料的过程。煤的液化可分为直接液化和间接液化两类。

煤的直接液化是采用加氢的方法使煤转化为液态烃,所以又称为煤的加氢液化。液化产物称为人造石油,可进一步加工成各种液体燃料。

煤的间接液化是将煤预先制成合成气,然后通过催化剂作用将合成气转化为烃类燃料、含氧化合物燃料（如低碳混合醇、二甲醚）。

1.4.5　生物质及其加工

农、林、牧、副、渔业的产品及其废物（壳、芯、秆、糠、渣）等生物质通过化学或生物化学方法可以转变为基础化学品或中间产品,如葡萄糖、乳酸、柠檬酸、乙醇、丙酮、高级脂肪酸（月桂酸、硬脂酸、油酸、亚油酸、肉豆蔻酸）等。加工过程涉及一系列化学工艺,如化学水解、酶水解、微生物水解、皂化、催化加氢、气化、裂解、萃取,有些还用到 DNA 技术。

1.4.6　再生资源的开发利用

工农业和生活废料在原则上都可以回收处理并加工成有用的产品,这些再生资源的利用不仅可以节约自然资源,而且是治理污染、保护环境的有效措施之一。

将废塑料重新炼制成液体燃料的工业装置已建成,重炼的方法也很多。含碳的废料也可通过部分氧化法转化为小分子气体化合物,然后再加工利用。

1.4.7　空气和水

空气的体积组成为 78.16% N_2、20.90% O_2 和 0.93% Ar,其余 0.01% 为 He、Ne、Kr、Xe 等稀有气体。空气中的 O_2 和 N_2 是重要化工原料;含碳物质在纯氧中燃烧,不会产生氮氧化物和碳粒烟尘,对环境有好处;氢氧燃料电池为清洁能源;从空气中提取的高纯度氩、氦、氖、氪等气体可广泛应用于高科技领域。

水在化工中应用更普遍。例如,作为溶剂溶解固体、液体,吸收气体;作为反应物参加水解、水合等反应;作为载体去加热或冷却物料和设备;可吸收反应热并汽化成具有做功本领的高压蒸汽。地球上水的面积占地球表面的70%以上,但是可供使用的淡水体积只有总水体积的0.3%,因此节约和保护淡水资源、提高水的循环利用率刻不容缓。

1.5 本书的主要内容和特点

本书根据化学工业的结构特点、内在关系和发展趋势,按化学反应过程分类讲述化工工艺原理和知识。其内容主要包括化学工业概貌和化工工艺有关基本概念的介绍;化工原料资源及其加工利用途径;化工基础原料的典型生产过程等。

本书内容丰富,知识面广,注意点面结合,重点内容深入细致地阐述,注意理论与实际的结合,也介绍了近年来化学工艺及有关方面的新成就和未来发展趋势,其中1~8章均有思考题,启发思考,便于自学。学生在学习时,应注意培养分析问题和解决问题的能力,对于典型反应过程,要求理解并掌握工艺原理、选定工艺条件的依据、流程的组织和特点、各类反应设备的结构特点和优缺点等;对典型产品的各种原料来源、不同工艺路线及其技术经济指标、能量回收利用方法、副产物回收利用和废料处理方法等,应进行分析比较,找出它们的优缺点。

对化工工艺的研究、开发和实施工业化,需要应用化学和物理等基础科学理论、化学工程原理和方法、相关工程学的知识和技术,通过分析和综合,进行实践,才能获得成功。因此,在化工工艺学课程的学习中,应该注意这些理论和知识的综合运用,特别强调理论与实践相结合,才能培养开拓创新能力。

思考题

1-1 为什么说石油、天然气和煤是现代化学工业的重要原料资源? 它们的综合利用途径有哪些?

1-2 生物质和再生资源的利用前景如何?

Chapter 1　Introduction

1.1　The research field of chemical technology

Chemical industry, also known as the chemical processing industry, generally refers to the process industry in which chemical methods play a major role in the production process. The conversion from raw materials to chemicals is achieved through chemical processes. Chemical technology refers to the method and process of transforming raw materials into products mainly through chemical reactions, including all chemical and physical measures to achieve such transformation.

Chemical process is a subject that studies the principle, method, process and equipment of synthetic utilization of various raw materials to produce chemical products on the basis of chemical, physical and other scientific achievements. The purpose is to create a production process that is technologically advanced, economically reasonable, safe and environmentally sound.

Chemical process has the characteristics of process industry, that is, different chemical processes are used to produce different chemical products. Even if the same product is produced, but the raw material route is not the same, different chemical processes are also used. However, the contents of chemical technology are the same, including the selection and pretreatment of raw materials; selection and principle of production method; selection, construction and operation of equipment (reactors and others); selection and use of catalysts; influence and selection of operating conditions; process organization; production control; product specifications and separation, utilization of by-products; energy recovery and utilization; technical and economic evaluation of different process routes and processes.

Chemical technology and chemical engineering are both basic sciences of chemical industry. Chemical process is a process for the purpose of research, focus on solving the entire production process (flow) organization and optimization; integrate each individual chemical engineering technology under the premise of taking the product as the target to solve the matching and link between each unit; under the condition of ensuring product quality, the optimization of energy, material and safety pollution factors of the whole system is realized. Chemical technology is a process technology that applies the advanced technology of chemical engineering to the specific production process and takes chemical products as the target. Chemical engineering mainly studies the common laws of chemical and physical processes carried out in chemical industry and other process industrial production. One of its important tasks is to study the effects of engineering factors on processes and devices, especially the effects in amplification. The combination of chemical process and chemical engineering can solve the problems of chemical process development, device design, flow organization, operation principle and method, etc. In addition, solving the practical problems in chemical production also needs the theoretical guidance of these two disciplines. The development of chemical industry has promoted the continuous development and perfection of these two disciplines, which in turn can further promote the rapid development and improvement of chemical industry.

1.2 The development position and function of chemical industry

Chemical industry developed on the basis of human life and production needs, and conversely, the development of chemical production also promoted the development of society.

Before the 18th century, chemical production was artisanal, such as pottery, brewing and smelting in the early days. In the early 18th century, the first typical chemical plant was built, that is, a lead-chamber sulphuric acid plant with sulphur-bearing ore and saltpeter as raw materials. In 1791, the Leblanc alkali process appeared, which met the large demand for alkali in textile, glass, soap and other industries, and strongly promoted the industrial revolution started in Britain at that time. From the 18th century to the early 20th century, sulfuric acid contact method replaced lead chamber method, Solvay alkali production (ammonia alkali method) replaced Leblanc alkali process, acid and alkali based inorganic chemical industry began to take shape. At the same time, with the development of iron and steel industry, a large amount of coke oven gas, crude benzene and coal tar produced in coking process have been paid attention to and applied. In Germany, the first fertilizer industry and coal chemical industry were developed. Mankind entered the era of chemical synthesis. Dyes, pesticides, spices, medicine and other organic chemicals developed rapidly, chemical fertilizers and pesticides in the production of crops play an important role. In the early 20th century, chemist Fritz Haber invented the synthesis of ammonia, and in 1913, chemical engineer C. Bosch helped building the world's first synthetic ammonia plant, spurring the rapid development of the nitrogenous fertilizer industry.

Since the beginning of the 20th century, oil and natural gas have been exploited and utilized in large quantities, providing mankind with various fuels and abundant chemical raw materials.

After the industrialization of neoprene rubber in 1931 and the synthesis of polyhexamethylene adipamide (nylon 66) in 1937, the polymer chemical industry developed vigorously. By the early 1950s, the large-scale production of plastics, synthetic rubber and synthetic fiber industry was formed, and mankind entered the era of synthetic materials. Further promoted the level of industrial and agricultural production, the development of science and technology, human living standard has been significantly improved.

With the development of petrochemical and polymer chemical industry, the synthesis of high value-added functional chemicals has become one of the development directions of modern chemical industry in order to meet the higher needs of people's life. Its product have small batch, many varieties, high technical content and fast-update. Improving the refinement rate of chemical production has become an important index for the development of chemical industry in the world.

With the development of biotechnology, the mutual penetration and combination of chemical industry and biotechnology is also the development direction of today's chemical industry. Biochemical industry with broad development prospects has been preliminarily formed, adding new vitality to the traditional chemical industry.

To sum up, chemical industry provides essential chemicals and energy for industry and agriculture, modern transportation, national defense and military, and cutting-edge science and technology, ensuring and promoting the development and technological progress of these sectors. Chemical industry is closely related to human's life. In modern human's life, chemical products cannot be separated from material life such as clothing, food, housing, transportation and overcoming diseases to spiritual life such as culture, art and entertainment. The development, production and application of some chemical products have played an epoch-making role in promoting the industrial revolution, agricultural development and human living standards.

1.3　Characteristics and development direction of modern chemical industry

1.3.1　Characteristics of modern chemical industry

(1) Diversity and complexity of raw material, production methods and products

Many different chemical products can be made from the same raw material. The same product can be produced using different raw materials, or different methods and process routes. One product can have different uses, and different products may have the same uses. Because of this diversity, the chemical industry is able to provide mankind with an increasing number of new substances, materials and energy sources. At the same time, the production process of most chemical products is multi-step. Some steps are very complicated, and the influencing factors are also complicated.

(2) Development towards large-scale, comprehensive and refined

With the increase of plant scale, the productivity per unit volume per unit time increases significantly. The integration of production can make full and rational use of resources and energy. By-products and "waste" can be used locally and converted into useful products with no or minimum waste discharge. Integration is not only limited to the union of different chemical plants, but also should be a comprehensive enterprise of chemical plants and other factories.

Refinement not only refers to the production of small batch chemical products, more mainly refers to the production of high technical content, high additional output value with excellent performance or function of the product, which can adapt to the rapid change of market demand, and constantly changing product varieties and models. The chemical process is also more refined, deep into the atomic level of the molecule to carry out chemical synthesis, so that the production of products is more efficient, energy saving, resources saving.

(3) Multi-disciplinary cooperation and technology-intensive production

Modern chemical industry is highly automated and mechanized production, and further towards intelligent development. The sustainable development of today's chemical industry relies more and more on the rapid transformation of scientific research achievements into productive forces by using high and new technologies. For example, the combination of biological and chemical engineering, microelectronics and chemistry, materials and chemical engineering and other different disciplines can create more excellent new substances and materials. The high level development of computer technology has made chemical production realize remote automatic control, and will provide powerful intelligent tools for chemical synthesis. The combination of combinatorial chemistry, computational chemistry and computer can accurately design and synthesize new molecules and materials, saving a lot of experimental time and manpower. Therefore, modern chemical industry needs high level, creative and pioneering technical experts in various disciplines and different majors, as well as well-educated and trained operators and managers who are familiar with production technology.

(4) Attach importance to rational utilization of energy and actively adopt energy-saving processes and methods

Chemical production is a process in which raw materials are transformed into product materials by chemical changes, accompanied by energy transfer and conversion, which must consume energy. Chemical production department is a large consumer of energy consumption. Rational use of energy and energy saving is very important. Many of the advanced production process is reflected in the use of low energy consumption process or energy saving process.

(5) Capital intensive, fast investment recovery and high profit

Modern chemical industry needs a lot of investment because of its complex equipment, high technology and rapid product renewal. However, the production value of chemical products is high, the cost is low, and the profit is high. Once the factory is completed and put into operation, the investment can be quickly recovered and the profit can be made. The output value of chemical industry has become an important part of the gross output value index of national economy.

(6) The problems of safety and environmental protection are becoming increasingly prominent

In chemical production, inflammable, explosive and toxic are still the primary problems to be solved by modern chemical enterprises. It is necessary to adopt safe production technology, have reliable safety technology guarantee, strict regulations and supervision institutions. It is one of the keys for the sustainable development of chemical industry to create a clean production environment, develop green chemical industry and produce environmentally friendly products by adopting non-toxic and harmless methods and processes.

1.3.2 The direction of chemical industry development

With the continuous development of human life and production, it also brings fierce market competition, the reduction of natural resources and energy, the aggravation of environmental pollution and other problems. The chemical industry is also facing the challenges of these problems. To take the road of sustainable development, we must do the following aspects.

(1) Facing the situation of fierce market competition, actively develop new and high technologies, shorten the cycle of industrialization of new technologies and new processes, and speed up product renewal and upgrading.

(2) Make the best and most thorough use of raw materials. In addition to the development of large integrated manufacturing enterprises for the comprehensive utilization of raw materials, products and by-products, the design and development of atomic economic reactions are advocated.

(3) We will vigorously develop green chemical industry. Including the use of non-toxic and harmless raw materials, solvents and catalysts; application of high reaction selectivity process and catalyst; converting by-products or waste into useful

substances; using atomic economic reaction to improve the utilization rate of atoms in raw materials and achieve zero emissions; eliminating products that pollute the environment and destroy ecological balance, developing and producing environmentally friendly products, etc.

(4) The chemical process should be efficient, energy saving and intelligent.

(5) Implementing waste recycling projects.

The realization of the above aspects requires the hard work of all chemists and chemical engineers, the sincere cooperation of multiple disciplines and departments, and more importantly, the continuous progress of science and the development of high and new technologies.

1.4　Chemical raw material resources and processing

1.4.1　Inorganic chemical minerals and processing

1. Major inorganic chemical minerals

Inorganic chemical minerals are mainly used for the production of inorganic compounds and smelting metals. The mining and beneficiation of mineral resources are called chemical mining industry, which belongs to one of the chemical industry in China. Chemical mine products are very diverse, only the main mineral products listed below.

(1) Salt mine

Rock salt, sea salt or lake salt, and so on, used for the manufacture of soda, caustic soda, hydrochloric acid and vinyl chloride, etc.

(2) Sulfur deposits

Sulfur (S), pyrite (FeS_2) and so on, used for the production of sulfuric acid and sulfur.

(3) Phosphate rock

Fluorapatite $[Ca_5F(PO_4)_3]$, chlorapatite $[Ca_5Cl(PO_4)_3]$, used for the production of phosphate fertilizer, phosphoric acid and phosphate.

(4) Potash deposit

Potassium salt (mixture of KCl and NaCl), carnallite ($KCl \cdot MgCl_2 \cdot 6H_2O$), potassium salt magnesium alum ($KCl \cdot MgSO_4 \cdot 3H_2O$).

(5) Bauxite

A mixture of hydrodiaspore ($\alpha - Al_2O_3 \cdot H_2O$) and trihydrate ($Al_2O_3 \cdot 3H_2O$).

Rare metal ore and precious metal are precious resources with little reserves but high practical value. For example, neodymium, lithium, europium, tantalum and other materials are needed for high-tech. Titanium is a competitor of steel because of its strong heat and corrosion resistance. It is also one of the main components of olefin polymerization catalyst. Some heterogeneous catalysts also contain TiO_2. Platinum, palladium, rhodium, rhenium and other precious metals are important catalyst materials. Without them, catalytic chemistry would be difficult to develop, many important chemicals and new materials would not be synthesized.

With the exception of a few minerals of high grade and quality that can be extracted without preliminary processing, most minerals need to be processed at the mining site to remove unwanted impurities and to be processed into a specified shape. The main methods of mineral preliminary processing include classification, crushing, lumping and sintering, cleaning, dehydration and dust removal, some or all of which should be selected according to the requirements of the user department for raw materials.

2. Phosphorite and pyrite processing

Phosphate and pyrite are the two most productive inorganic chemical minerals. Most of the phosphate rocks are fluorapatite $[Ca_5 F(PO_4)_3]$, after classification, water eluting, flotation and other methods to remove impurities, they become commercial phosphate rock. Pyrite includes pyrite (cubic system FeS_2), marcasite (orthorhombic system FeS_2) and magnetic pyrite (Fe_nS_{n+1}), of which pyrite is the main one.

Phosphate rock is the raw material for the production of phosphate fertilizer, phosphoric acid, elemental phosphorus, phosphate compounds and phosphate. More than 85% of phosphate rocks are used to make phosphate fertilizer. The ordinary superphosphate produced in the past has been eliminated due to its low phosphorus content. Now the phosphate fertilizer with the largest yield is ammonium phosphate, which belongs to nitrogen and phosphorus compound fertilizer. There are two main methods of producing phosphate fertilizers.

(1) Acid process (also known as wet process)

It uses sulfuric acid or nitric acid and other inorganic acids to treat phosphate ore, the most commonly used is sulfuric acid. Sulfuric acid reacts with phosphate rock to generate phosphoric acid and calcium sulfate crystals, and the main reaction formula is

$$Ca_5F(PO_4)_3 + 5H_2SO_4 + 5nH_2O \longrightarrow 3H_3PO_4 + 5CaSO_4 \cdot nH_2O + HF \quad (1-1)$$

(2) Hot method

High temperature is used to decompose phosphate rock and further make phosphate that can be absorbed by crops. Elemental phosphorus, phosphorus pentoxide and phosphoric acid can also be produced by thermal process. Thermal production consumes more electric energy and heat energy.

Pyrite is used to make sulphuric acid, and more than half of the world's total sulphuric acid production is used to make phosphate and chlorine fertilizers. The main reaction formula for producing sulfuric acid from pyrite is

$$4FeS_2 + 11O_2 \longrightarrow 2Fe_2O_3 + 8SO_2 \qquad (1-2)$$

$$SO_2 + \frac{1}{2}O_2 \longrightarrow SO_3 \qquad (1-3)$$

$$SO_3 + H_2O \longrightarrow H_2SO_4 \qquad (1-4)$$

1.4.2　Petroleum and its processing

Petrochemicals began to flourish in the 1950s. Up to now, about 90% of the products of basic organic chemicals, polymer chemicals, fine chemicals and nitrogen fertilizer industry are made from petroleum and natural gas. About 90% of the upstream raw materials of organic chemical products can be attributed to alkenes (ethylene, propylene, butadiene), benzol (benzene, toluene, xylene), acetylene, naphthalene and methanol. Among them, alkenes is mainly made from petroleum, while benzol, naphthalene and methanol can be made from petroleum, natural gas and coal.

1. Composition of petroleum

Oil is an odorous brown-black or yellow-brown viscous liquid with a relative density of about 0.75 to 1.0 depending on its composition. Some oil fields are often associated with field gas. Petroleum is a mixture of a large number of compounds with different relative molecular weights, compositions and structures, whose boiling points range from room temperature to more than 500℃. The two elements with the largest content in petroleum are C and H. Petroleum contain 83%~87% carbon and 11%~14% hydrogen respectively, and they mainly exist in the form of hydrocarbons. The content of other elements fluctuates greatly with different producing areas, such as sulfur 0.02%~5.5%, nitrogen 0.02%~1.7%, oxygen 0.08%~1.82%. Metallic elements such as Ni, V, Fe and Cu contain only trace amounts, ranging from parts per billion to a few parts per million. There are K, Na, Ca, Mg and other chloride dissolved in the water phase co-existing with oil underground, which is easy to remove. The compounds in petroleum can be divided into hydrocarbons, non-hydrocarbons, colloid and bitumen.

(1) Hydrocarbon compounds

Hydrocarbons, or hydrocarbon compounds, make up the vast majority of petroleum, numbering in the tens of thousands. They mainly include chain saturated hydrocarbons, cycloalkanes, aromatic hydrocarbons, and so on. These hydrocarbons are the basic raw materials of organic chemicals, and many hydrocarbons are also components of gasoline, aviation kerosene and diesel.

Olefins and alkynes are almost absent from petroleum, but they are important raw materials for petrochemical industry, especially olefins, which can only be obtained by chemical processing of petroleum.

(2) Non-hydrocarbon compounds

Organic compounds containing carbon, hydrogen and other heteroatom, such as sulfides, nitrides, oxygen-containing compounds, metal-organic compounds, etc.

Although the content of non-hydrocarbon compounds is very low, it has a great impact on the nature of oil and oil products processing. Some even make catalyst poisoning, some will corrosion of pipelines and equipment, some will pollute environment when using, etc. So we should advance its removal in oil processing and recycling. Desulfurization, denitrogenation and demetallization is the important processes of petroleum chemical processing.

(3) Colloid and asphaltene

After distillation of crude oil, the fraction with boiling point higher than 500℃ is residue, which contains a considerable amount of gum and asphaltene. They are mixtures composed of various compounds with different structures and large molecular weight, mostly viscous cyclic alkane, viscous cyclic aromatic hydrocarbon, cyclic compounds containing S, N and other heteroatom.

2. Atmospheric and vacuum distillation of petroleum

In order to make full use of the precious petroleum resources, it is necessary to carry out primary and secondary processing of petroleum, producing gasoline, aviation kerosene, diesel oil, boiler fuel and liquefied gasoline, and at the same time, making various chemical raw materials.

When oil is extracted and not processed, it is called crude oil, and a processing method is atmospheric distillation and vacuum distillation. Distillation is a method of separation by using the difference of volatilization between each component in liquid mixture (different boiling points). It is a physical process of mass and heat transfer without chemical reaction. The main equipment is a distillation tower. There are three types of normal and vacuum distillation processes.

(1) Fuel type

It mainly produces gasoline, kerosene, diesel oil and so on. It does not make full use of petroleum resources and is rarely used now.

(2) Fuel-lubricating oil type

In addition to the production of light and heavy fuel oil, it also produces a variety of lubricants and paraffins.

(3) Fuel-chemical type

In addition to the production of gasoline, kerosene, diesel oil and other fuel oil, aromatic hydrocarbons are also extracted from naphtha distillate, and important organic chemical basic raw materials such as olefins and aromatic hydrocarbons are prepared from naphtha or diesel pyrolysis. The gas produced by refining is also chemical raw materials. Some factories also use fuel-lubricating oil-chemical process, the main products are fuel and chemical products.

Schematic diagram of crude oil atmospheric and vacuum distillation process is shown in Figure 1 - 1.

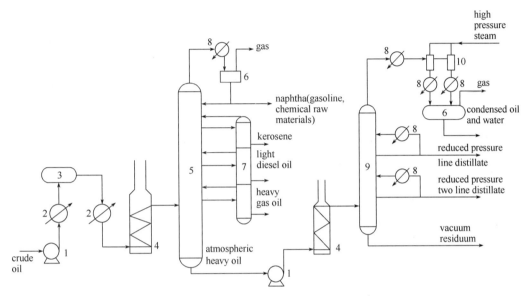

1 - Fuel delivery pump; 2 - Heat exchanger; 3 - Desalting tank; 4 - Heating furnace;
5 - Atmospheric distillation column; 6 - Liquid storage tank; 7 - Gas stripping column;
8 - Condensing cooler; 9 - Vacuum still; 10 - Steam-jet pump.

Figure 1 - 1   Schematic diagram of crude oil atmospheric and vacuum distillation process

3. Chemical processing of distillate oil

Atmospheric and vacuum distillation can only cut crude oil into a few fractions, the production of fuel is limited and cannot meet the demand, only gas from the top of the tower can be used directly as a chemical material. In order to produce more fuel and chemical feedstock, each distillate needs to be reprocessed. There are many processing methods, mainly chemical processing methods. The following is a brief introduction to the main several processing processes.

(1) Catalytic reforming

Catalytic reforming is a process in which gasoline fractions (naphtha) are heated under the action of platinum-containing catalyst to rearrange the hydrocarbon molecules to form new molecules. Catalytic reforming unit can provide gasoline with high octane number, aromatic raw materials such as benzene, toluene and xylene for chemical fiber, rubber, plastic and finechemicals, as well as liquefied gas, solvent oil and by-product hydrogen.

Schematic diagram of platinum-reforming process is shown in Figure 1 - 2.

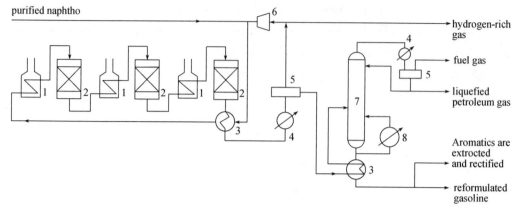

1 – Heating furnace; 2 – Reforming reactor; 3 – Heat exchanger; 4 – Cooler condenser;
5 – Oil-gas separator; 6 – Circulating hydrogen compressor; 7 – Fractionating tower;
8 – Reboiler.

Figure 1 – 2   Schematic diagram of platinum reforming process

(2) Catalytic cracking

Catalytic cracking is a process in which heavy distillate oil is heated under the action of catalyst, the large molecular hydrocarbon compounds are cracked and converted into high quality gasoline, and the by-products such as diesel oil, boiler fuel, liquefied gas and gas products are produced.

The raw material can be straight-run diesel oil, heavy diesel oil, vacuum diesel oil or lubricating oil distillate, or even residual oil coking after petroleum coke. Most of the hydrocarbons contain more than 18 carbons.

(3) Catalytic hydrocracking

Catalytic hydrocracking refers to the process in which heavy oil is heated to undergo various hydrogenation and cracking reactions in the presence of catalysts and high hydrogen pressure, and is converted into aviation kerosene, diesel, gasoline (or reforming raw materials) and gas products.

Hydrocracking process is shown in Figure 1 – 3.

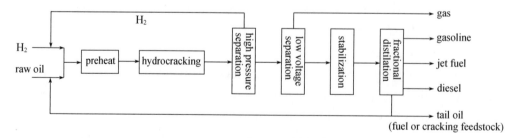

Figure 1 – 3   Hydrocracking process

(4) Pyrolysis of hydrocarbons

Hydrocarbon pyrolysis do not use catalyst. The hydrocarbon is heated to 750℃ ～

900℃ for pyrolysis. The reaction is quite complex, mainly high carbon alkanes crack to produce low carbon olefin and diolefin, accompanied by dehydrogenation, aromatization, coking and many other reactions. After cracking, the products were cooled and condensed to obtain two kinds of mixture of cracking gas and cracking gasoline.

The main purpose of hydrocarbon pyrolysis is to make ethylene and propylene, as well as aromatic hydrocarbons such as butene, butadiene, benzene, toluene, xylene, ethylbenzene and other chemical raw materials. It is an essential primary process in every petrochemical plant.

Secondary processing of petroleum includes alkylation, isomerization, coking and so on, which can obtain high-octane number gasoline and chemical raw materials.

The main way of making fuel and chemical raw materials from petroleum is shown in Figure 1 - 4.

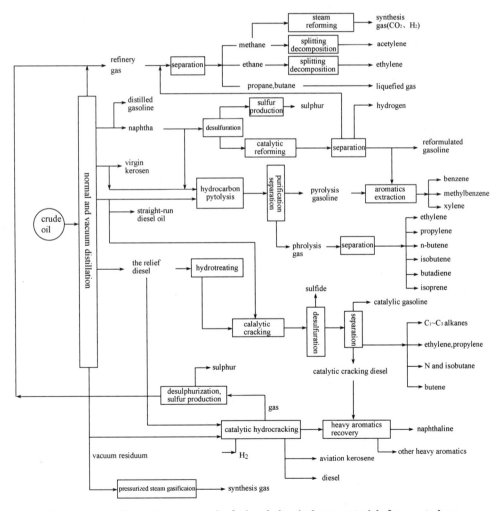

Figure 1 - 4   The main way to make fuel and chemical raw materials from petroleum

1.4.3　Natural gas and its processing

The main component of natural gas is methane. The natural gas with methane content above 90% is called dry gas, and the natural gas with $C_2 \sim C_4$ alkane content of $15\% \sim 20\%$ or more is called wet gas. Some natural gas is symbiotic with oil (associated gas in oil fields).

Natural gas is a kind of clean energy with high calorific value and less pollution, and its proportion in energy structure is increasing year by year. It is also an important raw material resource of petrochemical industry. Natural gas processing and utilization mainly includes the following aspects.

(1) Production of hydrogen and ammonia from natural gas.

(2) Catalytic conversion of natural gas to fuel and chemical products through syngas route.

(3) Natural gas is directly catalyzed into chemical products.

(4) Natural gas pyrolysis to produce chemical products.

(5) Chlorination, nitrification, ammonia oxidation and sulfurization of methane to produce chemical products.

(6) Utilization of $C_2 \sim C_4$ alkanes in wet natural gas.

1.4.4　Coal and its processing

Coal is composed of large molecular organic matter with various structures of carbon and hydrogen and a small amount of inorganic minerals of silicon, aluminum, iron, calcium and magnesium. Due to the different time of coal forming process, there are peat, lignite, bituminous and anthracite. According to the mass fraction, peat contains $60\% \sim 70\%$ carbon, lignite contains $70\% \sim 80\%$ carbon, bituminous coal contains $80\% \sim 90\%$ carbon, anthracite contains up to $90\% \sim 93\%$ carbon.

The known coal reserves are more than ten times larger than petroleum reserves. The comprehensive utilization of coal can provide valuable raw materials for energy, chemical industry and metallurgy at the same time. The processing route in coal chemical industry mainly has the following kinds.

(1) Coal carbonization

It is the process of heating coal under isolated air conditions to decompose it into coke, coal tar, crude benzene and coke oven gas.

(2) Coal gasification

It refers to the process in which solid fuel such as coal, coke or semi-coke reacts with gasification agent at high temperature ($900℃ \sim 1\,300℃$) and is converted into gases such as hydrogen and carbon monoxide.

(3) Coal liquefaction

It refers to the chemical process by which coal is converted into liquid fuel. Coal liquefaction can be divided into direct liquefaction and indirect liquefaction.

Direct liquefaction of coal is converted coal into liquid hydrocarbon by hydrogenation, so it is also called hydroliquefaction of coal. The liquefied products, also known as artificial petroleum, can be further processed into various liquid fuels.

Indirect liquefaction of coal is the production of syngas in advance, then which is converted to hydrocarbon fuels and oxygen-containing fuels (e. g. low carbon blended alcohols, dimethyl ether) through the action of catalysts.

1.4.5   Biomass and its processing

The products and wastes (core, shell, stalk, chaff, slag) of agriculture, forestry, animal husbandry and fishery can be changed to the basic chemicals or intermediate products by chemical or biological chemical method, such as glucose, lactic acid, citric acid, ethanol, acetone, senior fatty acid (lauric acid, stearic acid, oleic acid, linoleic acid, myristic acid), etc. The process involves a range of chemical processes, such as chemical hydrolysis, enzymatic hydrolysis, microbial hydrolysis, saponification, catalytic hydrogenation, gasification, cracking, extraction, and some of them using DNA technology.

1.4.6   Development and utilization of renewable resources

In principle, industrial, agricultural and domestic wastes can be recycled and processed into useful products. The utilization of these renewable resources can not only save natural resources, but also be one of the effective measures to prevent pollution and protect the environment.

Industrial plants have been built to reprocess waste plastics into liquid fuels, and there are many ways to reprocess them. The waste containing carbon can also be converted into small molecular gas compounds by partial oxidation and then reused.

1.4.7   Air and water

The volume composition of air is 78. 16% N_2, 20. 90% O_2, 0. 93% Ar, and the remaining 0. 01% is rare gases such as He, Ne, Kr and Xe. O_2 and N_2 in air are important chemical raw materials. Combustion of carbon containing materials in pure oxygen does not produce nitrogen oxides and carbon dust, which is good for the environment. Hydrogen and oxygen fuel cells can generate clean energy. Argon, helium, neon, krypton and other gases with high purity extracted from the air are widely used in high-tech fields.

Water is more commonly used in chemical industry, for example, as a solvent to dissolve solid and liquid, or absorb gas, as a reactant to participate in hydrolysis, hydration and other reactions, as a carrier to heat or cool materials and equipment. It can absorb the heat of reaction and vaporize into high pressure steam with the ability to do work. The area of water on the earth accounts for more than 70% of the earth's surface, but the volume of fresh water being available for use is only 0.3% of the total volume of water, so it is urgent to save and protect fresh water resources and improve the recycling efficiency of water.

1.5　The main content and features of this book

According to the structural characteristics, internal relations and development trend of chemical industry, this book describes the principles and knowledge of chemical process according to the classification of chemical reaction process. The content mainly includes the introduction of the general picture of chemical industry and the basic concept of chemical process, chemical raw material resources and their processing and utilization, typical production process of chemical basic raw materials.

This book is rich in content and has a wide range of knowledge. It pays attention to the combination of point and aspect. The key contents are elaborated thoroughly and carefully, and the combination of theory and practice is paid attention to. Students should pay attention to cultivate the ability to analyze and solve problems. For typical reaction process, it is required to understand and master the process principle, the basis for selecting process conditions, the organization and characteristics of the process, the structural characteristics, advantages and disadvantages of various reaction equipment, etc. Various raw material sources, different process routes and their technical and economic indicators, energy recovery and utilization methods, by-product recovery and utilization methods, waste disposal methods of typical products should be analyzed and compared to find out their advantages and disadvantages.

The research, development and industrialization of chemical process need to apply basic scientific theories such as chemistry and physics, chemical engineering principles and methods, and relevant engineering knowledge and technology, through analysis and synthesis to achieve success. Therefore, in the course of chemical technology, we should pay attention to the comprehensive application of these theories and knowledge, especially to the combination of theory and practice, so as to cultivate the ability of innovation.

Questions

1 - 1   Why petroleum, natural gas and coal are important raw materials for modern chemical industry? What are the ways of their comprehensive utilization?

1 - 2   What is the utilization prospect of biomass and renewable resources?

第2章 化工工艺学基础

2.1 化工生产过程及流程

2.1.1 化工生产过程

化工生产过程一般可概括为原料预处理、化学反应和产品分离及精制三大步骤。

（1）原料预处理

主要目的是使初始原料达到反应所需要的状态和规格。在多数生产过程中，原料预处理本身就很复杂，要用到许多物理和化学的方法和技术，有些原料预处理成本占总生产成本的大部分。

（2）化学反应

通过该步骤完成由原料到产物的转变，是化工生产过程的核心。反应温度、压力、浓度、催化剂（多数反应需要）或其他物料的性质以及反应设备的技术水平等各种因素对产品的数量和质量有重要影响，是化学工艺学研究的重点内容。

（3）产品的分离和精制

目的是获取符合规格的产品，并回收、利用副产物。在多数反应过程中，由于诸多原因，致使反应后产物是包括目的产物在内的许多物质的混合物，有时目的产物的浓度甚至很低，必须对反应后的混合物进行分离、提浓和精制，才能得到符合规格的产品。同时要回收剩余反应物，以提高原料利用率。

化工过程常常包括多步反应转化过程，因此除了起始原料和最终产品外，尚有多种中间产物生成，原料和产品也可能是多个，因此化工过程通常由上述三个步骤交替组成，以化学反应为中心，将反应与分离过程有机地组织起来。

2.1.2 化工生产工艺流程

1. 工艺流程和流程图

原料需要经过包括物质和能量转换的一系列加工，方能转变成所需产品。这些转换需要由相应的功能单元来完成，按物料加工顺序将这些功能单元有机地组合起来，则构筑成工艺流程。将原料转变成化工产品的工艺流程称为化工生产工艺流程。

化工生产中的工艺流程是丰富多彩的，不同产品的生产工艺流程不同；同一产品用不同原料来生产，工艺流程也大不相同；有时即使原料相同，产品也相同，若采用的工艺路线或加工方法不同，则在流程上也有区别。工艺流程多采用图示方法来表达，称为工艺流程图。

化学工艺学教科书中主要采用工艺流程示意图,它简明地反映出由原料到产品过程中各物料的流向和经历的加工步骤,从中可了解每个操作单元或设备的功能以及相互间的关系、能量的传递和利用情况、副产物和三废的排放及其处理方法等重要工艺和工程知识。

2. 化工生产工艺流程的组织

工艺流程的组织或合成是化工过程的开发和设计中的重要环节。组织工艺流程需要有化学、物理的理论基础以及工程知识,要结合生产实践,借鉴前人的经验。同时,可运用推论分析、功能分析、形态分析等方法论来进行流程的设计。

推论分析法是从"目标"出发,寻找实现此"目标"的"前提",将具有不同功能的单元进行逻辑组合,形成一个具有整体功能的系统。

功能分析法是缜密地研究每个单元的基本功能和基本属性,然后组成几个可以比较的方案以供选择。

形态分析法是对每种可供选择的方案进行精确的分析和评价,择优劣汰,选择其中的最优方案。

化学工业广泛地使用热能、电能和机械能,是耗能大户。在组织工艺流程时,不仅要考虑高产出、高质量,还要考虑合理地利用能源、回收能源,做到最大限度地节约能源,才能达到经济先进性。

热能有不同的温位,想要有高的利用率,应合理地安排相应的回收利用设备,能量回收利用的效率体现了工艺流程及技术水平的高低。应尽可能利用物料所带的显热,使之在离开系统时接近环境温度,以免热量损失到环境中。

2.2 化工过程主要效率指标

2.2.1 生产能力和生产强度

1. 生产能力

生产能力指一个设备、一套装置或一个工厂在单位时间内生产的产品量,或在单位时间内处理的原料量。其单位为 kg/h、t/d 或 kt/a、万吨/年等。

2. 生产强度

生产强度为设备单位特征几何量的生产能力,即设备的单位体积的生产能力,或单位面积的生产能力,其单位为 kg/(h·m³)、t/(d·m³) 或 kg/(h·m²)、t/(d·m²) 等。生产强度指标主要用于比较那些相同反应过程或物理加工过程的设备或装置的优劣。设备中进行的过程速率高,其生产强度就高。

3. 有效生产周期

$$开工因子 = \frac{全年开工生产天数}{365} \qquad (2-1)$$

开工因子通常在 0.9 左右,开工因子大意味着停工检修带来的损失小,即设备先进可靠,催化剂寿命长。

2.2.2　化学反应效率

1. 原子经济性

原子经济性由美国 Stanford 大学的 B. M. Trost 教授首次提出,获得了 1998 年美国"总统绿色化学挑战奖"的学术奖。

原子经济性 AE 定义为

$$AE = \frac{\sum_i P_i M_i}{\sum_j F_j M_j} \times 100\% \tag{2-2}$$

式中,P_i 为目的产物分子中各类原子数;F_j 为反应原料中各类原子数;M 为相应各类原子的相对质量。

2. 环境因子

环境因子由荷兰化学家 Sheldon 提出,定义为

$$E = \frac{废物质量}{目标产物质量} \tag{2-3}$$

上述指标从本质上反映了合成工艺是否最大限度地利用了资源,避免了废物的产生和由此而带来的环境污染。

2.2.3　转化率、选择性和收率

化工总过程的核心是化学反应,提高反应的转化率、选择性和收率是提高化工过程效率的关键。

1. 转化率

转化率指某一反应物参加反应而转化的数量占该反应物起始量的分率或百分率,用符号 X 表示。其定义式为

$$X = \frac{某一反应物的转化量}{该反应物的起始量} \tag{2-4}$$

2. 选择性

对于复杂反应体系,同时存在着生成目的产物的主反应和生成副产物的副反应,只用转化率来衡量是不够的。选择性是指体系中转化成目的产物的某反应物量与参加所有反应而转化的该反应物总量之比,用符号 S 表示。

$$S = \frac{转化为目的产物的某反应物的量}{该反应物的转化总量} \tag{2-5}$$

3. 收率(产率)

收率是从产物角度来描述反应过程的效率,用符号 Y 表示。

$$Y = \frac{转化为目的产物的某反应物的量}{该反应物的起始量} \tag{2-6}$$

2.2.4　平衡转化率和平衡产率

可逆反应达到平衡时的转化率称为平衡转化率,此时所得产物的产率为平衡产率。平衡转化率和平衡产率是可逆反应所能达到的极限值(最大值),但是,反应达平衡往往需要相当长的时间。随着反应的进行,正反应速率降低,逆反应速率升高,所以净反应速率不断下降直到零。在实际生产中应保持高的净反应速率,不能等待反应达平衡,故实际转化率和产率比平衡值低。若平衡产率高,则可获得较高的实际产率。工艺学的任务之一是通过热力学分析,寻找提高平衡产率的有利条件,并计算出平衡产率。

2.3　化工反应条件的影响

反应温度、压力、浓度、反应时间、原料的纯度和配比等众多条件是影响反应速率和平衡的重要因素,关系到生产过程的效率。在本书其他各章中均有具体过程的影响因素分析,此处仅简述以下几个重要因素的影响规律。

2.3.1　温度影响

1. 温度对化学平衡的影响

对于不可逆反应不需考虑化学平衡,而对于可逆反应,其平衡常数与温度的关系为

$$\log K = -\frac{\Delta H^{\ominus}}{2.303\,RT} + C \qquad (2-7)$$

式中,K 为平衡常数;$\Delta H^{\ominus}$ 为标准反应焓差;R 为气体常数;T 为反应温度;C 为积分常数。

对于吸热反应,$\Delta H^{\ominus} > 0$,K 值随着温度升高而增大,有利于反应,产物的平衡产率增加;对于放热反应,$\Delta H^{\ominus} < 0$,K 值随着温度升高而减小,平衡产率降低,故只有降低温度才能使平衡产率增高。

2. 温度对反应速率的影响

反应速率是指单位时间、单位体积某反应物组分的消耗量,或某产物的生成量。

对于不可逆反应,逆反应速率忽略不计,故产物生成速率总是随温度的升高而加快。

对于可逆反应而言,正、逆反应速率之差即为产物生成的净速率。温度升高时,正、逆反应速率常数都增大,所以正、逆反应速率都提高。理论上讲,放热可逆反应应在最佳反应温度下进行,此时净反应速率最大。

2.3.2　浓度影响

根据反应平衡移动原理,反应物浓度越高,越有利于平衡向产物方向移动。当有多种反应物参加反应时,往往使价廉易得的反应物过量,从而使价格贵或难得的反应物更多地转化为产物,提高其利用率。

2.3.3　压力影响

一般来说,压力对液相和固相反应的平衡影响较小。气体的体积受压力影响大,故压力对有气相物质参加的反应平衡影响很大,其规律为:

(1) 对分子数增加的反应,降低压力可以提高平衡产率。

(2) 对分子数减少的反应,压力升高,产物的平衡产率增大。

(3) 对分子数没有变化的反应,压力对平衡产率无影响。

在一定的压力范围内,加压可减小气体反应体积,且对加快反应速率有一定好处,但压力过高,能耗增大,对设备投资高,反而不经济。

惰性气体的存在可降低反应物的分压,对反应速率不利,但有利于分子数增加的反应的平衡产率。

2.4　催化剂

据统计,当今 90% 的化学反应中均包含催化(catalysis)过程,催化剂(catalyst)在化学工艺中占有相当重要的地位,其作用主要体现在以下几方面。

(1) 提高反应速率和选择性。有许多反应,虽然在热力学上是可能进行的,但反应速率太慢或选择性太低,不具有实用价值。对于这些反应,一旦发明和使用催化剂,则可能实现工业化,为人类生产出重要的化工产品。

(2) 改进操作条件。采用或改进催化剂可以降低反应温度和操作压力,可以提高化学加工过程的效率。

(3) 催化剂有助于开发新的反应过程,发展新的化工技术。

(4) 催化剂在能源开发和消除污染中可发挥重要作用。

2.4.1　催化剂基本特征

在一个反应系统中使化学反应速率明显加快,但在反应前后其数量和化学性质不变的物质称为催化剂。催化剂的作用是与反应物生成不稳定中间化合物,改变反应途径,活化能得以降低。

催化剂有以下三个基本特征。

(1) 催化剂是参与了反应的,但反应终了时,催化剂本身未发生化学性质和数量的变化。因此,催化剂在生产过程中可以在较长时间内使用。

(2) 催化剂只能缩短达到化学平衡的时间(即加速作用),但不能改变平衡。对于那些受平衡限制的反应体系,必须在有利于平衡向产物方向移动的条件下来选择和使用催化剂。

(3) 催化剂具有明显的选择性,特定的催化剂只能催化特定的反应。催化剂的这一特性在有机化学反应领域中起了非常重要的作用,因为有机反应体系往往同时存在许多反应,选用合适的催化剂,可使反应向需要的方向进行。

2.4.2 催化剂分类

按催化反应体系的物相均一性分类,有均相催化剂和非均相催化剂。

按反应类别分类,有加氢、脱氢、氧化、裂化、水合、聚合、烷基化、异构化、芳构化、羰基化、卤化等众多催化剂。

按反应机理分类,有氧化还原型催化剂、酸碱催化剂等。

按使用条件下的物态分类,有金属催化剂、氧化物催化剂、硫化物催化剂、酸催化剂、碱催化剂、配位化合物催化剂和生物催化剂等。

2.4.3 工业催化剂使用中的有关问题

在采用催化剂的化工生产中,正确地选择并使用催化剂是个非常重要的问题,关系到生产效率和效益。通常对工业催化剂的以下几种性能有一定的要求。

1. 工业催化剂使用性能

(1) 活性

催化剂活性是指在给定的温度、压力和反应物流量(或空间速度)下,催化剂使原料转化的能力。催化剂活性越高,则原料的转化率愈高;或者在转化率及其他条件相同时,催化剂活性愈高,则需要的反应温度愈低。工业催化剂应有足够高的活性。

(2) 选择性

催化剂的选择性是指反应所消耗的原料中有多少转化为目的产物。催化剂选择性愈高,生产单位量的产物的原料消耗定额愈低,也愈有利于产物的后处理,故工业催化剂的选择性应较高。当催化剂的活性与选择性难以两全时,若反应原料昂贵或产物分离很困难,宜选用选择性高的催化剂;若原料价廉易得或产物易分离,则可选用活性高的催化剂。

(3) 寿命

催化剂寿命是指其使用期限的长短,寿命的表征是生产单位量产品所消耗的催化剂量,或在满足生产要求的技术水平上催化剂能使用的时间长短,有的催化剂使用寿命可达数年,有的则只能使用数月。

2. 催化剂活化

许多固体催化剂在出售时的状态一般是较稳定的,但这种稳定状态不具有催化性能,催化剂使用时必须在反应前对其进行活化,使其转化成具有活性的状态。不同类型的催化剂要用不同的活化方法,有还原、氧化、硫化、酸化、热处理等,每种活化方法均有各自的活化条件和操作要求,应该严格按照操作规程进行活化,才能保证催化剂发挥良好的作用。如果活化操作失误,轻则使催化剂性能下降,重则使催化剂报废,造成经济损失。

3. 催化剂失活和再生

引起催化剂失活的原因较多,对于配位催化剂而言,主要是超温,大多数配位化合物在 250℃ 以上就分解而失活;对于生物催化剂而言,过热、化学物质和杂菌的污染、pH 失调等均是失活的原因。对于固体催化剂而言,其失活原因主要有:

（1）超温过热，使催化剂表面发生烧结、晶型转变或物相转变。

（2）原料气中混有毒物杂质，使催化剂中毒。

（3）有污垢覆盖催化剂表面。污垢可能是原料带入，也可能是设备内的机械杂质如油污、灰尘、铁锈等；有烃类或其他含碳化合物参加的反应往往易析碳，催化剂酸性过强或催化活性较低时析碳严重，发生积碳或结焦，覆盖催化剂活性中心，导致失活。

催化剂中毒有暂时性和永久性两种情况。暂时性中毒是可逆的，当原料中除去毒物后，催化剂可逐渐恢复活性；永久性中毒则是不可逆的。催化剂积碳可通过烧炭再生。但无论是暂时性中毒后的再生，还是积碳后的再生，通常均会引起催化剂结构不同程度的损伤，致使活性下降。

因此，应严格控制操作条件，采用结构合理的反应器，使反应温度在催化剂最佳使用温度范围内合理地分布，防止超温；反应原料中的毒物杂质应该预先加以脱除，使毒物含量低于催化剂耐受值；在有析碳反应的体系中，应采用有利于防止析碳的反应条件，并选用抗积碳性能高的催化剂。

4. 催化剂运输、贮存和装卸

催化剂一般价格较贵，要注意保护。在运输和贮藏中应防止其受污染和破坏；固体催化剂在装填于反应器时，要防止污染和破裂。装填要均匀，避免出现"架桥"现象，以防止反应工况恶化。许多催化剂使用后在停工卸出之前，需要进行钝化处理，尤其是金属催化剂一定要经过低含氧量的气体钝化后，才能暴露于空气，否则会遇空气剧烈氧化自燃，烧坏催化剂和设备。

2.5 反应过程物料衡算和热量衡算

物料衡算和热量衡算是化学工艺的基础之一。通过物料、热量衡算，计算生产过程的原料消耗指标、热负荷和产品产率等，可为设计、选择反应器和其他设备的尺寸、类型及台数提供定量依据；可以核查生产过程中各物料量及有关数据是否正常，是否有泄漏，以及热量回收、利用水平和热损失的大小，从而查找出生产上的薄弱环节和瓶颈部位，为改善操作和进行系统优化提供依据。在化工原理课程中已学习过除反应过程以外的化工单元操作过程的物料、热量衡算，所以本节只涉及反应过程的物料、热量衡算。

2.5.1 反应过程物料衡算

1. 物料衡算基本方程

物料衡算总是围绕一个特定范围来进行，可称此范围为衡算系统。衡算系统可以是一个总厂，一个分厂或车间，一套装置，一个设备，甚至一个节点等。物料衡算的理论依据是质量守恒定律，按此定律写出衡算系统的物料衡算通式为

输入物料的总质量＝输出物料的总质量＋系统内积累的物料质量　　（2-8）

2. 间歇操作过程的物料衡算

间歇操作属批量生产，即一次投料到反应器内进行反应，反应完成后一次出料，然

后再进行第二批生产。其特点是在反应过程中浓度等参数随时间而变化。分批投料和分批出料也属于间歇操作。

3. 稳定流动过程的物料衡算

生产中绝大多数化工过程为连续式操作,设备或装置可连续运行很长时间。除了开工和停工阶段外,在绝大多数时间内是处于稳定状态的流动过程,物料不断地流进和流出系统。其特点是系统中各点的参数如温度、压力、浓度和流量等不随时间而变化,系统中没有积累。当然,设备内不同点或截面的参数可相同,也可不同。稳定流动过程的物料衡算式为

$$输入系统的物料总质量=输出系统的物料总质量 \qquad (2-9)$$

4. 物料衡算步骤

化工生产的许多过程是比较复杂的,在对其做物料衡算时,应该按一定步骤进行,才能给出清晰的计算过程和正确的结果。通常遵循以下步骤:

第一步,绘出流程的方框图,以便选定衡算系统。图形表达方式宜简单,但代表的内容要准确,进、出物料不能有任何遗漏,否则衡算会造成错误。

第二步,写出化学反应方程式并配平。如果反应过于复杂,或反应不太明确、写不出反应式,则应用原子衡算法来进行计算,不必写反应式。

第三步,选定衡算基准。衡算基准是为进行物料衡算所选择的起始物理量,包括物料名称、数量和单位,衡算结果得到的其他物料量均是相对于该基准而言的。

第四步,收集或计算必要的各种数据,要注意数据的适用范围和条件。

第五步,设未知数,列方程组,联立求解。有几个未知数则应列出几个独立的方程式,这些方程式除物料衡算式外,有时尚需其他关系式,诸如组成关系约束式、化学平衡约束式、相平衡约束式、物料量比例等。

第六步,计算和核对。

第七步,报告计算结果。通常将已知及计算结果列成物料平衡表,表格可以有不同形式,但要全面反映输入及输出的各种物料和包含的组分的绝对量和相对含量。

2.5.2 反应过程热量衡算

根据能量守恒定律,进出系统的能量衡算式为

$$输入能量=输出能量+积累能量 \qquad (2-10)$$

在做热量衡算时应注意以下几点:

(1) 确定衡算对象。

(2) 选定物料衡算基准。

(3) 确定温度基准。

(4) 注意物质的相态。

思考题

2-1 何谓化工生产工艺流程? 举例说明工艺流程是如何组织的。

2-2 何谓循环式工艺流程？它有什么优缺点？

2-3 何谓转化率？何谓选择性？对于多反应体系，为什么要同时考虑转化率和选择性两个指标？

2-4 催化剂有哪些基本特征？它在化工生产中起到什么作用？在生产中如何正确使用催化剂？

Chapter 2　Fundamentals of chemical technology

2.1　Chemical production process and flow chart

2.1.1　Chemical production process

Chemical production process can be generally summarized as raw material pretreatment, chemical reaction, product separation and refining.

(1) Raw material pretreatment

The main purpose is to make the raw material to the required state and specification of the reaction. In most production processes, raw material pretreatment itself is very complicated, and many physical and chemical technologies are needed. Some raw material pretreatment costs account for most of the total production costs.

(2) Chemical reaction

It is the core of chemical production process to complete the transformation from raw material to product. Reaction temperature, pressure, concentration, catalyst (most reaction needs) or the nature of other materials, the technical level of reaction equipment and other factors have an important impact on the quantity and quality of products. It is the focus of chemical technology research content.

(3) Separation and refinement of products

The purpose is to obtain products in line with specifications, and recycle by-products. In most reaction process, due to many reasons, the product after reaction is a mixture of many substances including the target product. Sometimes the concentration of the target product is very low, it must be separated, concentrated and refined after the reaction of the mixture, in order to get the product in line with specifications. At the same time, the remaining reactants should be recovered to improve the utilization rate of raw materials.

Chemical processes often include a multi-step reaction transformation process, so in addition to the initial raw materials and final products, there is a variety of intermediates generated, raw materials and products may be multiple as well, so the chemical process is usually composed of the above three steps alternately, and the reaction and separation process are organically organized with the chemical reaction as the center.

2.1.2　Chemical production process flow

1. Process flow and flow chart

Raw materials need to go through a series of processing including material and energy conversion, and they can be transformed into the required products. These conversion needs to have the corresponding functional units to complete, combined these functional units organically according to the material processing sequence, then build a technological process. The process flow that transforms raw materials into chemical products is called chemical production process flow.

Chemical production process is colorful. The production process of different products is different; the process of the same product with different raw materials is also different. Sometimes both of the products and the ingredients are the same, but if the technological route or processing method is different, it will also have distinction on flow. The process flow is mostly expressed by graphical method, known as flowsheet or process flowsheet.

Mainly used in chemical technology textbook, the process flowsheet simply reflects the material flow and processing steps of each material from raw materials to products, from which we can understand the function of each operating unit or equipment as well as the relationship between each other, transmission and utilization of energy, discharge and processing methods of by-products and "three wastes", other important process and engineering knowledge.

2. Organization of chemical process flow

The organization or synthesis of process flows is an important step in the development and design of chemical processes. The organization of process flow requires theoretical basis of chemistry and physics as well as engineering knowledge, combining with production practice and previous experience. At the same time, inferential analysis, functional analysis, morphological analysis and other methodologies can be used to design the process.

Inferential analysis method starts from the "goal", seeks the "premise" to achieve this "goal", and logically combines units with different functions to form a system with overall functions.

Functional analysis method is to carefully study the basic functions and attributes of each unit, and then compose several comparable schemes for selection.

Morphological analysis method is to accurately analyze and evaluate each alternative scheme, focus on the best and eliminate the worst.

The chemical industry uses heat, electric and mechanical energy extensively and it is a heavy consumer of energy. In the organization of technological process, we should not only consider high yield and high quality, but also consider reasonable utilization and recovery of energy, so as to save energy to the maximum extent and achieve economic advancement.

There are different temperature levels of heat energy, so it is necessary to have high utilization rate. Corresponding recovery and utilization equipment should be rationally arranged. The efficiency of energy recovery and utilization reflects the technological process and technical level. The sensible heat of the material should be used as much as possible to make it close to the ambient temperature when it leaves the system to avoid heat loss.

2.2 Main efficiency index of chemical process

2.2.1 Production capacity and intensity

1. Production capacity

It refers to the amount of products produced or raw materials processed per unit of time by a piece of equipment, a set of equipment or a plant. Its unit is kg/h, t /d or kt /a, etc.

2. Production intensity

It is the production capacity of unit characteristic geometric quantity of equipment. That is, the production capacity of equipment per unit volume, or per unit area. The unit is $kg/(h \cdot m^3)$, $t /(d \cdot m^3)$, or $kg/(h \cdot m^2)$, or $t /(d \cdot m^2)$, etc. The production intensity index is mainly used to compare the advantages and disadvantages of equipment or devices in the same reaction process or physical process. The higher process rate in the equipment, the higher the production intensity.

3. Effective production cycle

$$\text{Construction factor} = \frac{\text{number of production days annual}}{365} \qquad (2-1)$$

The construction factor is usually about 0. 9. A large construction factor means that the loss caused by maintenance shutdown is small, that is, the equipment is advanced and reliable, and the catalyst life is long.

2.2.2 Efficiency of chemical reaction-synthesis efficiency

1. Atom Economy

It was first proposed by professor B. M. Trost of Stanford University, which won the 1998 Presidential Green Chemistry Challenge Award.

Atomic Economy (AE) is defined as:

$$AE = \frac{\sum\limits_{i} P_i M_i}{\sum\limits_{j} F_j M_j} \times 100\% \qquad (2-2)$$

P_i—the number of various atoms in the molecule of the target product; F_j—the number of atoms in the reaction materials; M—the relative mass of the corresponding class of atoms.

2. Environmental factors

Sheldon, a Dutch chemist, defines environmental factor as:

$$E = \frac{\text{quality of waste}}{\text{target product quality}} \qquad (2-3)$$

In essence, the above indexes reflect whether the synthetic process maximizes the use of resources and avoids the generation of waste, as well as the resulting environmental pollution.

2.2.3 Conversion, selectivity and yield

Chemical reaction is the core of chemical process. Improving the conversion rate, selectivity and yield of reaction is the key to improve the efficiency of chemical process.

1. Conversion rate

Conversion rate is the fraction or percentage of the amount of a reactant participating in a reaction to the initial amount of the reactant, indicated by the symbol X. Its definition is

$$X = \frac{\text{the amount of conversion of a reactant}}{\text{the initial amount of the reactant}} \qquad (2-4)$$

2. Selectivity

For complex reaction systems, there are main reactions to produce the target products and the side reactions to produce the by-product, so conversion rate alone is not enough to measure chemical process. Selectivity refers to the ratio of the amount of a reactant converted to the target product in the system to the total amount of the reactant transformed to participate in all reactions.

$$S = \frac{\text{the amount of one of the reactants converted to the target product}}{\text{the total amount of transformation of the reactant}} \qquad (2-5)$$

3. Yield

The efficiency of the reaction is described in terms of products.

$$Y = \frac{\text{the amount of one of the reactants converted to the target product}}{\text{the initial amount of the reactant}} \qquad (2-6)$$

2.2.4　Balance conversion rate and yield

The conversion rate when the reversible reaction reaches equilibrium is called equilibrium conversion rate, and the yield of the product obtained at this time is called equilibrium yield. The equilibrium conversion and yield are the maximum values that can be achieved by a reversible reaction, but it often takes a long time to reach equilibrium. As the reaction progresses, the forward reaction rate decreases and the reverse reaction rate increases, so the net reaction rate decreases continuously to zero. In actual production, the net reaction rate should be kept high, and the reaction cannot wait for equilibrium, so the actual conversion rate and yield are lower than the equilibrium value. If the equilibrium yield is high, the actual yield can be high. One of the tasks of technology is to find the favorable conditions to improve the equilibrium yield and calculate the equilibrium yield through thermodynamic analysis.

2.3　Effects of chemical reaction conditions

Reaction temperature, pressure, concentration, reaction time, purity, ratio of raw materials and many other conditions are important factors affecting the reaction rate and balance, which related to the efficiency of the production process. In other chapters of this book, there are specific process analysis of influencing factors, so only the following several important factors are briefly described here.

2.3.1　Influence of temperature

1. The influence of temperature on chemical equilibrium

For irreversible reactions, chemical equilibrium is not considered, while for reversible reactions, the relationship between the equilibrium constant and temperature is

$$\log K = -\frac{\Delta H^{\ominus}}{2.303\,RT} + C \tag{2-7}$$

K—equilibrium constant; $\Delta H^{\ominus}$—standard reaction enthalpy difference; R—gas constant; T—reaction temperature; C—integral constant.

For the endothermic reaction, $\Delta H^{\ominus} > 0$, K value increases with the increase of temperature, which is conducive to the reaction, and the equilibrium yield of the product increases.

For the exothermic reaction, $\Delta H^{\ominus} < 0$, K value decreases with the increasing temperature, and the equilibrium yield decreases. Therefore, the equilibrium yield can only be increased by decreasing temperature.

2. The influence of temperature on reaction rate

Reaction rate refers to the amount of a reactant component consumed per unit time or volume, or the amount of a product produced.

For the irreversible reaction, the reverse reaction rate is ignored, so the rate of product formation always increases with the increase of temperature.

For reversible reactions, the difference between the rates of the forward and reverse reactions is the net rate of product formation. As the temperature rises, the rate constants of both the forward and reverse reactions increase, so the rates of both the forward and reverse reactions increase. Theoretically, the exothermic reversible reaction should be carried out at the optimum reaction temperature, when the net reaction rate is maximum.

2.3.2　Influence of concentration

According to the principle of equilibrium movement of reaction, the higher the concentration of reactants, the more conducive to the shift of equilibrium toward the product. When various reactants participate in the reaction, the cheap and easily available reactants are usually excessive, so that the expensive or rare reactants can be more converted into products to improve their utilization rate.

2.3.3　Influence of pressure

In general, pressure has little effect on the equilibrium of liquid phase and solid phase. The volume of gas is greatly affected by pressure, so pressure has a great influence on the reaction equilibrium with gas phase, and its law is as follows:

(1) The equilibrium yield can be improved by decreasing the pressure in the reactions with increasing molecular number.

(2) When the number of molecules decreased, the pressure increased and the equilibrium yield of the product increased.

(3) For the reaction with no change in the number of molecules, pressure has no effect on the equilibrium yield.

In a certain pressure range, inflating can reduce the gas reaction volume, and has certain benefits to speed up the reaction rate. But if the pressure is too high, the energy consumption increases, the equipment investment is high, which is not economical.

The presence of an inert gas reduces the partial pressure of the reactants, which is not good for the reaction rate, but is good for the equilibrium yield of the reaction with an increase in the number of molecules.

2.4　Catalyst

According to statistics, 90% of today's chemical reactions contain catalytic process. Catalyst occupies a very important position in the chemical process. Its role is mainly reflected in the following aspects.

(1) Improve the reaction rate and selectivity

Many reactions, although thermally possible, are too slow or too selective to be of practical value. Once catalysts are invented and used, they can be industrialized and produce important chemical products for mankind.

(2) Improve operating conditions

Adopting or improving catalyst can reduce reaction temperature and operating pressure and can improve the efficiency of chemical processing.

(3) Catalysts contribute to the development of new reaction processes and new chemical technology.

(4) Catalysts can play an important role in energy development and pollution elimination.

2.4.1　Basic characteristics of catalysts

In a reaction system, the rate of chemical reaction is obviously accelerated by the addition of a substance, but the quantity and chemical properties of the substance remain unchanged before and after the reaction, the substance is called catalyst. The function of the catalyst is that it can form unstable intermediate compounds with the reactants, change the reaction pathway and reduce the activation energy.

Catalysts have the following three basic characteristics.

(1) The catalyst is involved in the reaction, but at the end of the reaction, the catalyst itself does not change in chemical properties and quantity. So the catalyst can be used for a long time in the production process.

(2) The catalyst can only shorten the time to reach chemical equilibrium (i. e. accelerate the action), but cannot change the equilibrium. For equilibria-bound reaction systems, catalysts must be selected and used in conditions conducive to equilibrium movement towards products.

(3) Catalysts have obvious selectivity, a specific catalyst can only catalyze a specific reaction. This characteristic of catalyst plays a very important role in the field of organic chemical reaction, as the organic reaction system often exists many reactions at the same time, the selection of appropriate catalyst can make the reaction to the required direction.

2.4.2　Classification of catalysts

According to the phase homogeneity of the catalytic reaction system, there are homogeneous catalysts and heterogeneous catalysts.

According to the reaction category, there are many catalysts such as hydrogenation, dehydrogenation, oxidation, cracking, hydration, polymerization, alkylation, isomerization, aromatization, carbonylation, halogenation and so on.

According to the reaction mechanism, there are prototype oxidation catalyst, acid and base catalyst, etc.

Under the conditions of use, there are metal catalyst, oxide catalyst, sulfide catalyst, acid catalyst, alkali catalyst, coordination compound catalyst and biocatalyst, etc.

2.4.3　Related problems in the use of industrial catalysts

It is very important to select and use catalyst correctly in chemical production, which is related to production efficiency and benefit. The following properties of industrial catalysts are usually required.

1. Performance of industrial catalysts

(1) Activity

It refers to the ability of a catalyst to convert raw materials at a given temperature, pressure and reaction flow rate (or space velocity). The higher the activity, the higher the conversion rate. Or at the same conversion rate and other conditions, the higher the catalyst activity, the lower the reaction temperature is required. Industrial catalysts should be sufficiently active.

(2) Selectivity

It refers to how much of the raw material consumed by the reaction is converted to the target product. The higher the selectivity, the lower the raw material consumption quota of the product per unit amount, and the more conducive to the post-treatment of the product, so the selectivity of industrial catalyst should be higher. When the activity and selectivity of catalyst are difficult to strike a balance, if the raw material is expensive or the product separation is difficult, it is appropriate to choose the catalyst with high selectivity; if the raw material is cheap and easy to obtain or the product is easy to separate, the catalyst with high activity can be selected.

(3) Life

It refers to the length of a catalyst's service life, which is characterized by the amount of catalyst consumed per unit of product, or the length of time the catalyst can be used at the technical level to meet the production requirements. Some catalysts can be used for several years, and some can only be used for several months.

2. Activation of catalyst

Many solid catalysts are generally sold in a stable state, but this stable state does not have catalytic performance. The catalyst must be activated before the reaction, so that it can be transformed into an active state. Different types of catalysts should be activated by different methods, such as reduction, oxidation, vulcanization, acidification, heat treatment, etc. Each activation method has its own conditions and operating requirements. It should be activated strictly in accordance with the operating procedures to ensure that the catalyst plays a good role. If the activation operation is wrong, the performance of the catalyst will decline, or the catalyst will be scrapped, resulting in economic losses.

3. Deactivation and regeneration of catalyst

There are many reasons for catalyst deactivation. For coordination catalyst, the main reason is overtemperature, most coordination compounds decompose and deactivate above 250℃. For biocatalysts, overheating, contamination of chemical substances and bacteria, and pH imbalance are the causes of inactivation. For solid catalysts, the main reasons for inactivation are:

(1) Over temperature makes the catalyst surface sintering, crystal transformation or phase transformation.

(2) Mixed with toxic impurities in raw material gas, which makes the catalyst poisoning.

(3) If there is dirt covering the surface of the catalyst, it will cause inactivation. The dirt may be brought from raw materials, or mechanical impurity in the equipment such as oil, dirt, rust. Hydrocarbon or other carbon compounds for reaction tends to precipitate carbon. When the catalyst is too acidic or low active, precipitating carbon is severer, carbon deposition or coking happen which cover the catalyst activity center and lead the catalyst deactivation.

There are two kinds of catalyst poisoning: temporary and permanent. The temporary poisoning is reversible. When the poison is removed from the raw material, the catalyst can gradually recover its activity, while the permanent poisoning is irreversible. Catalyst carbon deposition can be regenerated by burning carbon. However, either regeneration after temporary poisoning or regeneration after carbon deposition usually results in different degree of damage to the catalyst structure, resulting in decreased activity.

Therefore, the operating conditions should be strictly controlled, and the reactor with reasonable structure should be used to make the reaction temperature reasonably distributed within the optimum operating temperature range of catalyst to prevent over temperature. The toxic impurities in the reaction material should be removed in advance to make the toxic content below the catalyst tolerance value. In the system with carbon evolution reaction, the reaction conditions conducive to preventing carbon evolution should be adopted, and the catalyst with high resistance to carbon accumulation should be selected.

4. Transport, storage, loading and unloading of catalysts

Catalyst is generally expensive, protection should be paid attention. It should be prevented from being contaminated and damaged during transportation and storage. Solid catalysts must be protected from contamination and rupture when loaded into the reactor. The loading should be uniform to avoid the phenomenon of "bridging", so as to prevent the deterioration of reaction conditions. Many catalysts need to be passivated before they are unloaded after shutdown, especially the metal catalyst must be passivated by hypoxic gas before being exposed to air, otherwise it will be violently oxidized and spontaneous combustion, burning out the catalyst and equipment.

2.5 Material balance and heat balance of the reaction process

Material balance and heat balance is one of the basis of chemical process. Through material and heat balance, we can calculate the production process of raw material consumption index, heat load, product yield and so on to provide quantitative basis including size, type and number for the design and selection of reactor and other equipment. It can check whether the amount of materials and related data in the production process is normal, whether there is leakage, as well as heat recovery, utilization level and the size of heat loss, so as to find out the weak links and bottlenecks in the production, and provide the basis for improving the operation and the optimization of the system. We have learned material and heat balance calculation of chemical unit operation process other than reaction process in the course of chemical principle, so this section only deals with material and heat balance calculation of reaction process.

2.5.1 Material balance in the reaction process

1. Basic equation of material balance

Material accounting is always carried out around a specific range, which can be called the accounting system. A balance system can be a head plant, a branch plant or workshop, a set of devices, a piece of equipment, or even a node, etc. The theoretical basis of material balance is the law of mass conservation, according to which the general formula of material balance system is written as follow.

total mass of input material $=$ total mass of output material $+$

mass of accumulated material in the system

$$(2-8)$$

2. Material balance during intermittent operation

Intermittent operation is batch production, that is, one feed to the reactor, after the reaction is completed, one discharge, and then the second batch of production. Its characteristic is concentration and other parameters change with time in the reaction process. Batch feeding and batch discharging are also intermittent operations.

3. Material balance in steady flow process

In the production, the vast majority of chemical processes is continuous operation, equipment or devices can run for a long time. In addition to the start and stop phase, in the vast majority of time the production is in a stable state of the flow process, materials continue to flow into and out of the system. Its characteristic is that the parameters of each point in the system, such as temperature, pressure, concentration and flow rate, do not change with time, and there is no accumulation in the system. Of course, the parameters of different points or sections within the equipment may be the same or different. The material balance of the steady flow process is

total mass of material input system $=$ total mass of material output system

$$(2-9)$$

4. Material balance procedure

Many processes of chemical production are quite complicated, so when doing material balance, we should follow certain steps to give clear calculation process and correct results. The following steps are usually followed.

Step 1: Draw a block diagram of the process in order to select a balance system. Graphical expression should be simple, but the content of the representative should be accurate, there can be no omission of materials, otherwise the balance will be wrong.

Step 2: Write the chemical reaction equation and balance it. If the reaction is too complicated, or if the reaction is too vague to write down the equation, then you're going to use the atomic balance algorithm instead of writing down the equation.

Step 3: Select the balance base. The balance basis is the initial physical quantity selected for material balance, including the name, quantity and unit of the material. Other material quantities obtained from the balance result are relative to this basis.

Step 4: Collect or calculate the necessary data, pay attention to the scope of application and conditions of the data.

Step 5: Set up the unknown, set up the equation group, solve them simultaneously. If there are several unknowns, several independent equations should be listed. In addition to material balance formula, these equations also need other relations, such as composition relation constraint, chemical balance constraint, phase balance constraint, proportion of material quantity, etc.

Step 6: Calculate and check.

Step 7: Report the results. Generally, the known and calculated results are listed in the material balance sheet, which can be in different forms but should comprehensively all the materials and components imported and exported.

2.5.2　Heat balance of reaction process

According to the law of energy conservation, the energy balance in and out of the system is

$$\text{input energy} = \text{output energy} + \text{accumulated energy} \qquad (2-10)$$

The following points should be paid attention to when doing heat balance calculation:

(1) Determine the balance object.

(2) Select material balance basis.

(3) Determine the temperature reference.

(4) Pay attention to the phase states of substances.

Questions

2-1　What is chemical production process flow? Give an example of how the process is organized.

2-2　What is cyclic process flow? What are its advantages and disadvantages?

2-3　What is conversion rate? What is selectivity? For a multi-reaction system, why consider both conversion rate and selectivity?

2-4　What are the basic characteristics of catalysts? What role does it play in chemical production? How to use catalyst correctly in production?

第3章 烃类热裂解

乙烯、丙烯和丁二烯等低级烯烃分子中具有双键,化学性质活泼,能与许多物质发生加成、共聚或自聚等反应,生成一系列重要的产物,是化学工业的重要原料。工业上获得低级烯烃的主要方法是将烃类热裂解。烃类热裂解法是将石油系烃类燃料(天然气、炼厂气、轻油、柴油、重油等)经高温作用,使烃类分子发生碳链断裂或脱氢反应,生成相对分子质量较小的烯烃、烷烃和其他相对分子质量不同的轻质和重质烃类。

在低级不饱和烃中,乙烯最重要,产量也最大。乙烯产量常用来衡量一个国家基本化学工业的发展水平。

烃类热裂解制乙烯的生产工艺主要由两部分组成:原料烃的热裂解和裂解产物的分离。

3.1 热裂解过程基本反应

3.1.1 烃类裂解规律

1. 烷烃裂解反应

(1) 正构烷烃

正构烷烃的裂解反应主要有脱氢反应和断链反应,对于 C_5 以上的烷烃还可能发生环化脱氢反应。

脱氢反应是 C—H 键断裂的反应,生成碳原子数相同的烯烃和氢,其通式为

$$C_nH_{2n+2} \rightleftharpoons C_nH_{2n} + H_2 \qquad (3-1)$$

C_5 以上的正构烷烃可发生环化脱氢反应生成环烷烃,如正己烷脱氢生成环己烷。

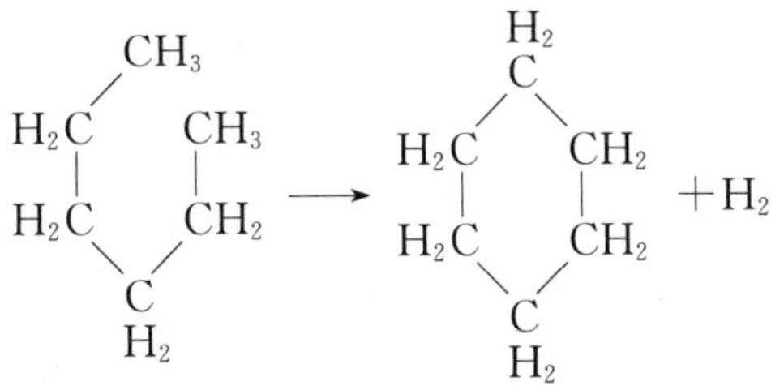

$$(3-2)$$

断链反应是 C—C 键断链的反应,反应产物是碳原子数较少的烷烃和烯烃,其通式为

$$C_nH_{2n+2} \longrightarrow C_mH_{2m} + C_kH_{2k+2} \qquad m+k=n \qquad (3-3)$$

相同烷烃脱氢和断链的难易,可以从分子结构中碳氢键和碳碳键的键能数值的大

小来判断。

(2) 异构烷烃裂解反应

异构烷烃结构各异,其裂解反应差异较大,与正构烷烃相比有如下特点。

① C—C 键或 C—H 键的键能较正构烷烃的低,故容易裂解或脱氢。

② 脱氢能力与分子结构有关,难易顺序为叔碳氢＞仲碳氢＞伯碳氢。

③ 异构烷烃裂解所得乙烯、丙烯收率远较正构烷烃裂解所得收率低,而氢、甲烷、C_4 及 C_4 以上烯烃收率较高。

④ 随着碳原子数的增加,异构烷烃与正构烷烃裂解所得乙烯和丙烯收率的差异减小。

2. 烯烃裂解反应

由于烯烃的化学活泼性,自然界石油系原料中,基本不含烯烃,但在炼厂气中和二次加工油品中含一定量烯烃。作为裂解过程中的目的产物,烯烃也有可能进一步发生反应,所以为了控制反应按人们所需的方向进行,必须了解烯烃在裂解过程中的反应规律。烯烃可能发生的主要反应有以下几种。

(1) 断链反应

较大分子的烯烃裂解可断链生成两个较小的烯烃分子,其通式为

$$C_{n+m}H_{2(n+m)} \longrightarrow C_nH_{2n} + C_mH_{2m} \tag{3-4}$$

例如:

$$H_2C=CH-CH_2-CH_2-CH_3 \longrightarrow H_2C=CH-CH_3 + H_2C=CH_2 \tag{3-5}$$

(2) 脱氢反应

烯烃可进一步脱氢生成二烯烃和炔烃。

$$C_4H_8 \longrightarrow C_4H_6 + H_2 \tag{3-6}$$

$$C_2H_4 \longrightarrow C_2H_2 + H_2 \tag{3-7}$$

(3) 歧化反应

两个同一分子烯烃可歧化为两个不同烃分子。

$$2C_3H_6 \longrightarrow C_2H_4 + C_4H_8 \tag{3-8}$$

$$2C_3H_6 \longrightarrow C_2H_6 + C_4H_6 \tag{3-9}$$

$$2C_3H_6 \longrightarrow C_5H_8 + CH_4 \tag{3-10}$$

(4) 双烯合成反应(Diels-Alder 反应)

二烯烃与烯烃进行双烯合成而生成环烯烃,进一步脱氢生成芳烃。通式为

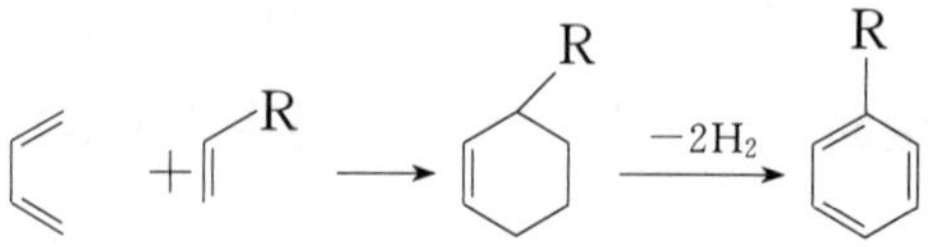

$$\left\langle\!\!\!\right| + \|^{R} \longrightarrow \bigcirc^{R} \xrightarrow{-2H_2} \bigcirc^{R} \tag{3-11}$$

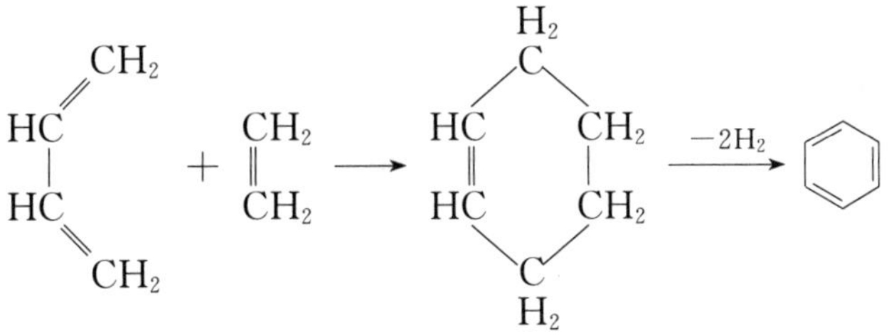

$$\text{(3-12)}$$

（5）芳构化反应

六个或更多碳原子数的烯烃,可以发生芳构化反应生成芳烃,通式如下:

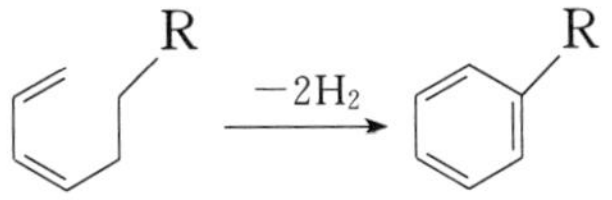

$$\text{(3-13)}$$

3. 环烷烃裂解反应

环烷烃较相应的链烷烃稳定,在一般裂解条件下可发生断链开环反应、脱氢反应、侧链断裂及开环脱氢反应,由此生成乙烯、丙烯、丁二烯、丁烯、芳烃、环烷烃、单环烯烃、单环二烯烃和氢气等产物。

例如环己烷:

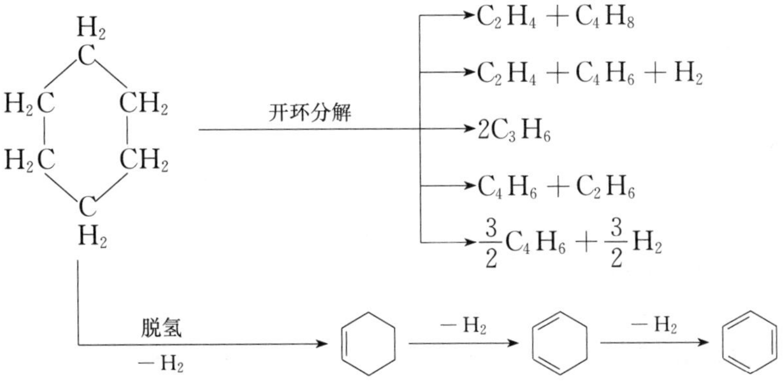

$$\begin{aligned}
&\rightarrow C_2H_4 + C_4H_8\\
&\rightarrow C_2H_4 + C_4H_6 + H_2\\
&\rightarrow 2C_3H_6\\
&\rightarrow C_4H_6 + C_2H_6\\
&\rightarrow \frac{3}{2}C_4H_6 + \frac{3}{2}H_2
\end{aligned}$$

开环分解

脱氢 $-H_2$ $\quad -H_2 \quad -H_2$

$$\text{(3-14)}$$

再如乙基环戊烷:

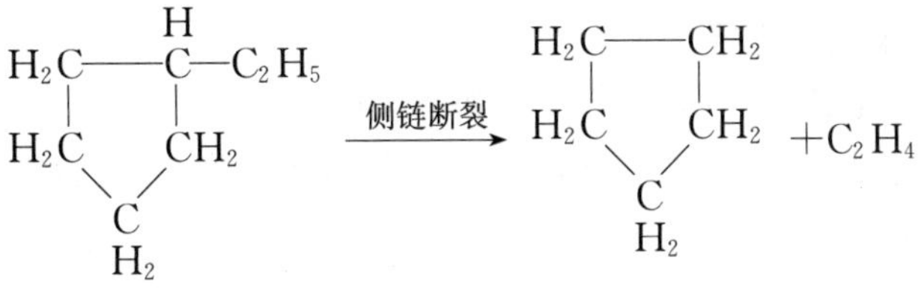

侧链断裂 $+C_2H_4$

$$\text{(3-15)}$$

4. 芳烃裂解反应

芳烃由于芳环的稳定性,不易发生裂开芳环的反应,而主要发生烷基芳烃的侧链断裂和脱氢反应,以及芳烃缩合生成多环芳烃,进一步成焦的反应。所以,含芳烃多的原料油不仅烯烃收率低,而且结焦严重,不是理想的裂解原料。

（1）烷基芳烃裂解

侧链脱烷基或断键反应：

$$Ar\text{-}C_nH_{2n+1} \begin{cases} \longrightarrow ArH + C_nH_{2n} \\ \longrightarrow Ar\text{-}C_kH_{2k+1} + C_mH_{2m} (m+k=n) \end{cases} \quad (3-16)$$

$$Ar\text{-}C_nH_{2n+1} \longrightarrow Ar\text{-}C_nH_{2n-1} + H_2 \quad (3-17)$$

（2）环烷基芳烃裂解

脱氢和异构脱氢反应：

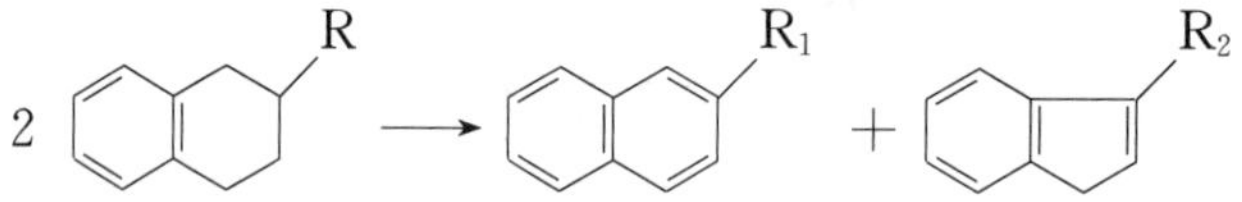

$$ (3-18) $$

缩合脱氢反应：

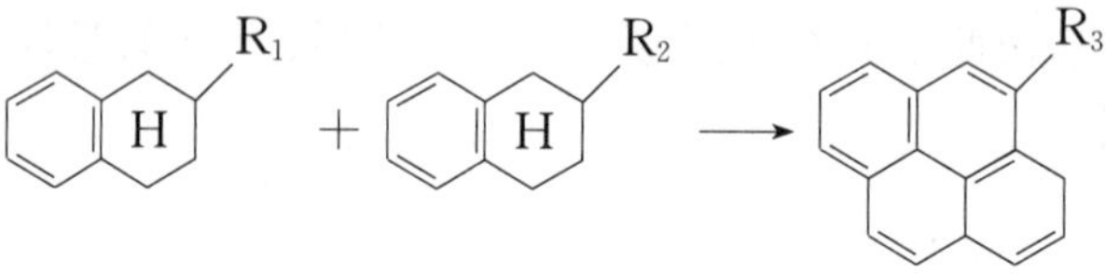

$$ (3-19) $$

（3）芳烃缩合反应

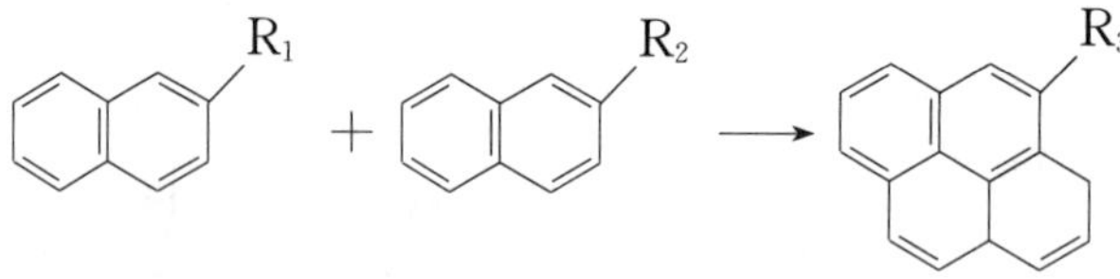

$$ (3-20) $$

5. 裂解过程中结焦生炭反应

（1）烯烃经过炔烃中间阶段而生碳

裂解过程中生成的乙烯在 900℃～1 000℃或更高的温度下经过乙炔阶段而生碳。

（2）经过芳烃中间阶段而结焦

高沸点稠环芳烃是馏分油裂解结焦的主要母体,裂解焦油中含大量稠环芳烃,裂解生成的焦油越多,裂解过程中结焦越严重。

6. 各族烃裂解反应规律

各族烃裂解生成乙烯、丙烯的能力有如下规律。

（1）烷烃:正构烷烃在各族烃中最利于乙烯、丙烯的生成。烷烃的相对分子质量愈小,其总产率愈高。异构烷烃的烯烃总产率低于同碳原子数的正构烷烃,但随着相对分子质量的增大,这种差别减小。

（2）烯烃:大分子烯烃裂解为乙烯和丙烯;烯烃能脱氢生成炔烃、二烯烃,进而生成芳烃。

（3）环烷烃:在通常裂解条件下,环烷烃生成芳烃的反应优于生成单烯烃的反应。相对于正烷烃来说,含环烷烃较多的原料丁二烯、芳烃的收率较高,而乙烯的收率较低。

（4）芳烃:无烷基的芳烃基本上不裂解为烯烃;有烷基的芳烃,主要是烷基发生断碳键和脱氢反应,而芳环保持不变,易脱氢缩合为多环芳烃,从而有结焦的倾向。

各族烃的裂解难易程度有下列顺序:

$$正烷烃>异烷烃>环烷烃(六碳环>五碳环)>芳烃$$

随着分子中碳原子数的增多,各族烃分子结构上的差别反映到裂解速度上的差异就逐渐减弱。

3.1.2　烃类裂解机理

烃类裂解反应机理研究表明,裂解时发生的基元反应大部分为自由基反应。

1. F. O. Rice 的自由基反应机理

大部分烃类裂解过程包括链引发反应、链增长反应和链终止反应三个阶段。链引发反应是自由基的产生过程;链增长反应是自由基的转变过程,在这个过程中一种自由基的消失伴随着另一种自由基的产生,反应前后均保持着自由基的存在;链终止是自由基消亡生成分子的过程。

链的引发是在热的作用下,一个分子断裂产生一对自由基。每个分子由于键的断裂位置不同可有多个可能发生的链引发反应,这取决于断裂处相关键的解离能大小,解离能小的反应更易于发生。

烷烃分子在引发反应中断裂 C—H 键的可能性较小,因为 C—H 键的解离能比 C—C 键大。故引发反应的通式为

$$R—R' \longrightarrow R \cdot + R' \tag{3-21}$$

链的增长反应包括自由基夺氢反应、自由基分解反应、自由基加成反应和自由基异构化反应,但以前两种为主。链增长反应的夺氢反应通式为

$$H \cdot + RH \longrightarrow H_2 + R \cdot \tag{3-22}$$

$$R' \cdot + RH \longrightarrow R'H + R \cdot \tag{3-23}$$

链增长反应中的自由基的分解反应是自由基自身进行分解,生成一个烯烃分子和一个碳原子数比原来要少的新自由基,而使其自由基传递下去。

这类反应的通式为

$$R \cdot \longrightarrow R' \cdot + 烯烃 \tag{3-24}$$

$$R \cdot \longrightarrow H \cdot + 烯烃 \tag{3-25}$$

自由基分解反应是生成烯烃的反应,而裂解的目的是生产烯烃,所以这类反应是很关键的。

2. 一次反应和二次反应

原料烃在裂解过程中所发生的反应是复杂的,一种烃可以平行地发生很多种反应,又可以连串地发生许多后继反应,所以裂解系统是一个平行反应和连串反应交叉的反应系统。从整个反应进程来看,裂解过程属于比较典型的连串反应。

随着反应的进行,不断分解出气态烃(小分子烷烃、烯烃)和氢;而液体产物的氢含量则逐渐下降,相对分子质量逐渐增大,以致结焦。

一次反应是指原料烃在裂解过程中首先发生的原料烃的裂解反应，二次反应则是指一次反应产物继续发生的后继反应。从裂解反应的实际反应历程看，一次反应和二次反应并没有严格的分界线，不同研究者对一次反应和二次反应的划分也不尽相同。

生成目的产物乙烯、丙烯的反应属于一次反应，这是希望发生的反应，在确定工艺条件、设计和生产操作中要设法促使一次反应充分进行。

乙烯、丙烯消失，生成相对分子质量较大的液体产物以至结焦生炭的反应是二次反应，是不希望发生的反应。这类反应的发生，不仅多消耗了原料，降低了主产物的产率，而且结焦生炭会恶化传热、堵塞设备，对裂解操作和稳定生产都带来极不利的影响，所以要设法抑制其进行。

3.1.3 裂解原料性质及评价

由于烃类裂解反应使用的原料是组成性质有很大差异的混合物，因此原料的特性无疑对裂解效果起着重要的决定作用，它是决定反应效果的内因，而工艺条件的调整、优化仅是其外部条件。

1. 族组成

裂解原料油中的各种烃按其结构可以分为四大族，即链烷烃族、烯烃族、环烷烃族和芳香族。这四大族的族组成以 PONA 值来表示，其含义为：P——烷烃（Paraffin）；N——环烷烃（Naphtene）；O——烯烃（Olefin）；A——芳烃（Aromatics）。

根据 PONA 值可以定性评价液体燃料的裂解性能，也可以根据族组成通过简化的反应动力学模型对裂解反应进行定量描述，因此 PONA 值是一个表征各种液体原料裂解性能的有实用价值的参数。

2. 氢含量

氢含量可以用裂解原料中所含氢的质量分数 $\omega(H_2)$ 表示，也可以用裂解原料中 C 与 H 的质量比（称为碳氢比）表示。

$$氢含量 \quad \omega(H_2) = \frac{n_H}{12n_C + n_H} \times 100 \quad\quad (3-26)$$

$$碳氢比 \quad \frac{m_C}{m_H} = \frac{12n_C}{n_H} \quad\quad (3-27)$$

式中，n_H、n_C 分别为原料烃中氢原子数和碳原子数。

氢含量顺序为烷烃＞环烷烃＞芳烃。

通过裂解反应，一定氢含量的裂解原料可以生成氢含量较高的 C_4 和 C_4 以下轻组分，以及氢含量较低的 C_5 和 C_5 以上的液体。从氢平衡可以断定，裂解原料氢含量愈高，获得的 C_4 和 C_4 以下轻烃的收率愈高，相应乙烯和丙烯收率一般也较高。显然，根据裂解原料的氢含量既可判断该原料可能达到的裂解深度，也可评价该原料裂解所得 C_4 和 C_4 以下轻烃的收率。

3.2 裂解过程参数和指标

3.2.1 裂解原料

烃类裂解反应使用的原料对裂解工艺过程及裂解产物起着决定性作用。裂解原料氢含量越高，获得 C_4 以下烯烃的收率越高，因此，烷烃尤其是低碳烷烃是首选的原料。

国外烃类裂解原料以轻烃（C_4 以下）和石脑油为主，占 90% 左右。而国内重质油、柴油的比例还高达 20% 以上，有待于进一步优化。

3.2.2 裂解温度和停留时间

1. 裂解温度

从自由基反应机理分析，在一定温度内，提高裂解温度有利于提高一次反应所得乙烯和丙烯的收率。

根据裂解反应动力学，使裂解反应控制在一定裂解深度范围内，就是使转化率控制在一定范围内。不同裂解原料的反应速率常数大不相同，因此，在相同停留时间的条件下，不同裂解原料所需裂解温度也不相同。裂解原料相对分子质量越小，其活化能和频率因子越高，反应活性越低，所需裂解温度越高。

在控制一定裂解深度的条件下，可以有各种不同的裂解温度-停留时间组合。因此，对于生产烯烃的裂解反应而言，裂解温度与停留时间是一组相互关联、不可分割的参数。而高温-短停留时间则是改善裂解反应产品收率的关键。

2. 停留时间

管式裂解炉中物料的停留时间是裂解原料经过辐射盘管的时间。由于裂解管中裂解反应是在非等温变容的条件下进行，很难计算其真实停留时间。

3.2.3 烃分压与稀释剂

1. 压力对裂解反应的影响

烃裂解的一次反应是分子数增多的过程。对于脱氢可逆反应，降低压力对提高乙烯平衡组成有利（断链反应因是不可逆反应，压力对其无影响）。烃聚合的二次反应是分子数减少的过程，降低压力对提高二次反应产物的平衡组成不利，可抑制结焦过程。

所以，降低压力可以促进生成乙烯的一次反应，抑制发生聚合的二次反应，从而减轻结焦的程度。

2. 稀释剂

裂解是在高温下操作的，不宜用抽真空减压的方法降低烃分压，这是因为高温密封困难，一旦空气漏入负压操作的裂解系统，与烃气体形成爆炸混合物就有爆炸的危险，而且减压操作对以后分离工序的压缩操作也不利，会增加能量消耗。所以，添加稀释剂以降低烃分压是一个较好的方法。这样，设备仍可在常压或正压操作，而烃分压则可降低。理论上可采用水蒸气、氢或任一种惰性气体作稀释剂，但目前较为成熟的裂解方法均采用水蒸气作稀释剂。

3.2.4 裂解深度

裂解深度是指裂解反应的进行程度。由于裂解反应的复杂性,很难以一个参数准确地对其进行定量的描述。根据不同情况,常常采用如下参数衡量裂解深度。

(1)原料转化率

原料转化率 X 反映了裂解反应时裂解原料的转化程度。因此,常用原料转化率衡量裂解深度。

(2)甲烷收率

裂解所得甲烷收率 $y(C_1)$ 随着裂解深度的提高而增加。由于甲烷比较稳定,基本上不因二次反应而消失,因此,裂解产品中甲烷收率可以在一定程度上衡量反应的深度。

(3)乙烯对丙烯收率比

在一定裂解深度范围内,随着裂解深度的增大,乙烯收率增高,而丙烯收率增加缓慢。到一定裂解深度后,乙烯收率尚进一步随裂解深度增加而上升,丙烯收率将由最高值开始下降。

(4)甲烷对乙烯或丙烯收率比

由于甲烷收率随反应进程的加深总是增大的,而乙烯或丙烯收率随裂解深度的增加在达到最高值后开始下降。

3.3 裂解气预分馏及净化

3.3.1 裂解气预分馏的目的与任务

裂解炉出口的高温裂解气经急冷换热器的冷却,再经急冷器进一步冷却后,温度可以降到 200℃～300℃。将急冷后的裂解气进一步冷却至常温,并在冷却过程中分馏出裂解气中的重组分(如燃料油、裂解汽油、水分),这个环节称为裂解气的预分馏。经预分馏处理的裂解气再送至裂解气压缩系统并进一步进行深冷分离。显然,裂解气的预分馏过程在乙烯装置中起着十分重要的作用。

3.3.2 预分馏工艺过程概述

1. 轻烃裂解装置裂解气的预分馏过程

轻烃裂解装置所得裂解气的重质馏分甚少,尤其乙烷和丙烷裂解时,裂解气中的燃料油含量甚微。此时,裂解气的预分馏过程主要是在裂解气进一步冷却过程中分馏水分和裂解汽油馏分。

轻烃裂解装置裂解气预分馏流程示意图见图 3-1。

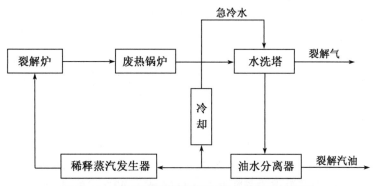

图 3 - 1 轻烃裂解装置裂解气预分馏流程示意图

2. 馏分油裂解装置裂解气预分馏过程

馏分油裂解装置所得裂解气中含相当量的重质馏分,这些重质燃料油馏分与水混合后会因乳化而难于进行油水分离。因此,在馏分油裂解装置中,必须在冷却裂解气的过程中先将裂解气中的重质燃料油馏分分馏出来,分馏重质燃料油馏分之后的裂解气再进一步送至水洗塔冷却,并分馏其中的水和裂解汽油。

馏分油裂解装置裂解气预分馏过程示意图见图 3 - 2。

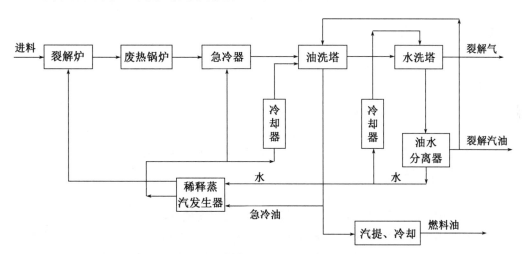

图 3 - 2 馏分油裂解装置裂解气预分馏过程示意图

3.3.3 裂解汽油与裂解燃料油

1. 裂解汽油

烃类裂解副产的裂解汽油包括 C_5 至沸点 204℃ 以下的所有裂解副产物。作为乙烯装置的副产品,其典型规格通常如下:

C_4 馏分　　　　　　0.5%(最大质量分数)

终馏点　　　　　　204℃

裂解汽油经一段加氢可作为高辛烷值汽油组分,如需经芳烃抽提分离芳烃产品,则应进行两段加氢,脱出其中的硫、氮,并使烯烃全部饱和。

可以将裂解汽油全部进行加氢,加氢后分为加氢 C_5 馏分、$C_6 \sim C_8$ 中心馏分、$C_9 \sim$ 204℃馏分。此时,加氢 C_5 馏分可返回循环裂解,而 $C_6 \sim C_8$ 中心馏分则是芳烃抽提的原料,C_9 馏分可作为歧化生产芳烃的原料。也可以将裂解汽油先分为 C_5 馏分、C_9 馏分、$C_6 \sim C_8$ 中心馏分,然后仅对 $C_6 \sim C_8$ 中心馏分进行加氢处理,由此,可使加氢处理量减少。

2. 裂解燃料油

烃类裂解副产的裂解燃料油是指沸点在 200℃ 以上的重组分。其中沸程在 200℃ ~ 360℃ 的馏分称为裂解轻质燃料油,相当于柴油馏分,其中,烷基萘含量较高,可作为脱烷基制萘的原料。沸程在 360℃ 以上的馏分称为裂解重质燃料油,相当于常压重油馏分。除作燃料外,由于裂解重质燃料油的灰分低,是生产炭黑的良好原料。

轻烃和轻质油裂解时,裂解燃料油较少,通常不再对轻质燃料油和重质燃料油进行分离。一般在柴油裂解时,需分出轻质燃料油,并以其作为裂解炉燃料,以此平衡柴油裂解厂气体燃料的不足。

裂解燃料油需要控制的规格主要是油品的闪点。通常,轻质裂解燃料油的闪点应控制在 70℃ 以上,重质裂解燃料油的闪点应控制在 100℃ 以上。裂解燃料油的硫含量取决于裂解原料的硫含量。石脑油裂解时,裂解原料总硫的 10% ~ 20% 聚积于裂解燃料油。柴油裂解时,则有 50% ~ 60% 的硫富集于裂解燃料油。

3.3.4 裂解气净化

裂解气中含 H_2S、CO_2、H_2O、C_2H_2、CO 等气体杂质,主要来源有三方面:一是原料中带来;二是裂解反应过程中生成;三是裂解气处理过程中引入。

这些杂质的含量虽不大,但对深冷分离过程是有害的。而且这些杂质若不脱除,进入乙烯、丙烯产品,会使产品达不到规定的标准,尤其是生产聚合级乙烯、丙烯,其杂质含量的控制是很严格的,为了达到产品所要求的规格,必须脱除这些杂质,对裂解气进行净化。

1. 酸性气体脱除

(1) 酸性气体杂质的来源和危害

裂解气中的酸性气体主要是 CO_2、H_2S 和其他气态硫化物。

裂解气中含有的酸性气体对裂解气分离装置以及乙烯和丙烯衍生物加工装置都会有很大危害。对裂解气分离装置而言,CO_2 会在低温下结成干冰,造成深冷分离系统设备和管道堵塞,H_2S 将造成加氢脱炔催化剂和甲烷化催化剂中毒。对于下游加工装置而言,当氢气、乙烯、丙烯产品中的酸性气体含量不合格时,可使下游加工装置的聚合过程或催化反应过程的催化剂中毒,也可能严重影响产品质量。因此,在裂解气精馏分离之前,需将裂解气中的酸性气体脱除干净。

(2) 碱洗法脱除酸性气体

碱洗法是用 NaOH 为吸收剂,通过化学吸收使 NaOH 与裂解气中的酸性气体发生化学反应,以达到脱除酸性气体的目的。

(3) 乙醇胺法脱除酸性气体

用乙醇胺作吸收剂除去裂解气中的 CO_2 和 H_2S,是一种物理吸收和化学吸收相结合的方法,所用的吸收剂主要是一乙醇胺(MEA)和二乙醇胺(DEA)。

（4）醇胺法与碱洗法比较

醇胺法与碱洗法相比,其主要优点是吸收剂可再生、循环使用。当酸性气含量较高时,从吸收液的消耗和废水处理量来看,醇胺法明显优于碱洗法。

2. 脱水

裂解气经预分馏处理后进入裂解气压缩机,压缩机入口裂解气中的水分为入口温度和压力条件下的饱和水含量。在裂解气压缩过程中,随着压力的升高,可在段间冷凝过程中分离出部分水分。

裂解气中的水含量不高,但要求脱水后物料的干燥度很高,因而,均采用吸附法进行干燥。

3. 炔烃脱除

（1）炔烃来源、危害及处理方法

在裂解气分离过程中,裂解气中的乙炔将富集于 C_2 馏分中,甲基乙炔和丙二烯(简称 MAPD)将富集于 C_3 馏分。通常 C_2 馏分中乙炔的摩尔分数为 $0.3\% \sim 1.2\%$,MAPD 富集于 C_3 馏分的摩尔分数为 $1\% \sim 5\%$。在 Kellogg 毫秒炉高温超短停留时间的裂解条件下,C_2 馏分中富集的乙炔摩尔分数可高达 $2.0\% \sim 2.5\%$,C_3 馏分中 MAPD 的摩尔分数可达 $5\% \sim 7\%$。

乙烯生产中常用的脱除乙炔的方法是溶剂吸收法和催化加氢法。溶剂吸收法是使用溶剂吸收裂解气中的乙炔以达到净化目的,该法同时也能回收一定量的乙炔。催化加氢法是将裂解气中乙炔加氢成为乙烯或乙烷,由此达到脱除乙炔的目的。溶剂吸收法和催化加氢法各有优缺点。目前,在不需要回收乙炔时,一般采用催化加氢法;在需要回收乙炔时,则采用溶剂吸收法。实际生产中,建有回收乙炔的溶剂吸收系统的工厂,往往同时设有催化加氢脱炔系统。两个系统并联,以具有一定的灵活性。

（2）催化加氢脱炔

① 炔烃的催化加氢

裂解气中的乙炔进行选择催化加氢时有如下反应发生。

正反应

$$C_2H_2 + H_2 \xrightarrow{k_1} C_2H_4 + \Delta H_1 \tag{3-28}$$

副反应

$$C_2H_2 + 2H_2 \xrightarrow{k_2} C_2H_6 + \Delta H_2 \tag{3-29}$$

$$C_2H_4 + H_2 \longrightarrow C_2H_6 + (\Delta H_2 - \Delta H_1) \tag{3-30}$$

$$mC_2H_2 + nC_2H_4 \longrightarrow 低聚物(绿油) \tag{3-31}$$

当反应温度升高到一定程度时,还可能发生生成 C、H_2 和 CH_4 的裂解反应。

② 前加氢和后加氢

前加氢过程是指在裂解气中氢气未分离出来之前,利用裂解气中的氢对炔烃进行选择性加氢,以脱除其中炔烃。所以,又称为自给氢催化加氢过程。

后加氢过程是指裂解气分离出 C_2 馏分和 C_3 馏分后,再分别对 C_2 和 C_3 馏分进行催化加氢,以脱除乙炔、甲基乙炔和丙二烯。

③ 加氢工艺流程

以后加氢过程为例,进料中乙炔的摩尔分数高于 0.7%,一般采用多段绝热床或等温反应器。

两段绝热床加氢工艺流程见图 3-3。

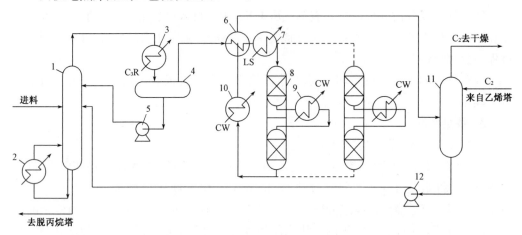

1—脱乙烷塔;2—再沸器;3—冷凝器;4—回流罐;5—回流泵;6—换热器;7—加热器;
8—加氢反应器;9—段间冷却器;10—冷却器;11—绿油吸收塔;12—绿油泵。

图 3-3　两段绝热床加氢工艺流程

(3) 溶剂吸收法脱除乙炔

溶剂吸收法是使用选择性溶剂将 C₂ 馏分中的少量乙炔选择性地吸收到溶剂中,从而实现脱除乙炔的方法。由于使用选择性吸收乙炔的溶剂,可以在一定条件下再把乙炔解吸出来,因此,溶剂吸收法脱除乙炔的同时,可回收到高纯度的乙炔。

DMF 溶剂吸收法脱乙炔工艺流程(Lummus)见图 3-4。

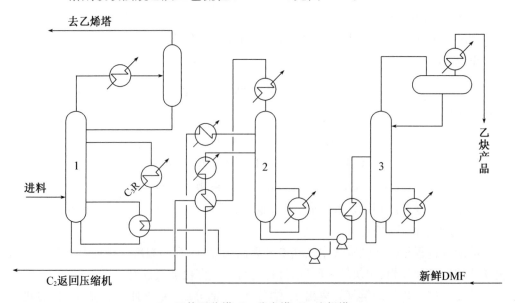

1—乙炔吸收塔;2—稳定塔;3—汽提塔。

图 3-4　DMF 溶剂吸收法脱乙炔工艺流程(Lummus)

3.4　压缩和制冷系统

3.4.1　裂解气压缩

裂解气中许多组分在常压下都是气体,其沸点很低。若在常压下进行各组分精馏分离,则分离温度很低,需要大量冷量。为了使分离温度不太低,可适当提高分离压力,裂解气分离中温度最低处是甲烷和氢气的分离,即甲烷塔塔顶。

3.4.2　裂解装置中的制冷系统

深冷分离过程需要制冷剂。制冷是利用制冷剂压缩和冷凝得到制冷剂液体,再于不同压力下蒸发,以获得不同温度级位的冷冻过程。

3.5　裂解气精馏分离系统

3.5.1　分离流程组织

在多组分系统的精馏分离中,能否合理地组织好流程,对于建设投资、能量消耗、操作费用、运转周期、产品的产量和质量、生产安全都有极大影响。

裂解气通过预分离被降至常温,并分馏出燃料油馏分、大部分水分、部分裂解汽油。

3.5.2　分离流程主要评价指标

1. 乙烯回收率

现代乙烯工厂中,乙烯回收率高低对工厂的经济性有很大影响,它是评价分离装置是否先进的一项重要技术经济指标。

2. 能量的综合利用水平

能量的综合利用水平决定了单位产品(乙烯、丙烯……)所需的能耗,为此要针对主要能耗设备加以分析,不断改进,降低能耗,提高能量综合利用水平。

综合上述原因,甲烷塔和乙烯塔既是保证乙烯回收率和乙烯产品质量(纯度)的关键设备,又是消耗冷量的主要设备(消耗冷量占总数的 88%)。因此,需重点讨论脱甲烷塔和乙烯塔。

3.5.3　脱甲烷塔

脱除裂解气中的氢和甲烷,是裂解气分离装置中投资最大、能耗最多的环节。在深冷分离装置中,需要在 $-90℃$ 以下的低温条件下进行氢和甲烷的脱除,其冷冻功耗约占全装置冷冻功耗的 50% 以上。

对于脱甲烷塔而言,其轻关键组分为甲烷,重关键组分为乙烯。应使塔顶分离出的甲烷轻馏分中乙烯含量尽可能低,以保证乙烯的回收率。而塔釜产品则应使甲烷含量尽可能低,以确保乙烯产品质量。

3.5.4　乙烯塔

C_2馏分经过加氢脱炔之后,到乙烯塔进行精馏,塔顶得产品乙烯,塔釜液为乙烷。要求塔顶乙烯的纯度达到聚合级。乙烯塔设计和操作的好坏,与乙烯产品的产量和质量有直接关系。由于乙烯塔温度仅次于脱甲烷塔,所以冷量消耗占总制冷量的比例也较大,约为38%～44%,对产品的成本有较大的影响。乙烯塔在深冷分离装置中是一个比较关键的塔。

思考题

3-1　在原料确定的情况下,从裂解过程的热力学和动力学出发,为了获取最佳裂解效果,应选择什么样的工艺参数(停留时间、温度、压力……),为什么?

3-2　提高反应温度的技术关键在何处? 应解决什么问题才能最大限度提高裂解温度?

3-3　为了降低烃分压,常加入稀释剂,试分析稀释剂加入量确定的原则是什么?

3-4　试讨论影响热裂解的主要因素有哪些? 评价裂解过程优劣的目标函数(指标)是什么?

3-5　裂解气出口的急冷操作的目的是什么? 可采取的方法有几种? 你认为哪种好,为什么? 若设计一个间接急冷换热器,其关键指标是什么? 如何评价一个急冷换热器的优劣?

3-6　裂解气进行预分离的目的和任务是什么? 裂解气中要严格控制的杂质有哪些? 这些杂质存在的害处是什么? 用什么方法除掉这些杂质,这些处理方法的原理是什么?

3-7　裂解气分离流程各有不同,其共同点是什么? 试绘出顺序分离流程、前脱乙烷后加氢流程、前脱丙烷后加氢流程简图,指出各流程的特点、适用范围和优缺点。

3-8　甲烷塔操作压力的不同对甲烷塔的操作参数(温度、回流比……)、塔设计(理论板数、材质……),以及未来的操作费用和投资有什么影响?

3-9　对于已有的甲烷塔,H_2/CH_4比值对乙烯回收率有何影响? 采用前冷工艺对甲烷塔分离有何好处?

3-10　何为非绝热精馏? 何种情况下采用中间冷凝器或中间再沸器? 分析其利弊。

Chapter 3   Hydrocarbon cracking

Low level olefin molecules such as ethylene, propylene and butadiene have double bonds and active chemical properties. They can react with many substances, such as addition, copolymerization or self-polymerization to generate a series of important products. They are important raw materials in the chemical industry. The main industrial method for obtaining lower olefin is thermal cracking of hydrocarbons. Hydrocarbon pyrolysis is making the petroleum hydrocarbon fuels (natural gas, refinery gas, light oil, diesel oil, heavy oil, etc.) through high temperature, so that the carbon chain fracture or have dehydrogenation reaction of hydrocarbon molecules, forming alkenes, alkanes and other relatively small molecular weight of light and heavy hydrocarbons.

Among the low unsaturated hydrocarbons, ethylene is the most important and has the largest yield. Ethylene production is often used to measure the development level of a country's basic chemical industry.

The production process of hydrocarbon pyrolysis to ethylene mainly consists of two parts: the pyrolysis of raw hydrocarbon and the separation of pyrolysis products.

3.1   Basic reactions during pyrolysis

3.1.1   Reaction law of hydrocarbon cracking

1. Cracking reaction of alkanes

(1) Normal-alkane

The cracking reactions of n-alkanes mainly include dehydrogenation and chain breaking reaction, and cyclization dehydrogenation may occur for alkanes above C_5.

Dehydrogenation is a reaction in which C—H bond is broken, alkenes and hydrogen with the same number of carbon atoms are generated. The general formula is

$$C_nH_{2n+2} \rightleftharpoons C_nH_{2n} + H_2 \qquad (3-1)$$

N-alkanes above C_5 can undergo cyclic dehydrogenation to form cycloalkanes, such as n-hexane dehydrogenation to cyclohexane.

$$\begin{array}{c} \text{CH}_3 \\ \diagup \\ \text{H}_2\text{C} \quad \text{CH}_3 \\ | \qquad | \\ \text{H}_2\text{C} \quad \text{CH}_2 \\ \diagdown \diagup \\ \text{C} \\ \text{H}_2 \end{array} \longrightarrow \begin{array}{c} \text{H}_2 \\ \text{C} \\ \diagup \diagdown \\ \text{H}_2\text{C} \quad \text{CH}_2 \\ | \qquad | \\ \text{H}_2\text{C} \quad \text{CH}_2 \\ \diagdown \diagup \\ \text{C} \\ \text{H}_2 \end{array} + \text{H}_2 \qquad (3-2)$$

Chain breaking reaction is about C—C bond, and the reaction products are alkanes and alkenes with fewer carbon atoms. The general formula is

$$C_nH_{2n+2} \longrightarrow C_mH_{2m} + C_kH_{2k+2} \quad m+k=n \qquad (3-3)$$

The difficulty of dehydrogenation and chain breaking of the same alkanes can be judged from the value of the bond energy of C—H bond and C—C bond in the molecular structure.

(2) Cracking reaction of isomeric alkanes

Isoalkanes have different structures and different cracking reactions. Compared with normal alkanes, they have the following characteristics.

① The bond energy of C—C bond or C—H bond is lower than that of n-alkanes, so it is easy to crack or dehydrogenate.

② The dehydrogenation ability is related to molecular structure, and the order of difficulty is tertiary hydrocarbon>secondary hydrocarbon>primary hydrocarbon.

③ The yields of ethylene and propylene from cracking isomerized alkanes are much lower than those from cracking normal alkanes, while the yields of hydrogen, methane, C_4 and above C_4 alkenes are higher.

④ With the increase of carbon atom number, the difference of ethylene and propylene yield from isomeric and n-alkanes pyrolysis decreases.

2. Olefins cracking reaction

Due to the chemical activity of olefins, there is almost no olefin in natural petroleum raw materials. But there are a certain amount of olefins in refinery gas and secondary processing oil. As the target product in pyrolysis process, olefins are also likely to react further, so in order to control the reaction to go in the direction people want it to go, it is necessary to understand the reaction rules of olefins in the pyrolysis process. The main reactions that may occur in olefins have the following kinds.

(1) Chain breaking reaction

The olefin cracking of the larger molecule can break the chain to generate two smaller olefin molecules, whose general formula is

$$C_{n+m}H_{2(n+m)} \longrightarrow C_nH_{2n} + C_mH_{2m} \qquad (3-4)$$

For instance

$$H_2C{=}CH{-}CH_2{-}CH_2{-}CH_3 \longrightarrow H_2C{=}CH{-}CH_3 + H_2C{=}CH_2 \qquad (3-5)$$

(2) Dehydrogenation

Alkenes may be further dehydrogenated to form dialkenes and alkynes.

$$C_4H_8 \longrightarrow C_4H_6 + H_2 \qquad (3-6)$$

$$C_2H_4 \longrightarrow C_2H_2 + H_2 \qquad (3-7)$$

(3) Disproportionation reaction

Two same olefins can be dismutated into two different hydrocarbon molecules.

$$2C_3H_6 \longrightarrow C_2H_4 + C_4H_8 \qquad (3-8)$$

$$2C_3H_6 \longrightarrow C_2H_6 + C_4H_6 \qquad (3-9)$$

$$2C_3H_6 \longrightarrow C_5H_8 + CH_4 \qquad (3-10)$$

(4) Diene synthesis (Diels-Alder reaction)

Diolefin and alkene undergo a diene synthesis to generate cycloolefin, further dehydrogenation to aromatic hydrocarbons. General formula is

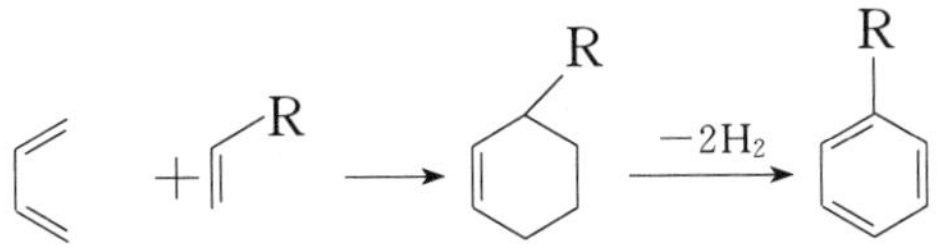

$$(3-11)$$

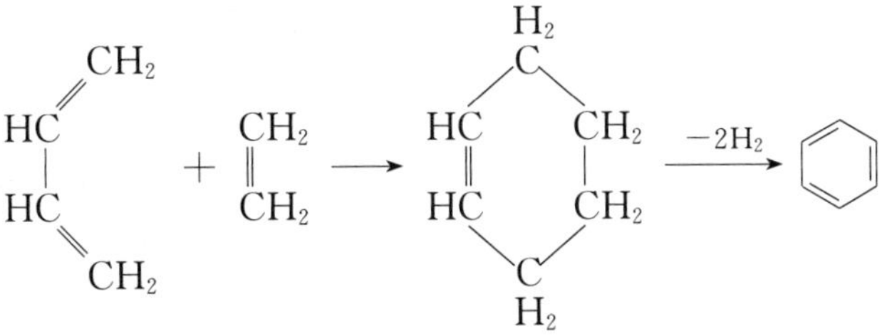

$$(3-12)$$

(5) Aromatization reaction

Alkenes with six or more carbon atoms can undergo aromatization reactions to form aromatics as follows.

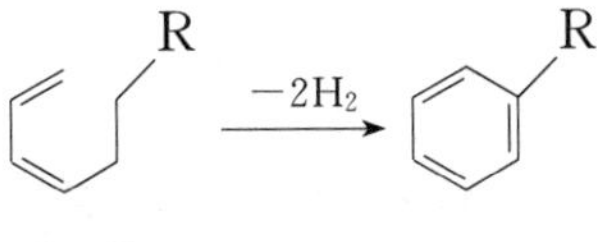

$$(3-13)$$

3. Cracking reaction of cycloalkanes

Cycloalkanes are more stable than their corresponding alkanes. Under general cracking conditions, chain breaking and ring-opening reaction, dehydrogenation, side chain breaking and ring-opening dehydrogenation can take place, from which products such as ethylene, propylene, butadiene, butene, aromatic hydrocarbons, cycloalkanes, monocyclic olefin, monocyclic diolefin and hydrogen are generated.

For example, cyclohexane

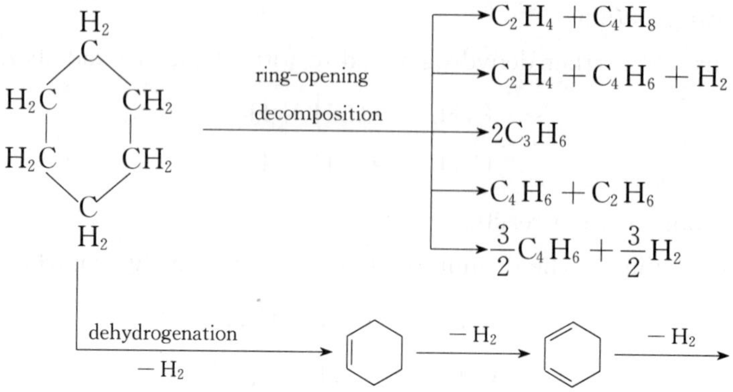

$$\text{(cyclohexane)} \quad
\begin{array}{l}
\xrightarrow{\text{ring-opening}}
\end{array}
\begin{cases}
\rightarrow C_2H_4 + C_4H_8 \\
\rightarrow C_2H_4 + C_4H_6 + H_2 \\
\rightarrow 2C_3H_6 \\
\rightarrow C_4H_6 + C_2H_6 \\
\rightarrow \frac{3}{2}C_4H_6 + \frac{3}{2}H_2
\end{cases}$$

ring-opening decomposition

dehydrogenation
$-H_2$

$\xrightarrow{\text{dehydrogenation} \atop -H_2}$ ⬡ $\xrightarrow{-H_2}$ ⬡ $\xrightarrow{-H_2}$ ⬡

$$(3-14)$$

Ethyl cyclopentane

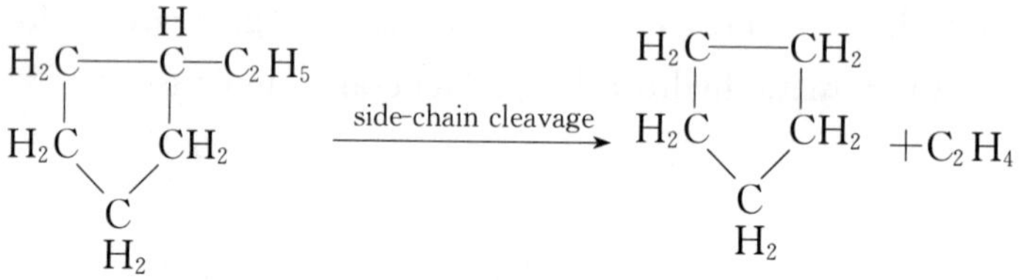

$$\text{(ethyl cyclopentane)} \xrightarrow{\text{side-chain cleavage}} \text{(cyclopentane)} + C_2H_4 \qquad (3-15)$$

4. Aromatics cracking reaction

Due to the stability of aromatic rings in aromatic hydrocarbons, aromatic ring splitting reaction is not easy to occur, there are mainly side chain breaking and dehydrogenation reaction of alkyl aromatic hydrocarbons, condensation of aromatic hydrocarbons to generate polycyclic aromatic hydrocarbons, and further coking reaction. Therefore, the raw oil containing more aromatic hydrocarbons not only has low olefin yield, but also has serious coking, which is not an ideal raw material for cracking.

(1) Cracking of alkyl aromatics

Side chain decalkyl or bond breaking reaction

$$\text{Ar-}C_nH_{2n+1}
\begin{cases}
\rightarrow \text{ArH} + C_nH_{2n} \\
\rightarrow \text{Ar-}C_kH_{2k+1} + C_mH_{2m} \ (m+k=n)
\end{cases} \qquad (3-16)$$

$$\text{Ar-}C_nH_{2n+1} \rightarrow \text{Ar-}C_nH_{2n-1} + H_2 \qquad (3-17)$$

(2) Cracking of naphthenic aromatic hydrocarbons

Dehydrogenation and isomerization dehydrogenation

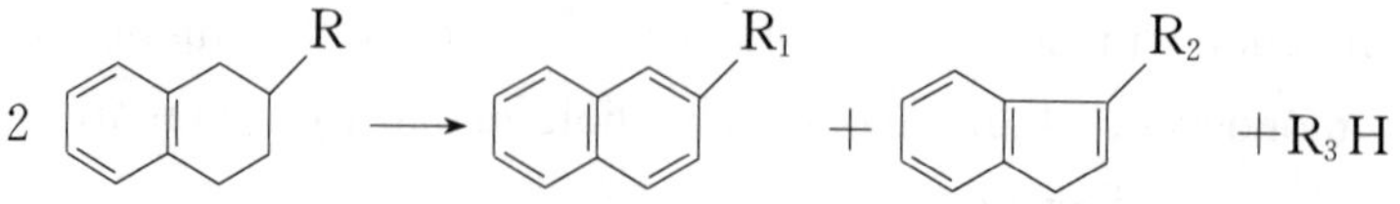

$$2 \ \text{(tetralin-R)} \longrightarrow \text{(naphthalene-}R_1) + \text{(indane-}R_2) + R_3H \qquad (3-18)$$

Condensation dehydrogenation

$$+R_4H+H_2 \quad (3-19)$$

(3) Condensation of aromatic hydrocarbons

$$+R_4H \quad (3-20)$$

5. Coking and charring reaction in cracking process

(1) Carbon is generated from olefin through the intermediate stage of alkynes

The ethylene produced in the cracking process passes through the acetylene stage to produce carbon at 900℃~1 000℃ or higher.

(2) Coking after the aromatic intermediate stage

High boiling point polycyclic aromatic hydrocarbons are the main parent of distillate pyrolysis coking. The pyrolysis tar contains a large number of polycyclic aromatic hydrocarbons. The more tar generated by pyrolysis, the more serious coking in the pyrolysis process.

6. Cracking reaction rule of hydrocarbons

The ability of hydrocarbon cracking to produce ethylene and propylene leads out the following rules.

(1) Alkanes—N-alkanes are the most conducive to the formation of ethylene and propylene. The lower the relative molecular weight of alkanes, the higher the total yield. The total olefin yield of isomeric alkanes is lower than that of normal alkanes with the same number of carbon atoms, but the difference decreases with the increase of relative molecular weight.

(2) Olefin—Macromolecular alkenes cracking into ethylene and propylene. Olefins can be dehydrogenated to alkynes, diolefins, and then produce aromatics.

(3) Cycloalkane—Under normal cracking conditions, the reaction of cycloalkanes to aromatics is better than that of monoolefines. Compared with n-alkanes, the yield of butadiene and aromatic hydrocarbon containing more cycloalkanes is higher, while the yield of ethylene is lower.

(4) Aromatic hydrocarbons—The aromatics without alkyl basically do not crack into olefin, but the aromatics with alkyl mainly undergo carbon bond breaking and dehydrogenation reaction, while the aromatic rings remain unchanged and are easily dehydrogenated and condensated to be polycyclic aromatic hydrocarbons, thus tending to coking.

The degree of difficulty in hydrocarbon cracking is in the following order:

N-alkanes $>$ isoalkanes $>$ cycloalkanes（six-carbon ring $>$ five-carbon ring）$>$ aromatics

With the number of carbon atoms in the molecule increases, the influence of hydrocarbon molecular structure on the cracking rate gradually weakens.

3.1.2　Reaction mechanism of hydrocarbon cracking

The study on the mechanism of hydrocarbon pyrolysis reaction shows that most of the elementary reactions occurred during pyrolysis are free radical reactions.

1.　Free radical reaction mechanism of F. O. Rice

Most hydrocarbon pyrolysis processes include chain initiation, chain growth and chain termination. Chain initiation is the production process of free radicals. Chain growth reaction is the transformation process of free radicals. In this process, the disappearance of one free radical is accompanied by the generation of another free radical, and the existence of free radicals is maintained before and after the reaction. Chain termination is the process by which free radicals disappear to form molecules.

Chain initiation is a molecule breaks to produce a pair of free radicals under the action of heat. Each molecule may have multiple possible chain initiation due to the different fracture locations of bonds, which depends on the dissociation energy of related bonds at the fracture location. Reactions with small dissociation energy are more likely to occur.

Alkane molecules are less likely to break the C—H bond in initiation reactions because the dissociation energy of the C—H bond is greater than that of the C—C bond. Therefore, the general formula for the initiation reaction is

$$R—R' \longrightarrow R\cdot +R'\cdot \tag{3-21}$$

Chain growth reactions include free radical hydrogen abstraction reaction, free radical decomposition reaction, free radical addition reaction and free radical isomerization reaction, but the former two are dominant. The general formula for the hydrogen abstraction reaction is as follows.

$$H\cdot +RH \longrightarrow H_2 +R\cdot \tag{3-22}$$

$$R'\cdot +RH \longrightarrow R'H+R\cdot \tag{3-23}$$

The decomposition reaction of free radicals in chain growth reaction is that the free radicals themselves decompose to form an olefin molecule and a new free radical with fewer carbon atoms than the original, and the free radicals are passed on.

The general formula for this type of reaction is as follows.

$$R\cdot \longrightarrow R'\cdot +Olefin \tag{3-24}$$

$$R\cdot \longrightarrow H\cdot +Olefin \tag{3-25}$$

Free radical decomposition reactions are reactions that produce alkenes, and the purpose of cracking is to produce alkenes, so such reactions are critical.

2. Primary reaction and secondary reaction

The reaction of raw hydrocarbon in the cracking process is complex. A hydrocarbon can take place in many kinds of reactions parallelly, and can take place many subsequent reactions in succession. So the pyrolysis system is a reaction system where parallel reaction and consecutive reaction cross. From the perspective of the whole reaction process, it is a typical consecutive reaction.

As the reaction goes on, gaseous hydrocarbons (molecule, alkenes) and hydrogen are decomposed continuously. The hydrogen content of liquid product decreased gradually, and the relative molecular weight increased accordingly, resulting in coking.

Primary reaction refers to the cracking reaction of raw hydrocarbon which occurs first in the cracking process of raw hydrocarbon, and secondary reaction refers to the subsequent reaction of the first reaction product. From the actual reaction process of pyrolysis reaction, there is no strict dividing line between primary reaction and secondary reaction, and different researchers have different division of primary reaction and secondary reaction.

The reaction that produces the target product such as ethylene, propylene belongs to the primary reaction, this is the reaction that people want to happen. In the determination of technological conditions/design and production operation, every effort should be made to promote sufficient primary reaction.

The disappearance of ethylene and propylene, the formation of liquid products with relatively large molecular weight and coking, charcoal is a secondary reaction, which is not expected to occur. The occurrence of this kind of reaction, not only consumes more raw materials, reduces the yield of the main product, but also deteriorates heat transfer, blocks equipment, and brings very adverse effects to cracking operation and stable production. So we should try every means to restrain it.

3.1.3　Properties and evaluation of pyrolysis raw materials

Due to the raw material used in hydrocarbon pyrolysis reaction is a mixture of very different composition properties, the characteristics of raw materials undoubtedly plays an important role in the pyrolysis effect. It is the internal reason that determines reaction effect, while the adjustment and optimization of process conditions is only its external conditions.

1. Group composition

All kinds of hydrocarbon in pyrolysis raw oil can be divided into four groups according to their structure, that is, paraffin group, olefin group, cycloparaffin group and aromatic group. The composition of the four groups is represented by PONA values, with the following meanings:

P—Paraffin, N—Naphtene,O—Olefin, A—Aromatics.

According to PONA value, the cracking performance of liquid fuel can be qualitatively evaluated, and the cracking reaction can also be quantitatively described by simplified reaction kinetics model according to the group composition. Therefore, PONA value is a useful parameter to characterize the cracking performance of various liquid feedstocks.

2. Hydrogen content

Hydrogen content can be expressed by the mass fraction ω (H_2) of hydrogen contained in the cracking raw material, or by the mass ratio of C to H in the cracking raw material (called hydrocarbon ratio).

$$\text{Hydrogen content} \quad \omega(H_2) = \frac{n_H}{12n_C + n_H} \times 100 \quad (3-26)$$

$$\text{Hydrocarbon ration} \quad \frac{m_C}{m_H} = \frac{12n_C}{n_H} \quad (3-27)$$

n_H and n_C are the number of hydrogen and carbon atoms in the hydrocarbon respectively.

Sequence of hydrogen content is P>N>A.

Through the pyrolysis reaction, the cracking raw material with certain hydrogen content can be used to produce light components (C_4 and below) with higher hydrogen content and liquid (C_5 and above) with lower hydrogen content. From the hydrogen balance, it can be concluded that the higher the hydrogen content of pyrolysis raw material, the higher the yield of light hydrocarbons (C_4 and below) obtained, the higher the yield of ethylene and propylene. Obviously, according to the hydrogen content of the cracking raw material, the possible cracking depth of the raw material can be determined, and the yield of light hydrocarbon (C_4 and below) obtained from the cracking can be evaluated.

3.2 The parameters and indexes of cracking process

3.2.1 Cracking raw material

The raw materials used in hydrocarbon pyrolysis play an important and decisive role in the pyrolysis process and pyrolysis products. The higher the hydrogen content of cracking raw materials, the higher the yield of alkenes below C_4. So alkanes, especially low carbon alkanes, are the preferred raw materials.

Foreign hydrocarbon cracking raw materials are mainly light hydrocarbons (below C_4) and naphtha, accounting for about 90%. However, the proportion of heavy oil and diesel oil in China is more than 20%, which needs to be further optimized.

3.2.2　Cracking temperature and residence time

1. Cracking temperature

Based on the analysis of free radical reaction mechanism, the yield of ethylene and propylene from primary reaction can be improved by increasing cracking temperature within a certain temperature.

According to the kinetics of cracking reaction, it is necessary to control the conversion rate within a certain range in order to control the cracking reaction within a certain range of cracking depth. As the reaction rate constants of different cracking materials are very different, the temperature required by different cracking materials is also different at the same residence time. The lower the relative molecular weight of the raw material, the higher the activation energy and frequency factor, the lower the reactivity, and the higher the cracking temperature.

At a certain cracking depth, there can be various cracking temperature-residence time combinations. Therefore, cracking temperature and residence time are a set of interrelated and indivisible parameters for olefins production. High temperature and short residence time is the key to improve the yield of pyrolysis products.

2. Residence time

The residence time of material in tubular cracking furnace is the time of cracking material passing through radiant coil. It is difficult to calculate the real residence time because the cracking reaction in the cracking tube is carried out under the condition of non-isothermal variability.

3.2.3　Hydrocarbon partial pressure and diluents

1. Influence of pressure on cracking reaction

The primary reaction of hydrocarbon cracking is the process which increases the number of molecules. For the reversible dehydrogenation reaction, reducing the pressure is beneficial to improve the equilibrium composition of ethylene (chain breaking reaction is irreversible, pressure has no effect). The secondary reaction of hydrocarbon polymerization is the process reducing the number of molecules. Reducing the pressure is not conducive to improving the equilibrium composition of the secondary reaction products, but can inhibit the coking process.

Therefore, reducing the pressure can promote the primary reaction of ethylene formation and inhibit the second reaction of polymerization, thus reducing the degree of coking.

2. Diluent

The pyrolysis is operating at high temperature, unfavorable to use the method of vacuuming decompression to reduce hydrocarbon partial pressure. This is because the high-temperature sealing is difficult. Once the air leaks into the negative pressure operated cracking system and forms explosive mixture with hydrocarbon gases, there is a risk of explosion. Moreover, the decompression operation against later compression operation in the separation process, which will increase energy consumption. Therefore, it is a better method to add diluent to reduce hydrocarbon partial pressure. In this way, the equipment can still be operated at normal or positive pressure, while the hydrocarbon partial pressure can be reduced. In theory, water vapor, hydrogen or any inert gas can be used as diluent, but at present, most of the mature cracking methods use water vapor as diluent.

3.2.4　Cracking depth

Cracking depth refers to the degree of cracking reaction. Because of the complexity of the pyrolysis reaction, it is difficult to describe quantitatively by one parameter. According to different situations, the following parameters are often used to measure the cracking depth.

(1) Conversion rate of raw materials

The conversion rate of raw materials (X) reflects the conversion degree of the raw material during the cracking reaction. Therefore, raw material conversion rate is commonly used to measure the cracking depth.

(2) Methane yield

The yield of methane $[y(C_1)]$ from pyrolysis increases with the increase of pyrolysis depth. Because methane is relatively stable, it will not disappear due to secondary reactions. Therefore, methane yield in pyrolysis products can measure the depth of reaction to a certain extent.

(3) Yield ratio of ethylene to propylene

In a certain range of cracking depth, the yield of ethylene increases with the increase of cracking depth, and the yield of propylene increases slowly. At a certain cracking depth, the yield of ethylene still increased with the increase of cracking depth, but the yield of propylene decreased from the maximum value.

(4) Yield ratio of methane to ethylene or propylene

The yield of methane always increases with the deepening of the reaction process, while the yield of ethylene or propylene begins to decrease after reaching the maximum value with the increase of the cracking depth.

3.3   Pre-fractionation and purification of cracked gas

3.3.1   Objective and task of cracked gas prefractionation

The high-temperature cracking gas at the outlet of the cracking furnace is cooled by the quench heat exchanger and further cooled by the quench cooler, and the temperature can be reduced to 200℃～300℃. After quenching, the cracking gas is further cooled to room temperature, and in the process of cooling, the heavy components (such as fuel oil, cracking gasoline, water) in the cracking gas are fractionated, this link is called the pre-fractionation of cracking gas. The cracking gas treated by pre-fractionation is then sent to cracked gas compressor for further cryogenic separation. Obviously, the pre-fractionation process of cracking gas plays an important role in ethylene plant.

3.3.2   Summary of pre-fractionation process

1. Pre-fractionation process of cracking gas in light hydrocarbon cracking unit

The heavy fraction of cracking gas obtained from light hydrocarbon cracking unit is very small, especially for ethane and propane cracking, the content of fuel oil in cracking gas is very small. At this point, the pre-fractionation process of cracking gas is mainly to fractionate water and cracking gasoline fraction in the further cooling process of cracking gas.

The schematic diagram of the pre-fractionation process of cracking gas in light hydrocarbon cracking unit is shown in Figure 3 – 1.

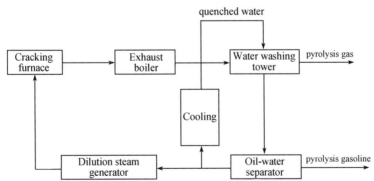

Figure 3 – 1   Pre-fractionation flow diagram of cracking gas in light hydrocarbon cracking unit

2. Pre-fractionation process of cracking gas in distillate oil cracking device

The cracking gas produced by the distillate oil cracking device contains a considerable amount of heavy fractions, which are difficult to separate oil and water because of emulsification after mixing with water. Therefore, in the distillate oil cracking device, it is necessary to fractionate the heavy fuel oil fraction in the cracking in the process of cooling, and the fractionated cracking gas is further sented to the water washing tower for cooling, and the water and cracking gasoline will be fractionated.

The schematic diagram of the pyrolysis gas pre-fractionation process in the distillate cracking unit is shown in Figure 3 – 2.

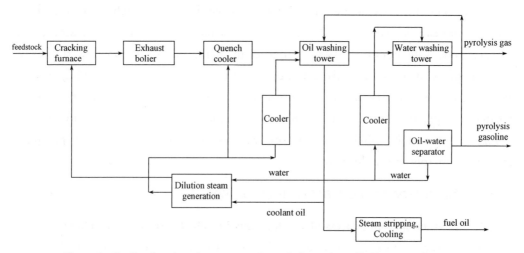

Figure 3 – 2　Pre-fractionation process of pyrolysis gas in a distillate cracking unit

3.3.3　Cracking gasoline and cracking fuel oil

1. Cracking gasoline

Cracking gasoline as a by-product of hydrocarbon pyrolysis consists of all pyrolysis by-products from C_5 to boiling point below 204℃. As a by-product of ethylene plant, its typical specifications usually as follows:

　　C_4 fraction　　　　　　　　0. 5% (Maximum mass fraction)
　　End boiling point　　　　　　204℃

Cracking gasoline can be used as high octane gasoline component after one-stage hydrogenation. If aromatics products are seperated by aromatics extraction, two stages of hydrogenation should be carried out to remove sulfur and nitrogen, as well as saturate all olefin.

All cracking gasoline can be hydrogenated, and can be divided into hydrogenated C_5 fraction, $C_6 \sim C_8$ central fraction and $C_9 \sim 204$℃ fraction after hydrogenation. At this point, the hydrogenated C_5 fraction can be returned to cyclic cracking, while the $C_6 \sim C_8$ central fraction is the raw material for aromatics extraction, and the C_9 fraction can be used as the raw material for disproportionation production of aromatics. Cracking gasoline can also be first divided into C_5 fraction, C_9 fraction, $C_6 \sim C_8$ central fraction. Then only $C_6 \sim C_8$ central fraction is hydrotreated, so the hydrotreating amount can be reduced.

2. Cracking fuel oil

Cracking fuel oil as a by-product of hydrocarbon pyrolysis refers to the heavy fraction with the boiling point above 200℃. The fraction with boiling range of 200℃ ~ 360℃ is called light pyrolysis fuel oil, which is equivalent to diesel oil fraction. Among them, the content of alkyl naphthalene is high, which can be used as the raw material of dealkylation to produce naphthalene. The fraction with boiling range above 360℃ is called heavy cracking fuel oil and is equivalent to atmospheric heavy oil fraction. In addition to being used as fuel, the heavy fuel oil is a good raw material for producing carbon black due to its low ash content.

When light hydrocarbons and light oils are cracked, there is less cracking fuel oil and usually no separation between light and heavy fuel oil is performed. When diesel fuel is cracked, light fuel oil should be separated and used as fuel in the cracking furnace to balance the shortage of gas fuel in the diesel cracking plant.

The mainly specification needed to be controlled of the cracking fuel oil is the flash point of oil. Usually, the flash point of light cracking fuel oil should be controlled above 70℃, and the flash point of heavy cracking fuel oil should be controlled above 100℃. The sulfur content of cracking fuel oil depends on the sulfur content of pyrolysis feedstock. During naphtha cracking, $10\% \sim 20\%$ of the total sulfur of the cracking feedstock accumulates in the cracking fuel oil. When diesel fuel is cracked, $50\% \sim 60\%$ of sulfur is enriched in the cracked fuel oil.

3.3.4 Purification of cracked gas

Cracking gas contains H_2S, CO_2, H_2O, C_2H_2, CO and other gaseous impurities. There are three main sources: the raw material, the pyrolysis reaction process, and the pyrolysis gas treatment process.

The amount of these impurities is small, but they are harmful to the cryogenic separation process. If these impurities are not removed, the ethylene or propylene products will not reach the specified standards. Especially in the production of polymerization grade ethylene and propylene, the control of the content of impurities is very strict, in order to meet the requirements of the product specifications, it is necessary to remove these impurities and purify the cracking gas.

1. Removal of acidic gases

(1) The source and harm of acidic gas impurities

The acidic gases in the cracking gas are mainly CO_2, H_2S and other gaseous sulfides.

The acidic gas contained in cracking gas will do great harm to cracking gas separation unit and the processing unit of ethylene and propylene derivative. For pyrolysis gas separation device, CO_2 will form dry ice at low temperature, resulting in clogging of cryogenic separation system equipment and pipeline, and H_2S will cause poisoning of hydrodealkyne catalyst and methanation catalyst. For downstream processing units, when the content of acidic gas in hydrogen, ethylene and propylene products is not up to standard, the catalyst in the polymerization process or catalytic reaction process of downstream processing units can be poisoned, and the product quality may be seriously affected. Therefore, the acidic gas in cracking gas should be removed before rectification and separation.

（2）Alkali washing method to remove acidic gas

Alkali washing method is to use NaOH as absorbent, through chemical absorption to make NaOH reacts with acidic gas in cracking gas, in order to achieve the purpose of removing acidic gas.

（3）Removal of acidic gas by ethanolamine method

Using ethanolamine as absorbent to remove CO_2 and H_2S from pyrolysis gas is a method combining physical absorption and chemical absorption. The absorbent is MEA and DEA.

（4）Comparison between alcohol-amine method and alkali washing method

Compared with alkali wash, the main advantage of alcohol-amine method is that the absorbent can be recycled. When the acidic gas content is high, the alcohol-amine method is obviously superior to alkaline washing method in terms of absorption liquid consumption and wastewater treatment capacity.

2. Dehydration

The cracking gas enters the cracking gas compressor after pre-fractionation treatment. The water content in the cracking gas at the compressor inlet is the saturated water content at the inlet temperature and pressure. In the process of cracking gas compression, with the increase of pressure, part of water can be separated in the process of intersegment condensation.

The water content in the cracking gas is not high, but it requires high dryness of the material after dehydration, so the adsorption method is used for drying.

3. Removal of alkynes

（1）Source, harm and treatment of alkynes

During the separation process of cracking gas, acetylene in cracking gas will be enriched in C_2 fraction, while methyl acetylene and propylene diene （MAPD for short）will be enriched in C_3 fraction. In general, the mole fraction of acetylene in C_2 fraction is $0.3\% \sim 1.2\%$, and the mole fraction of MAPD enriched in C_3 fraction is $1\% \sim 5\%$. Under the pyrolysis condition of ultra-short residence time at high temperature in Kellogg millisecond furnace, the acetylene mole fraction enriched in C_2 fraction can reach $2.0\% \sim 2.5\%$, the mole fraction of MAPD in C_3 fraction can reach

5%~7%.

Solvent absorption and catalytic hydrogenation are commonly used to remove acetylene in ethylene production. Solvent absorption is using solvent to absorb acetylene in cracking gas to achieve purified gas, and recover a certain amount of acetylene as well. Catalytic hydrogenation is hydrogenating acetylene to ethylene or ethane so as to remove acetylene. Solvent absorption and catalytic hydrogenation have their advantages and disadvantages. At present, catalytic hydrogenation is generally used when there is no need to recycle acetylene. When acetylene recovery is required, the solvent absorption method is used. In the actual production, the plant with a solvent absorption system for acetylene recovery, often has catalytic hydrogenation of acetylene system. The two systems are in parallel to provide flexibility.

(2) Catalytic hydrogenation

① Catalytic hydrogenation of alkynes

The following reactions occured during the selective catalytic hydrogenation of acetylene in cracking gas.

Main reaction

$$C_2H_2 + H_2 \xrightarrow{k_1} C_2H_4 + \Delta H_1 \qquad (3-28)$$

Side reaction

$$C_2H_2 + 2H_2 \xrightarrow{k_2} C_2H_6 + \Delta H_2 \qquad (3-29)$$

$$C_2H_4 + H_2 \longrightarrow C_2H_6 + (\Delta H_2 - \Delta H_1) \qquad (3-30)$$

$$mC_2H_2 + nC_2H_4 \longrightarrow \text{Oligomer (Green oil)} \qquad (3-31)$$

When the reaction temperature rises to a certain extent, the pyrolysis reaction of C, H_2 and CH_4 may also occur.

② Front-end and back-end hydrogenation

Front-end hydrogenation is the selective hydrogenation of alkynes by hydrogen in cracking gas before hydrogen separation in order to remove alkynes. Therefore, it is also called self-contained hydrogen catalytic hydrogenation process.

The back-end hydrogenation process refers to the separation of C_2 and C_3 fractions by cracking gas, and then the catalytic hydrogenation of C_2 and C_3 fractions to remove acetylene, methylacetylene and allylene.

③ Hydrogenation process

Take the back-end hydrogenation process as an example, when the mole fraction of acetylene in the feed is higher than 0.7%, multi-stage adiabatic bed or isothermal reactor is generally adopted.

The hydrogenation process of two stage adiabatic bed is shown in Figure 3-3.

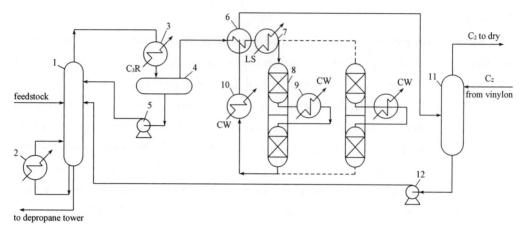

1 – Deethane column; 2 – Reboiler; 3 – Condenser; 4 – Reflux tank; 5 – Reflux pump;

6 – Heat exchanger; 7 – Heater; 8 – Hydrogenation reactor; 9 – Intersection cooler;

10 – Cooler; 11 – Green oil absorption tower; 12 – Green oil pump.

Figure 3 – 3   Two stage adiabatic bed hydrogenation process

(3) Acetylene removed by solvent absorption

Solvent absorption is using a selective solvent to selectively absorb a small amount of acetylene in the C_2 fraction into the solvent, so as to achieve the removal of acetylene. Due to the use of solvent that selectively absorbs acetylene, acetylene can be desorbed under certain conditions. Therefore, solvent absorption can remove acetylene, and can recover high purity acetylene at the same time.

DMF solvent absorption process for acetylene removal (Lummus) is shown in Figure 3 – 4.

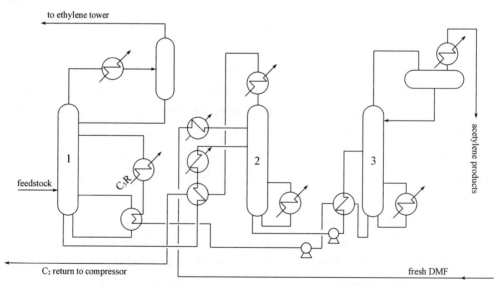

1 – Acetylene absorber; 2 – Stabilization tower; 3 – Stripping tower.

Figure 3 – 4   DMF solvent absorption process for acetylene removal (Lummus)

3.4　Compression and refrigeration system

3.4.1　Compression of cracking gas

Many components of pyrolysis gas are gases under atmospheric pressure, and their boiling point are very low. If the separation of each component is distilled under atmospheric pressure, the separation temperature is very low, requiring a large amount of cooling capacity. In order to keep the separation temperature not too low, the separation pressure can be appropriately increased. The lowest temperature part of the pyrolysis gas separation is the separation of methane and hydrogen, that is, the top of the methane tower.

3.4.2　Refrigeration system in cracking unit

Cryogenic separation process requires refrigerant. Refrigeration is using refrigerant compression and condensation to obtain refrigerant liquid, and then evaporating under different pressure to obtain different temperature level of the freezing process.

3.5　Distillation separation system for cracking gas

3.5.1　Organization of separation process

In the distillation separation of multi-component system, reasonable organization of the process has great influence on the construction investment, energy consumption, operation cost, operation cycle, product yield and quality, and production safety.

The pyrolysis gas is reduced to normal temperature by pre-separation. The fuel oil fraction, most of the water and part of the pyrolysis gasoline are fractionated from it.

3.5.2　Main evaluation indicators of separation process

1. Ethylene recovery rate

In modern ethylene plant, the rate of ethylene recovery has a great influence on the economy of the plant. It is an important technical and economic index to evaluate whether the separation unit is advanced or not.

2. Comprehensive utilization level of energy

This determines the amount of energy required per unit of product (ethylene, propylene ...). Therefore, it is necessary to analyze the main energy consumption equipment and improve it continuously to reduce energy consumption and improve the comprehensive utilization level of energy.

Combined with the above reasons, methane tower and ethylene tower are not only the key equipments to ensure ethylene recovery rate and ethylene product quality (purity), but also the main source of cold consumption (consumption of cold accounted for 88% of the total). Therefore, the demethanizer and ethylene tower are mainly discussed.

3.5.3 Demethanizer

Removing hydrogen and methane from cracking gas is the most invested and energy consuming part in cracking gas separation unit. In cryogenic separation unit, hydrogen and methane are removed at low temperature below $-90℃$, and its freezing power consumption accounts for more than 50% of the whole unit's freezing power consumption.

For demethanation tower, the light critical component is methane, and the heavy critical component is ethylene. The ethylene content in the light fraction of methane separated from the top of the tower should be as low as possible to ensure ethylene recovery rate. The methane content of the products should be as low as possible to ensure the quality of ethylene products.

3.5.4 Ethylene tower

After hydrogenation and deacetylene, the C_2 fraction goes to ethylene tower for rectification. Ethylene produced from the tower top, and tower kettle liquid is ethane. Ethylene purity at the top of the tower is required to reach polymer grade. The design and operation of this tower are directly related to the quantity and quality of ethylene products. Because the temperature of ethylene tower is only second to that of demethanation tower, the cold energy consumption accounts for a larger proportion of the total cooling capacity, about $38\%\sim44\%$, which has a greater impact on the cost of the product. Ethylene tower is a key tower in cryogenic separation unit.

Questions

3 - 1　What process parameters (residence time, temperature, pressure ...) should be selected in order to obtain the best cracking effect based on the thermodynamics and kinetics of the cracking process when the raw materials are determined? Why?

3 - 2　What are the technical keys to improve the reaction temperature, and what

problems should be solved to maximize the pyrolysis temperature?

3 – 3 In order to reduce hydrocarbon partial pressure, diluent is usually added. What is the principle for determining the amount of diluent?

3 – 4 What are the main factors affecting thermal cracking? What is the objective function (index) to evaluate the performance of pyrolysis process?

3 – 5 What is the purpose of quench operation at cracking gas outlet? There are several methods you can take. Which one do you think is better and why? What are the key indicators of designing an indirect quench heat exchanger? How to evaluate the advantages and disadvantages of a quench heat exchanger?

3 – 6 What is the purpose and task of cracking gas pre-separation? What impurities should be strictly controlled in cracking gas? What is the harm of these impurities? What methods are used to remove these impurities, and what are the principles of these treatments?

3 – 7 The cracking gas separation process is different, what is the common ground? Try out the sequential separation process, pre-dethane and post-hydrogenation process, pre-depropane and post-hydrogenation process, and point out the characteristics, application scope, advantages and disadvantages of each process.

3 – 8 What is the influence of the different operating pressure of methane column on the operating parameters (temperature, reflux ratio ...), tower design (number of theoretical plates, material ...), the future operating costs and investments?

3 – 9 What is the effect of H_2/CH_4 ratio on ethylene recovery rate in existing methane column? What are the advantages of using pre-cooling process for methane column separation?

3 – 10 What is non-adiabatic rectification and when to use intermediate condenser or intermediate reboiler? Analyze the advantages and disadvantages.

第 4 章 芳烃转化

4.1 概　述

芳烃是含苯环结构的碳氢化合物的总称。芳烃中的"三苯"(苯、甲苯和二甲苯,简称 BTX)和烯烃中的"三烯"(乙烯、丙烯和丁二烯)是化学工业的基础原料,具有重要地位。芳烃中以苯、甲苯、二甲苯、乙苯、异丙苯、十二烷基苯和萘最为重要,这些产品广泛应用于合成树脂、合成纤维、合成橡胶、合成洗涤剂、增塑剂、染料、医药、农药、炸药、香料、专用化学品等工业,对发展国民经济、改善人民生活起着极为重要的作用。

化学工业所需的芳烃主要是苯、甲苯及二甲苯。苯可用来合成苯乙烯、环己烷、苯酚、苯胺及烷基苯等;甲苯不仅是有机合成的优良溶剂,而且可以合成异氰酸酯、甲酚,或通过歧化和脱烷基制苯;二甲苯和乙苯同属 C_8 芳烃,二甲苯异构体分别为对二甲苯、邻二甲苯和间二甲苯。

4.1.1　芳烃来源与生产方法

近 20 年,芳烃生产得到迅速发展。1986 年全世界 BTX 的生产能力仅为 3 291 万吨,2003 年世界 BTX 生产能力已达到 8 738.7 万吨;中国 2003 年 BTX 生产能力达到 614.8 万吨。

芳烃最初全部来源于煤焦化工业。由于有机合成工业的迅速发展,芳烃的需求量上升,煤焦化工业生产的焦化芳烃在数量和质量上都不能满足需求,因此许多工业发达的国家开始以石油为原料生产石油芳烃,以弥补不足。石油芳烃发展至今,已成为芳烃的主要来源,约占全部芳烃的 80%。

石油芳烃主要来源于石脑油重整生成油及烃裂解生产乙烯副产的裂解汽油。由于各国资源不同,裂解汽油生产的芳烃在石油芳烃中比重也不同。美国乙烯生产大部分以天然气凝析液为原料,副产芳烃很少,故美国的石油芳烃主要来自催化重整油。日本与西欧各国乙烯生产主要以石脑油为原料,副产芳烃较多,且从裂解汽油中回收芳烃的投资与操作费用比重整生成油生产芳烃低。因此,日本与西欧各国从裂解汽油中回收芳烃量较大。随着乙烯工业的发展和乙烯原料由轻烃转向石脑油与柴油,预计来自裂解汽油生产的芳烃在世界芳烃产量中的比重将呈上升趋势,石脑油蒸汽裂解副产芳烃汽油量约为原料量的 25%(质量计)。焦化芳烃生产受冶金工业的限制,其产量将维持现状或略有增加。

1. 焦化芳烃生产

在高温作用下,煤在焦炉炭化室内进行干馏时,煤质发生一系列的物理、化学变化。

除生成 75％的焦炭外,还副产粗煤气约 25％,其中粗苯约占 1.1％,煤焦油约占 4.0％。粗煤气中含有多种化学品,其组成与数量随炼焦温度和原料配煤不同而有所波动。粗煤气经初冷、脱氨、脱萘、终冷后,进行粗苯回收。粗苯由多种芳烃和其他化合物组成,其主要组分是苯、甲苯和二甲苯。

用洗油从粗煤气中吸收粗苯后,经蒸馏脱吸,得到粗苯,粗苯回收率约 90％。

2. 石油芳烃生产

石油芳烃生产过程见图 4-1。

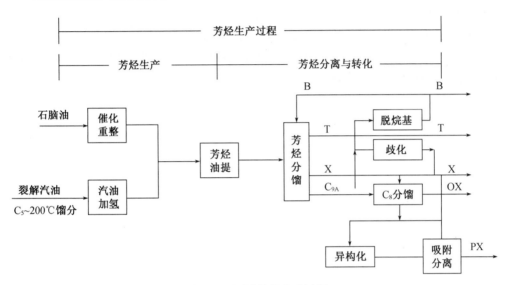

图 4-1　石油芳烃生产过程

以石脑油和裂解汽油为原料生产芳烃的过程,可分为反应、分离和转化三部分。不同国家的石油芳烃生产模式有所不同。芳烃资源丰富的美国,苯的需要量较大,需通过甲苯脱烷基制苯补充苯的不足,而对二甲苯与邻二甲苯主要从催化重整油中分离而得,很少采用烷基转移与二甲苯异构化等工艺过程。西欧与日本芳烃资源不够丰富,因而采用芳烃转化工艺过程较多。中国芳烃资源比较少,需充分利用有限的芳烃资源,因而采用甲苯、C_9 芳烃的烷基转移,甲苯歧化,二甲苯异构化等工艺过程,很少采用甲苯脱烷基工艺。

（1）催化重整生产芳烃

催化重整是炼油工业主要的二次加工装置之一,它用于生产高辛烷值汽油或 BTX 等芳烃,其中约 10％的装置用于生产芳烃产品。

① 催化重整的基本化学反应

催化重整包括环烷脱氢、五元环异构脱氢、烷烃脱氢环化、烷烃异构加氢裂解等反应,从而生成芳烃,同时也伴有加氢裂解、烯烃聚合等副反应。实际生产中要采取一定的措施抑制副反应的发生。

② 催化重整的原料

催化重整的原料为石脑油馏分,石脑油的烃族组成和馏程对重整生成油中芳烃含量、组成和生产的各项技术、经济指标有着决定性的影响。一般应尽可能选用含环烷烃多的石脑油为原料;对于馏程,一般根据生产目的适当选取。就生产 BTX 来说,原料的实际沸点馏程取 65℃~145℃,如要利用 C_9 芳烃增产 C_8 芳烃,则实际沸点馏程取 70℃~177℃ 为宜。

③ 重整催化剂

重整催化剂由一种或多种贵金属元素高度分散在多孔载体上制成,其主金属为铂,质量分数在 0.3%~0.7%;还有卤族元素(氟或氯),其质量分数在 0.5%~1.5%。

④ 催化重整工艺

催化重整过程是在临氢条件下进行,一般反应温度为 425℃~525℃,反应压力为 0.7 MPa~3.5 MPa,空速 1.5~3.0 h^{-1},氢油摩尔比为 3~6,重整油的收率为 75.85%(即 C_5 以上烃的收率),芳烃含量为 30%~70%,重整油的研究法辛烷值(RON)可高达 100。

连续重整反应原理流程见图 4-2。

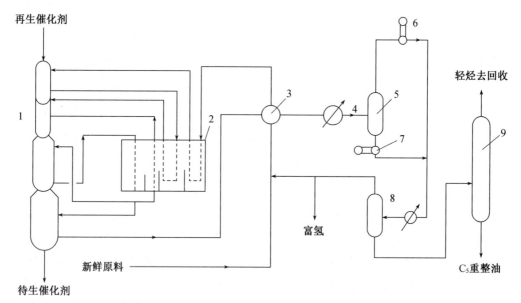

1—反应器;2—加热炉;3—换热器;4—冷却器;5—高压分离器;
6—压缩器;7—泵;8—低压分离器;9—稳定塔。

图 4-2 连续重整反应原理流程图

(2) 裂解汽油生产芳烃

乙烯是石油化工最重要的基础原料之一。随着乙烯工业的发展,副产的裂解汽油已是石油芳烃的重要来源。1984 年从裂解汽油生产的苯量已达 794 万吨,相当于当年苯产量的三分之一以上。裂解汽油除含 40%~60%的 C_6~C_9 芳烃外,还含有相当数量的二烯烃与单烯烃,少量的烷烃与微量氧、氮、硫及砷的化合物。裂解汽油中烯烃与各项杂质远远超过芳烃生产后续工序所能允许的标准,必须经过预处理,加氢精制后,才能作为芳烃抽提的原料。

① 裂解汽油预处理

裂解汽油为 $C_5 \sim 200℃$ 馏分。C_5 馏分中含有较多异戊二烯、间戊二烯与环戊二烯，它们是合成橡胶和精细化工的重要原料。C_5 馏分中二烯烃加氢生成的烯烃是很好的汽油组分。C_5 馏分烯烃进一步加氢生成 C_5 烷烃，可作为烃类裂解原料。

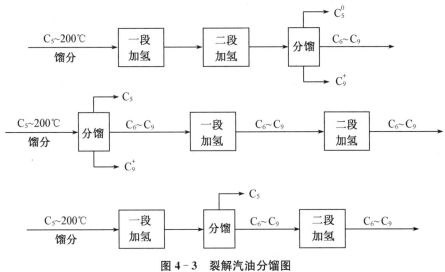

图 4-3 裂解汽油分馏图

② 裂解汽油加氢

裂解汽油加氢是目前普遍采用的精制方法。由于从裂解汽油中除去双烯烃、单烯烃和氧、氮、硫等有机化合物的工艺条件不同，一般采用二段加氢精制工艺。第一段加氢的目的是使易生胶的二烯烃加氢转化为单烯烃以及使烯基芳烃转化为芳烃。第二段加氢的目的主要是使单烯烃饱和并脱除硫、氧、氮等有机化合物。

（3）轻烃芳构化与重芳烃轻质化

催化重整和高温裂解的原料主要都是石脑油，而石脑油同时也是生产汽油的重要原料。由于汽油需求日益增长，人们不得不寻找石脑油以外的生产芳烃的原料。目前正在开发的一是利用液化石油气和其他轻烃进行芳构化，二是使重芳烃进行轻质化。这两种原料路线的工业化和基础研究均取得了重要进展，液化石油气芳构化制芳烃已初步实现工业化。

4.1.2 芳烃馏分分离

由催化重整和加氢精制裂解汽油得到的含芳烃馏分都是由芳烃与非芳烃组成的混合物。由于碳数相同的芳烃与非芳烃的沸点非常接近，有时还会形成共沸物，因此用一般的蒸馏方法难以将它们分离。为了满足对芳烃纯度的要求，目前工业上实际应用的主要是溶剂萃取法和萃取蒸馏法。前者适用于从宽馏分中分离苯、甲苯、二甲苯等；后者适用于从芳烃含量高的窄馏分中分离纯度高的单一芳烃。

1. 溶剂萃取

溶剂萃取法分离芳烃是利用一种或两种以上的溶剂（萃取剂）对芳烃和非芳烃选择

溶解,从而分离出芳烃。溶剂的性能与芳烃收率、芳烃质量、公用工程消耗及装置投资有直接关系。对溶剂性能的基本要求是对芳烃的溶解选择性好、溶解度高、与萃取原料密度差大、蒸发潜热与热容小、蒸汽压小、腐蚀性小,并有良好的化学稳定性与热稳定性等。

2. 萃取蒸馏

萃取蒸馏是利用极性溶剂与烃类混合时,能降低烃类蒸汽压使混合物初沸点提高的原理而设计的工艺过程。由于此种效应对芳烃的影响最大,对环烷烃的影响次之,对烷烃影响最小,因此有助于芳烃和非芳烃的分离。萃取蒸馏塔把溶剂萃取和蒸馏两种过程结合起来。待分离的物料预热后进入萃取蒸馏塔中部,溶剂进入塔的顶部,进行逆流传质过程。含微量芳烃的非芳烃呈气态从塔顶蒸出,冷凝后部分回流,其余出装置。溶剂和芳烃从塔底排出,进入汽提塔,汽提出芳烃后,溶剂循环使用。萃取蒸馏法是从富含芳烃的馏分中直接提取某种高纯芳烃的一种工艺过程。原料需首先进行预分馏,切除轻、重馏分,留下中心馏分送去萃取蒸馏。萃取蒸馏可用于从重整油或裂解汽油(加氢后)提取苯、甲苯或二甲苯,也可用于从未加氢的裂解汽油中直接提取苯乙烯。

4.1.3 芳烃转化

不同来源的各种芳烃馏分组成是不同的,能得到的各种芳烃的产量也不同。各种芳烃组分中用途最广、需求量最大的是苯与对二甲苯,其次是邻二甲苯。甲苯、间二甲苯及 C_9 芳烃迄今尚未获得重大的化工利用,因而有所过剩。

值得说明的是,由于苯的毒性,各国对汽油中苯含量的限制越来越严格,其他用途如化工原料和溶剂等也尽量使用代用品,苯的需求日趋降低。另外,最近由于石化工业的发展,邻、间二甲苯在合成树脂、染料、药物、增塑剂和各种中间体上发现了有其独特优点的新用途,促使邻、间二甲苯的需求增加。

1. 芳烃转化反应的化学过程

芳烃的转化反应主要有异构化、歧化与烷基转移、烷基化和脱烷基化等几类反应。主要转化反应及其反应机理如下。

异构化反应

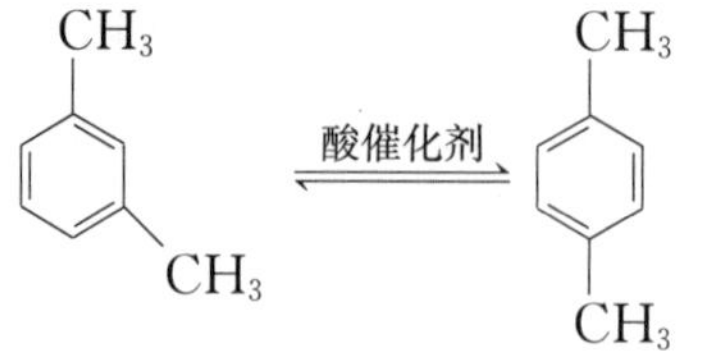

$$(4-1)$$

$$(4-2)$$

歧化反应

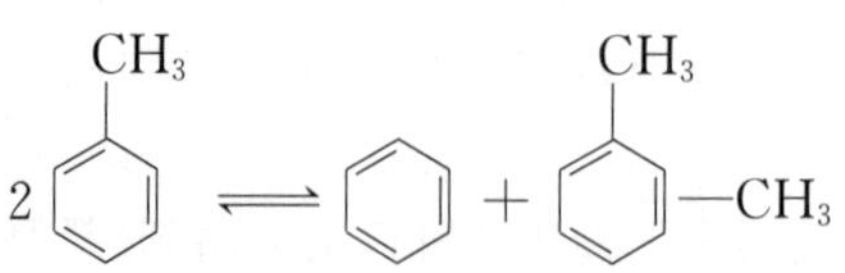

$$(4-3)$$

烷基化反应

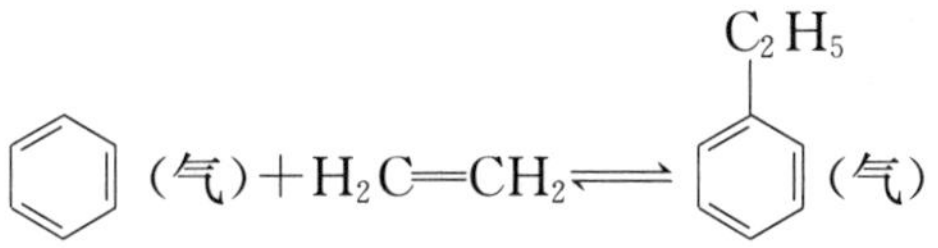

$$\text{苯（气）} + H_2C = CH_2 \rightleftharpoons \text{乙苯（气）} \tag{4-4}$$

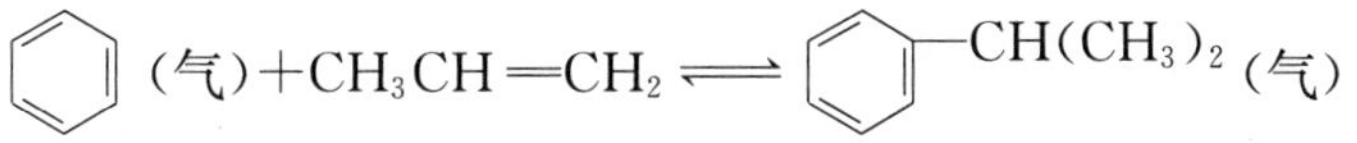

$$\text{苯（气）} + CH_3CH = CH_2 \rightleftharpoons \text{异丙苯} CH(CH_3)_2 \text{（气）} \tag{4-5}$$

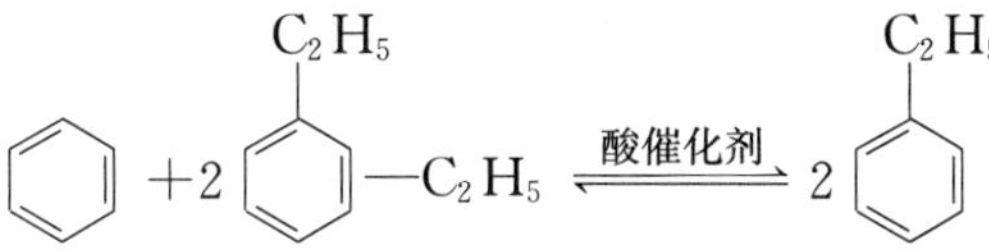

烷基转移反应

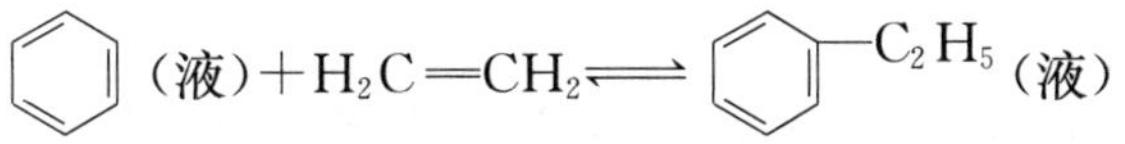

$$\text{苯（液）} + H_2C = CH_2 \rightleftharpoons \text{乙苯} C_2H_5 \text{（液）} \tag{4-6}$$

$$\text{苯} + 2 \text{二乙苯} - C_2H_5 \xrightarrow{\text{酸催化剂}} 2 \text{乙苯} \tag{4-7}$$

脱烷基化反应

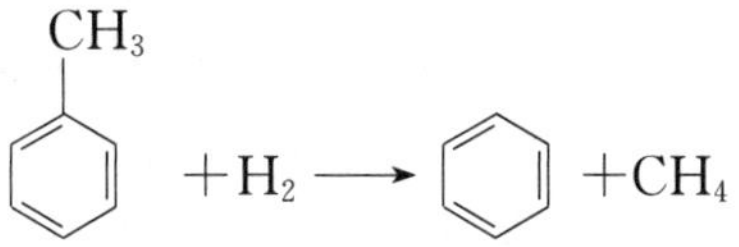

$$\text{甲苯} CH_3 + H_2 \longrightarrow \text{苯} + CH_4 \tag{4-8}$$

$$\text{甲基萘} CH_3 + H_2 \longrightarrow \text{萘} + CH_4 \tag{4-9}$$

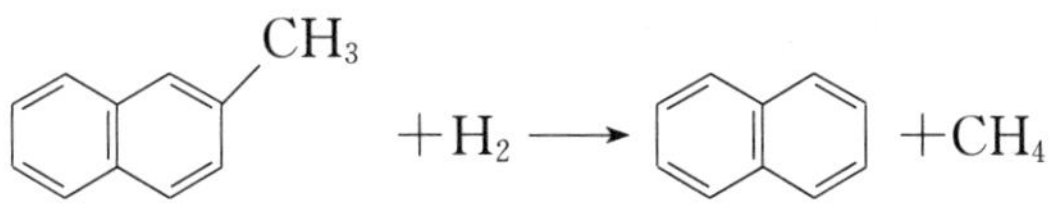

芳烃的转化反应(脱烷基反应除外)都是在酸性催化剂存在下进行的,具有相同的离子反应机理(但在特殊条件下,如自由基引发或高温条件下也可发生自由基反应),其反应历程包括正烃离子的生成及正烃离子的进一步反应。芳烃的异构化、歧化与烷基转移和烷基化都是按离子型反应机理进行的反应,而正烃离子是非常活泼的,在其寿命内可以参加多方面的反应,因此造成各类芳烃转化反应产物的复杂化。至于不同转化反应之间的竞争,主要决定于离子的寿命和它在有关反应中的活性。

2. 催化剂

芳烃转化反应是酸碱型催化反应,其反应速度不仅与芳烃(和烯烃)的碱性有关,也与酸性催化剂的活性有关。而酸性催化剂的活性与酸浓度、酸强度和酸存在的形态均有关。

芳烃转化反应所采用的催化剂主要有以下两类。

(1) 酸性卤化物

酸性卤化物分子如 $AlBr_3$、$AlCl_3$、BF_3 等都具有接受一对电子对的能力,是路易斯酸。

(2) 固体酸

① 浸附在适当载体上的质子酸

例如,载于载体上的 H_2SO_4、H_3PO_4、HF 等。

② 浸附在适当载体上的酸性卤化物

例如，载于载体上的 $AlCl_3$、$AlBr_3$、BF_3、$FeCl_3$、$ZnCl_2$ 和 $TiCl_4$。

③ 混合氧化物催化剂

常用的是 $SiO_2 - Al_2O_3$ 催化剂，亦称硅酸铝催化剂，主要用于异构化和烷基化反应。

④ 贵金属-氧化硅-氧化铝催化剂

主要是 $Pt/SiO_2 - Al_2O_3$ 催化剂，这类催化剂不仅具有酸功能，也具有加氢脱氢功能。

⑤ 分子筛催化剂

经改性的 Y 型分子筛、丝光沸石(亦称 M 型分子筛)和 ZSM 系列分子筛是广泛用于芳烃歧化与烷基转移、异构化和烷基化等反应的催化剂。

4.2 芳烃转化

各种芳烃组分中用途最广、需求量最大的是苯与对二甲苯，其次是邻二甲苯。甲苯、间二甲苯及 C_9 芳烃迄今尚未获得重大的化工利用，而有所过剩。为解决苯与对二甲苯的迫切需求，在 20 世纪 60 年代初发展了脱烷基制苯工艺；在 20 世纪 60 年代后期又发展了甲苯歧化、甲苯、C_9 芳烃烷基转移及二甲苯异构化等芳烃转化工艺。这些工艺是增产苯与对二甲苯的有效手段，从而得到较快的发展。

4.2.1 芳烃脱烷基化

烷基芳烃分子中与苯环直接相连的烷基，在一定条件下可以被脱去，此类反应称为芳烃的脱烷基化，工业上主要应用于甲苯脱甲基制苯、甲基萘脱甲基制萘。

1. 脱烷基反应的化学过程——以甲苯加氢脱烷基制苯为例

甲苯加氢脱烷基制苯是 20 世纪 60 年代以后，由于对苯的需要量增长很快，为了调整苯的供需平衡而发展起来的增产苯的途径之一。

主反应如式(4-8)所示。

副反应

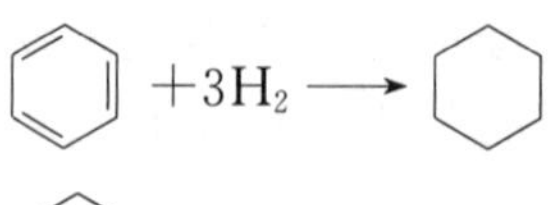

$$+3H_2 \longrightarrow \qquad\qquad (4-10)$$

$$+6H_2 \longrightarrow 6CH_4 \qquad\qquad (4-11)$$

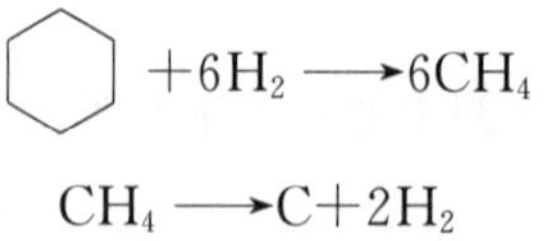

$$CH_4 \longrightarrow C+2H_2 \qquad\qquad (4-12)$$

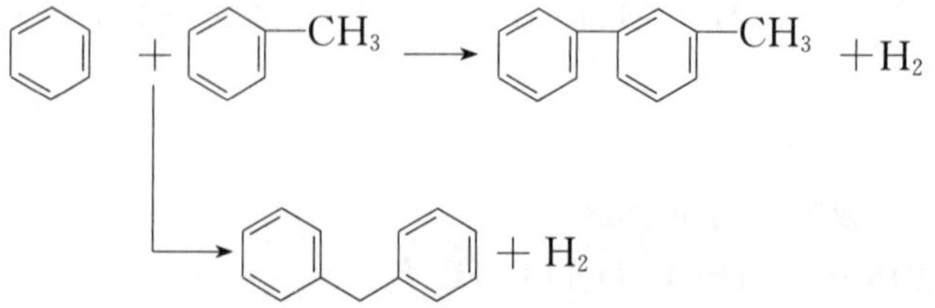

$$+ \quad -CH_3 \longrightarrow \quad -CH_3 \quad +H_2 \qquad\qquad (4-13)$$

$$\longrightarrow \qquad +H_2$$

2. 脱烷基化的方法

（1）烷基芳烃的催化脱烷基

烷基苯在催化裂化的条件下可以发生脱烷基反应生成苯和烯烃。此反应为苯烷基化的逆反应，是强吸热反应。例如，异丙苯在硅酸铝催化剂作用下于 $350℃\sim550℃$ 催化脱烷基成苯和丙烯。

$$C_6H_5CH(CH_3)_2 \longrightarrow C_6H_6+CH_3CHCH_2 \tag{4-14}$$

（2）烷基芳烃的催化氧化脱烷基

烷基芳烃在某些氧化催化剂作用下，用空气氧化可发生氧化脱烷基生成芳烃母体及二氧化碳和水。

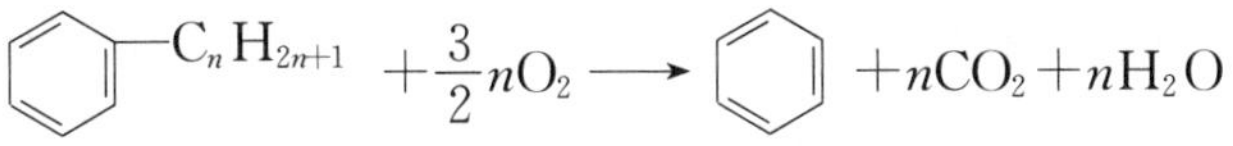

$$\bigcirc\!-\!C_nH_{2n+1}+\frac{3}{2}nO_2 \longrightarrow \bigcirc+nCO_2+nH_2O \tag{4-15}$$

（3）烷基芳烃的加氢脱烷基

在加压及大量氢气存在的条件下，使烷基芳烃发生氢解反应脱去烷基生成母体芳烃和烷烃。

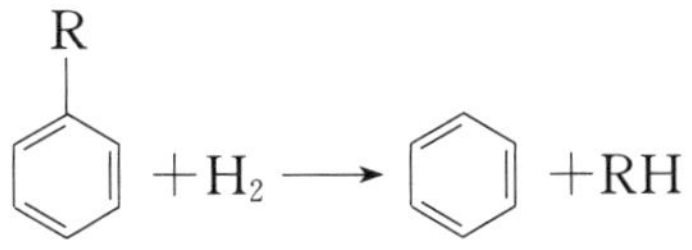

$$\overset{R}{\bigcirc}+H_2 \longrightarrow \bigcirc+RH \tag{4-16}$$

（4）烷基苯的水蒸气脱烷基法

本法与加氢脱烷基的反应条件相同，但用水蒸气代替氢气进行脱烷基反应。

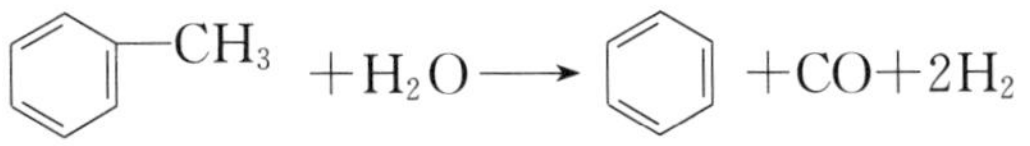

$$\bigcirc\!-\!CH_3+H_2O \longrightarrow \bigcirc+CO+2H_2 \tag{4-17}$$

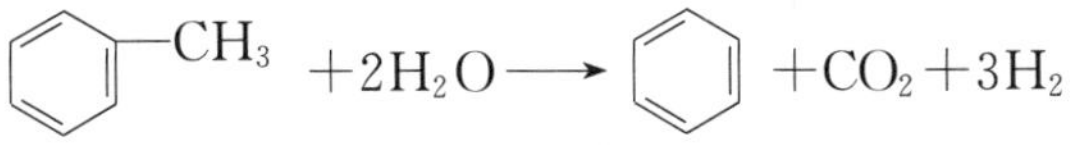

$$\bigcirc\!-\!CH_3+2H_2O \longrightarrow \bigcirc+CO_2+3H_2 \tag{4-18}$$

3. 工业生产方法

脱烷基制苯是甲苯最大的化工利用途径，也是苯的主要来源之一，但随着苯的使用受到限制，此类装置发展趋于停滞，甚至有的已经关闭。我国仅有少量甲苯用于脱烷基制苯。脱烷基制苯工艺分为催化脱烷基与热脱烷基两种。

4.2.2　芳烃歧化与烷基转移

芳烃的歧化是指两个相同芳烃分子在酸性催化剂作用下，一个芳烃分子上的侧链烷基转移到另一个芳烃分子上去的反应，如式（4-3）。烷基转移反应是指两个不同芳烃分子之间发生烷基转移的过程，如式（4-7）。从以上两式可以看出歧化和烷基转移反应互为逆反应。在工业中应用最广的是甲苯的歧化反应，通过甲苯歧化反应可使用途较少而过剩的甲苯转化为苯和二甲苯两种重要的芳烃原料。若同时进行 C_9 芳烃的烷基转移反应，还可增产二甲苯。

1. 甲苯歧化的主、副反应

甲苯歧化的主反应如式（4-3）所示，是可逆吸热反应，但反应热效应甚小。

甲苯歧化反应的副反应有如下几种。

(1) 产物二甲苯的二次歧化

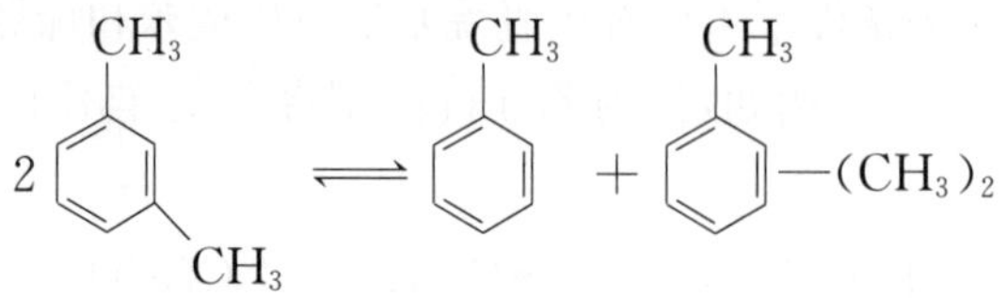

$$2 \text{[苯环-CH}_3\text{,CH}_3\text{]} \rightleftharpoons \text{[苯环-CH}_3\text{]} + \text{[苯环-CH}_3\text{]}-(CH_3)_2 \tag{4-19}$$

$$2 \text{[苯环-CH}_3\text{]}-(CH_3)_2 \rightleftharpoons \text{[苯环-CH}_3\text{]}-CH_3 + \text{[苯环-CH}_3\text{]}-(CH_3)_3 \tag{4-20}$$

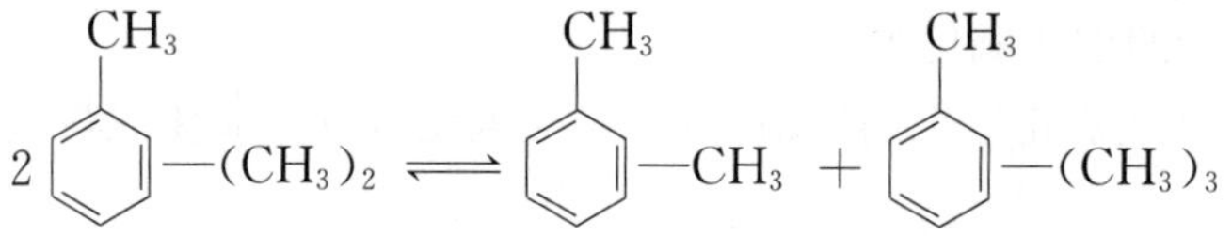

(2) 产物二甲苯与原料甲苯或副产物多甲苯之间的烷基转移反应

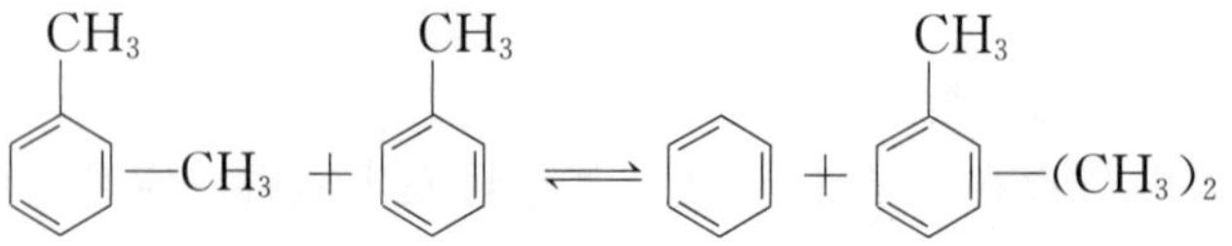

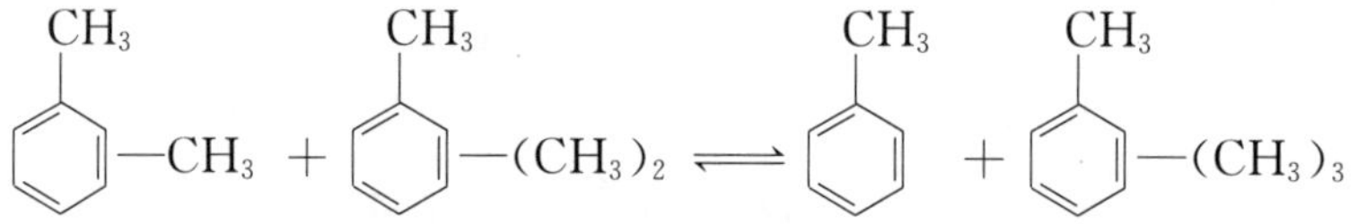

$$\text{[苯环-CH}_3\text{]}-CH_3 + \text{[苯环-CH}_3\text{]} \rightleftharpoons \text{[苯环]} + \text{[苯环-CH}_3\text{]}-(CH_3)_2 \tag{4-21}$$

$$\text{[苯环-CH}_3\text{]}-CH_3 + \text{[苯环-CH}_3\text{]}-(CH_3)_2 \rightleftharpoons \text{[苯环-CH}_3\text{]} + \text{[苯环-CH}_3\text{]}-(CH_3)_3 \tag{4-22}$$

$$\text{[苯环-CH}_3\text{]} + \text{[苯环-CH}_3\text{]}-(CH_3)_2 \rightleftharpoons 2\text{[苯环-CH}_3\text{]}-CH_3 \tag{4-23}$$

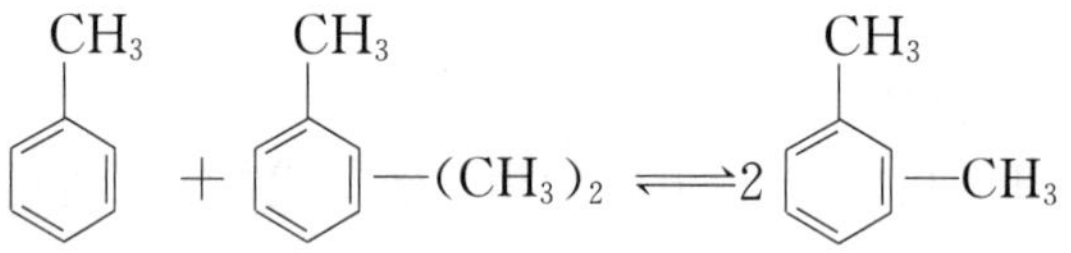

(3) 甲苯的脱烷基反应

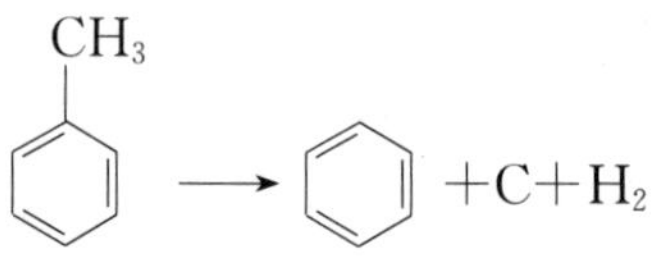

$$\text{[苯环-CH}_3\text{]} \rightarrow \text{[苯环]} + C + H_2 \tag{4-24}$$

(4) 芳烃脱氢缩合生成稠环芳烃和焦

此副反应的发生会使催化剂表面迅速结焦而活性下降,为了抑制焦的生成和延长催化剂的寿命,工业生产上采用临氢歧化法。在氢存在下进行甲苯歧化反应,不仅可抑制焦的生成,也能阻抑副反应式(4-24)的进行,避免碳的沉积。但在临氢条件下会增加甲苯加氢脱甲基转化为苯和甲烷以及苯环氢解为烷烃的副反应,尤其后者会使芳烃的收率降低,应尽量减少发生。

2. 甲苯歧化产物的平衡组成

甲苯歧化是可逆反应,反应热效应小,温度对平衡常数的影响不大。

甲苯歧化反应过程比较复杂,除所生成的二甲苯会发生异构化反应外,还会发生一系列歧化和烷基转移反应,故所得歧化产物是多种芳烃的平衡混合物。

4.2.3 C$_8$芳烃异构化

工业上 C$_8$ 芳烃的异构化是以不含或少含对二甲苯的 C$_8$ 芳烃为原料,通过催化剂的作用,转化成浓度接近平衡浓度的 C_8 芳烃,从而达到增产对二甲苯的目的。

1. 主副反应及热力学分析

C_8 芳烃异构化时,可能进行的主反应是三种二甲苯异构体之间的相互转化和乙苯与二甲苯之间的转化;副反应是歧化和芳烃的加氢反应等。

2. 动力学分析

(1) 二甲苯的异构化过程

对于二甲苯异构化的反应图式有两种看法。

一种是三种异构体之间的相互转化:

$$间二甲苯$$
$$\Updownarrow \qquad \Updownarrow \qquad (4-25)$$
$$邻二甲苯 \Longleftrightarrow 对二甲苯$$

另一种是连串式异构化反应:

$$邻二甲苯 \Longleftrightarrow 间二甲苯 \Longleftrightarrow 对二甲苯 \qquad (4-26)$$

(2) 乙苯的异构化过程

根据在 Pt/Al_2O_3 催化剂上对乙苯的气相临氢异构化的研究,整个异构化过程包括了加氢、异构和脱氢等反应。而低温有利于加氢,高温有利于异构和脱氢,故只有协调好各种条件才能使乙苯异构化得到较好的结果。

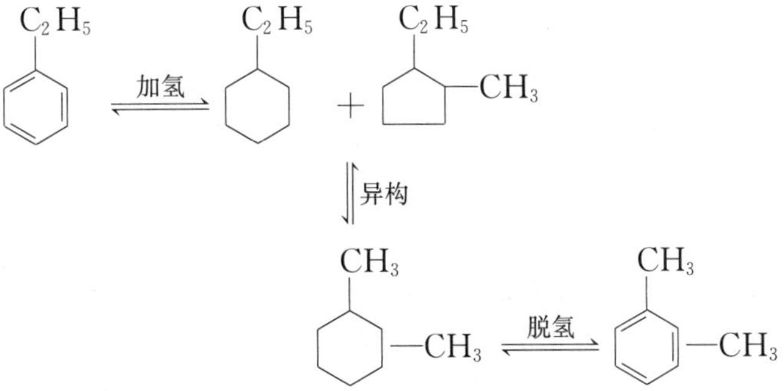

$$(4-27)$$

3. 催化剂

异构化催化剂主要有无定型 $SiO_2 - Al_2O_3$ 催化剂、负载型铂催化剂、ZSM 分子筛催化剂和 $HF - BF_3$ 催化剂等。

4.2.4　芳烃烷基化

芳烃的烷基化是芳烃分子中苯环上的一个或几个氢被烷基取代而生成烷基芳烃的反应,在芳烃的烷基化反应中以苯的烷基化最为重要。这类反应在工业中主要用于生产乙苯、异丙苯和十二烷基苯等。能为烃的烷基化提供烷基的物质称为烷基化剂,可采用的烷基化剂有多种,工业上常用的有烯烃和卤代烷烃。烯烃如乙烯、丙烯、十二烯,不仅具有较好的反应活性,而且比较容易得到。由于烯烃在烷基化过程中形成的正烃离子会发生骨架重排而取得最稳定的结构,所以乙烯以上烯烃与苯进行烷基化反应时,只能得到异构烷基苯而不能得到正构烷基苯。烯烃的活泼顺序为异丁烯＞正丁烯＞乙烯;卤代烷烃主要是氯代烷烃,如氯乙烷、氯代十二烷等。此外,醇类、酯类、醚类等也可作为烷基化剂。

4.3 C₈芳烃分离

4.3.1 C₈芳烃组成与性质

各种来源的 C_8 芳烃都是三种二甲苯异构体与乙苯的混合物。邻二甲苯与间二甲苯的沸点差为 5.3℃,工业上可采用精馏法分离;乙苯与对二甲苯的沸点差为 2.2℃,工业上尚可用 300~400 块板的精馏塔进行分离,但绝大多数的加工流程都不采用耗能大的精馏法回收乙苯,而是在异构化装置中将其转化;间二甲苯与对二甲苯沸点接近,借助普通的精馏法进行分离是非常困难的。所以 C_8 芳烃分离的技术难点主要在于间二甲苯与对二甲苯的分离。最早工业上主要利用凝固点差异采用深冷结晶分离法,后又开发了吸附分离和络合分离的工艺,其中吸附分离占有越来越重要的地位。

4.3.2 C₈芳烃单体分离

1. 邻二甲苯和乙苯的分离

(1) 邻二甲苯的分离

C_8 芳烃中邻二甲苯的沸点最高,与关键组分间二甲苯的沸点相差 5.3℃,可以用精馏法分离,精馏塔需 150~200 块塔板,两塔串联,回流比为 7~10,可得产品纯度为 98%~99.6%。

(2) 乙苯的分离

C_8 芳烃中乙苯的沸点最低,与关键组分对二甲苯的沸点仅差 2.2℃,可以用精馏法分离,但较困难。工业上分离乙苯的精馏塔实际塔板数达 300~400(相当于理论塔板数 200~250),三塔串联,塔釜压力 0.35 MPa~0.4 MPa,回流比为 50~100,可得纯度在 99.6% 以上的乙苯。其他分离方法有络合萃取法(如日本三菱瓦斯化学公司的 Pomex 法)以及吸附法(如美国 UOP 公司的 Ebex 法)。

2. 对二甲苯、间二甲苯的分离

由于对二甲苯与间二甲苯的沸点差只有 0.75℃,难以采用精馏方法进行分离。目前工业上分离对二甲苯的方法主要有深冷结晶分离法、络合分离法和模拟移动床吸附分离法三种。

(1) 深冷结晶分离法

C_8 芳烃深度冷却至 -60℃~-75℃时,熔点最高的对二甲苯首先被结晶出来。在对二甲苯结晶过程中,晶体内不可避免地包含一部分 C_8 芳烃混合物,从而影响对二甲苯的纯度。为提高对二甲苯纯度,工业上多采用二段结晶工艺。第一段结晶,对二甲苯纯度约为 85%~90%;第二段结晶,对二甲苯纯度可达 99.2%~99.5%。另外,受共熔温度的限制,如再降低温度,邻二甲苯、间二甲苯将同时被结晶出来,而此时,未结晶的C_8芳烃液体中仍含有对二甲苯的量约为 6.2%～6.9%。因此,结晶分离的单程收率较低,仅为 60%左右。

Amoco 对二甲苯结晶分离流程见图 4-4。

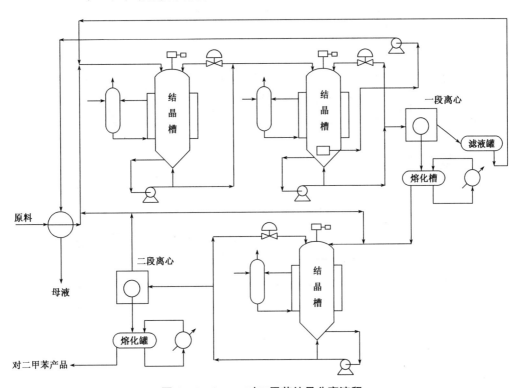

图 4-4　Amoco 对二甲苯结晶分离流程

(2) 络合萃取分离法

利用一些化合物能与二甲苯异构体形成配位化合物的特性可以达到分离各异构体的目的。络合分离法中最成功的工业实例是日本三菱瓦斯化学公司发展的 MGCC 法。此法是有效分离间二甲苯的唯一工业化方法,同时也使其他 C_8 芳烃分离过程大为简化。MGCC 络合萃取分离法分离二甲苯流程见图 4-5。

C_8 芳烃四个异构体与 HF 共存于一个系统时,形成两个互相分离的液层:上层为烃层,下层为 HF 层。当加入 BF 后,发生下列反应而生成在 HF 中溶解度大的配位化合物。

$$X+HF+BF_3 \longrightarrow XHBF_4 \tag{4-28}$$

$$MX+PXHBF_4 \longrightarrow PX+MXHBF_4 \tag{4-29}$$

$$MX+OXHBF_4 \longrightarrow OX+MXHBF_4 \tag{4-30}$$

(3) 吸附分离法

吸附分离是利用固体吸附剂吸附二甲苯各异构体的能力不同进行的一种分离方法。吸附分离法由美国 UOP 公司解决了三个关键问题而实现了工业化;一是研制成

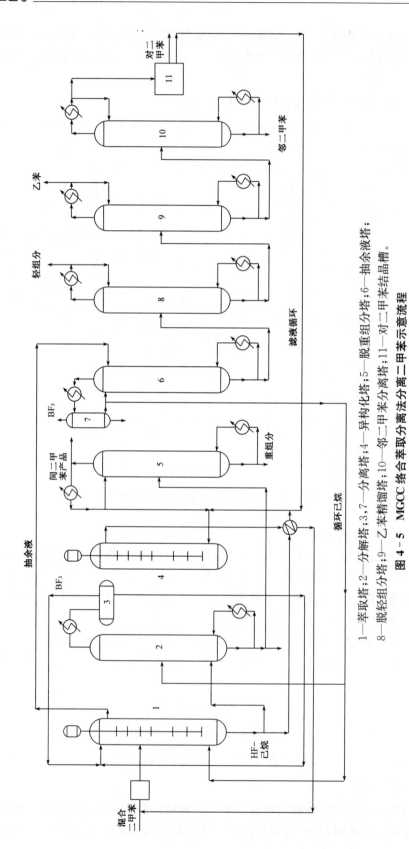

图 4-5 MGCC 络合萃取法分离二甲苯示意流程

1—萃取塔；2—分解塔；3,7—分离塔；4—异构化塔；5—脱重组分塔；6—抽余液塔；
8—脱轻组分塔；9—乙苯精馏塔；10—邻二甲苯分离塔；11—对二甲苯结晶槽。

功一种对各种二甲苯异构体有较高选择性的固体吸附剂;二是研制成功以24通道旋转阀进行切换操作的模拟移动床技术;三是选到一种与对二甲苯有相同吸附亲和力的脱附剂。吸附分离比结晶分离有较多的优点,如工艺过程简单,单程回收率达98%,生产成本较低,因此已取代深冷结晶,成为一种广泛采用的二甲苯分离技术。

4.4 芳烃生产技术发展方向

由于芳烃在化工原料中所占有的重要地位,其生产技术的发展受到了广泛重视。近年来,在开辟新的原料来源、开发新一代更高水平的催化剂和工艺、提高芳烃收率和选择性、降低能耗和操作费用、现有装置的技术改造、提高生产方案的灵活性等方面取得了长足的技术进步,大大提高了芳烃生产的技术水平。有的已经在工业生产中发挥重要作用,有的已显示出巨大的潜力。芳烃生产技术的发展主要有如下几个方面:

（1）扩大芳烃原料来源。

（2）工艺革新,提高技术水平。

（3）产品新用途促进了产品的结构调整。

（4）化学工程新技术发挥重要作用。

（5）新、老技术共同发展。

思考题

4-1 简述芳烃的主要来源及主要生产过程。

4-2 芳烃的主要产品有哪些? 各有何用途?

4-3 试论芳烃转化的必要性与意义。主要的芳烃转化反应有哪些?

4-4 试分析我国与美国、日本的芳烃生产各有何特点及其原因。

4-5 简述苯、甲苯和各种二甲苯单体的主要生产过程,并说明各自的特点。

4-6 简述芳烃生产技术的新进展及其主要特征。

4-7 如何理解芳烃生产、转化与分离过程之间的关系? 试组织两种不同的芳烃生产方案。

Chapter 4　Aromatic Conversion

4.1　Introduction

Aromatics are the general term for hydrocarbons containing phenyl structure. The "triphenyl" in aromatic hydrocarbons (benzene, toluene and xylene, referred to as BTX) and the "triene" in olefin (ethylene, propylene and butadiene) are the basic raw materials of the chemical industry with an important position. Benzene, toluene, xylene, ethylbenzene, isopropylbenzene, dodecyl benzene and naphthalene of aromatics are the most important, these products are widely used in synthetic resin, synthetic fiber, synthetic rubber, synthetic detergent, plasticizer, dye, medicine, pesticides, explosives, spices, specialty chemicals and other industries. It plays an extremely important role in developing national economy and improving people's life.

The aromatic hydrocarbons needed by chemical industry are benzene, toluene and xylene. Benzene can be used to synthesize styrene, cyclohexane, phenol, aniline and alkylbenzene, etc. Toluene is not only an excellent solvent for organic synthesis, but also can synthesize isocyanate, cresol, or prepare benzene by disproportionation and dealkylation. Xylene and ethylbenzene are both C_8 aromatic hydrocarbons. Xylene isomers are p-xylene, o-xylene and m-xylene respectively.

4.1.1　Sources and production methods of aromatic hydrocarbons

In the past 20 years, aromatic hydrocarbon production has developed rapidly. In 1986, the world's BTX production capacity was only 32.91 million tons, and in 2003, the world's BTX production capacity reached 87.387 million tons. China's BTX production capacity reached 6.148 million tons in 2003.

Aromatic hydrocarbons originally from coal coking industry. Due to the rapid development of organic synthesis industry, the demand for aromatic hydrocarbons increases, but the quantity and quality of coking aromatic hydrocarbons produced by coal coking industry cannot meet the demand. Therefore, many industrial-developed countries began to develop petroleum as raw materials to produce petroleum aromatic hydrocarbons to make up for the shortage. Up to now, petroleum aromatic hydrocarbons have become the main source of aromatic hydrocarbons, accounting for about 80% of all aromatic hydrocarbons.

Petroleum aromatics are mainly derived from naphtha reformate oil and cracking gasoline produced by pyrolysis of hydrocarbon to produce ethylene. Due to different resources in different countries, the proportion of aromatic hydrocarbons produced by pyrolysis gasoline in petroleum aromatic hydrocarbons is also different. Most of the ethylene production in the United States uses natural gas liquids as raw material, and the by-products of aromatic hydrocarbons are few, so the petroleum aromatic hydrocarbons in the United States mainly come from catalytic reforming oil. Ethylene production in Japan and Western European countries mainly uses naphtha as raw material, and there are many by-products of aromatics. The investment and operation cost of aromatics recovery from pyrolysis gasoline is lower than that from reformate oil. Therefore, the amount of aromatic hydrocarbons recovered from pyrolysis gasoline is large in Japan and Western Europe. With the development of ethylene industry and the shift of ethylene raw material from light hydrocarbon to naphtha and diesel oil, it is expected that the proportion of aromatics from pyrolysis gasoline production in the world's production of aromatic hydrocarbon will increase, and the mass of aromatic gasoline produced by naphtha steam cracking is about 25% of the raw material (mass). Coking aromatic hydrocarbon production is limited by metallurgical industry, and its output will maintain the current status or slightly increase.

1. Coking aromatics production

Under the action of high temperature, a series of physical and chemical changes of coal quality occur when coal is retorted in coke oven carbonization chamber. In addition to 75% coke, the by-product of raw gas is about 25%, of which crude benzene is about 1.1%, coal tar is about 4.0%. There are many kinds of chemicals in raw gas, whose composition and quantity fluctuate with coking temperature and raw coal blending. Crude benzene is recovered from raw gas after primary cooling, deamination, removing naphthalene and final cooling. Crude benzene is composed of a variety of aromatic hydrocarbons and other compounds, its main components are benzene, toluene and xylene.

The crude benzene was absorbed from the raw gas by washing oil and desorbed by distillation. The recovery of crude benzene was about 90%.

2. Production of petroleum aromatics

The production process of petroleum aromatic hydrocarbons is shown in Figure 4-1.

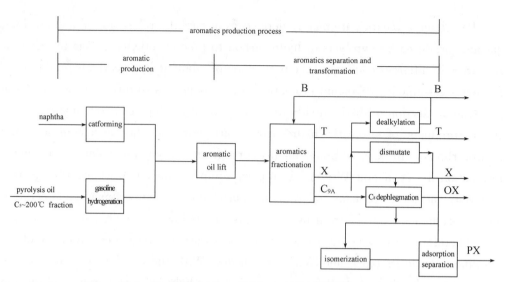

Figure 4 - 1　**Production process of petroleum aromatic hydrocarbons**

The process of aromatics production from naphtha and pyrolysis gasoline can be divided into three parts: reaction, separation and transformation. Different countries have different production modes of petroleum aromatics. In the United States, which is rich in aromatic hydrocarbon resources, there is a large demand for benzene, which needs to supplement the deficiency of benzene through toluene dealkylation to produce benzene. P-xylene and o-xylene are mainly separated from catalytic reforming oil, and few processes such as alkyl transfer and xylene isomerization are rarely used. In Western Europe and Japan, aromatic hydrocarbon resources are not rich, so aromatic hydrocarbon conversion processes are often used. China has few aromatic hydrocarbon resources, which means it is necessary to make full use of the limited aromatic hydrocarbon resources, so toluene and C_9 aromatic alkyl transfer, toluene disproportionation, xylene isomerization and other processes are used, but toluene dealkylation process is rarely used.

（1）Production of aromatic hydrocarbons by catalytic reforming

Catalytic reforming is one of the main secondary processing units in oil refining industry. It is used to produce high-octane gasoline or BTX and other aromatic hydrocarbons. About 10% of the units are used to produce aromatic hydrocarbons.

① Basic chemical reaction of catalytic reforming

Catalytic reforming includes naphthenic dehydrogenation, five-membered ring isomerization dehydrogenation, dehydrogenation and cyclization of paraffins, isomerization hydrocracking of paraffins to generate aromatic hydrocarbons. At the same time, there are also side reactions such as hydrocracking and olefin polymerization. In actual production, certain measures should be taken to suppress the occurrence of side reactions.

② Raw materials for catalytic reforming

Naphtha fraction is the raw material for catalytic reforming. The composition of hydrocarbon group and distillation range of naphtha have a decisive influence on the content and composition of aromatic hydrocarbons in the reformating oil. Naphtha with more naphthene is generally selected as raw materials. For distillation, the selection of range generally according to the production purpose. For the production of BTX, the actual boiling range of raw materials is 65℃~145℃. If C_9 aromatics are used to increase the production of C_8 aromatics, the actual boiling range is 70℃~177℃.

③ Reforming catalyst

Reforming catalyst is made of one or more noble metal elements highly dispersed on the porous carrier. The main metal is platinum, its mass fraction is 0.3%~0.7%. There are also halogen elements (fluorine or chlorine), with a mass fraction between 0.5% and 1.5%.

④ Catalytic reforming process

The catalytic reforming process is carried out in the presence of hydrogen, the general reaction temperature is 425℃~525℃, the reaction pressure is 0.7 MPa~3.5 MPa, the space velocity is 1.5~3.0 h^{-1}, the hydrogen-oil molar ratio is 3~6, the yield of reformate oil is 75.85%(that is, the yield of hydrocarbons above C_5), the aromatic content is 30%~70%. The research octane number (RON) of reforming oil can be as high as 100.

The principle flow of continuous reforming reaction is shown in Figure 4-2.

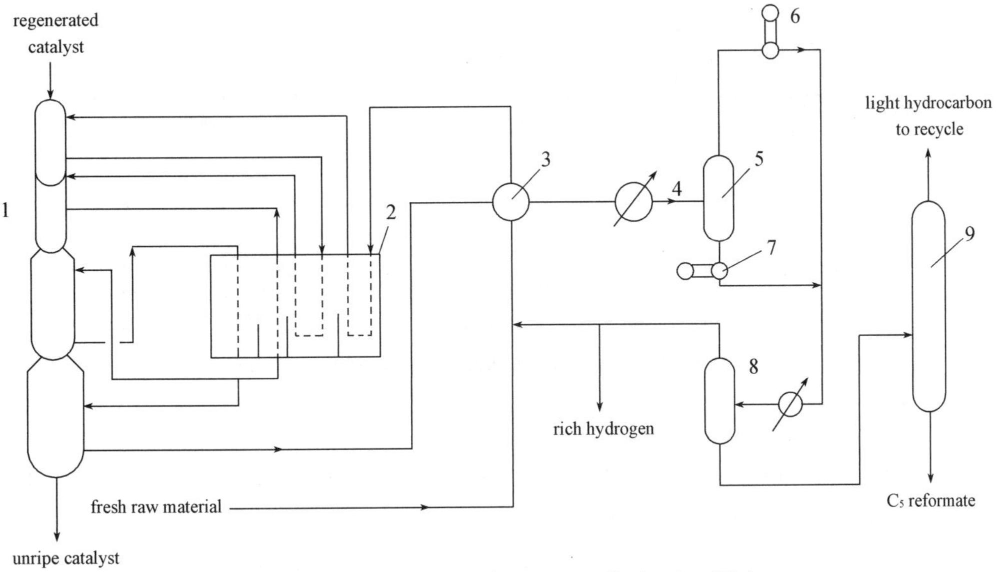

1 – Reactor; 2 – Heating furnace; 3 – Heat exchanger; 4 – Cooler; 5 – High-pressure separator;

6 – Compressor; 7 – Pump; 8 – Low pressure separator; 9 – Stabilizer.

Figure 4-2   Flow chart of the principle of continuous reforming reaction

（2）Pyrolysis gasoline to produce aromatics

Ethylene is one of the most important basic raw materials in petrochemical industry. With the development of ethylene industry, pyrolysis gasoline as a by-product has become an important source of petroleum aromatics. Benzene produced from pyrolysis gasoline reached 7.94 million tons in 1984, which equivalented to more than a third of benzene production that year. In addition to $40\% \sim 60\%$ $C_6 \sim C_9$ aromatics, cracking gasoline also contains a considerable amount of diolefin and monoolefin, a small amount of alkanes and trace amounts of oxygen, nitrogen, sulfur and arsenic compounds. Olefins and all the impurities in pyrolysis gasoline far exceed the standards allowed by the subsequent process of aromatics production. It must be pretreated and hydrorefined before it can be used as raw material for aromatics extraction.

① Pyrolysis gasoline pretreatment

The pyrolysis gasoline is $C_5 \sim 200℃$ fraction. The C_5 fraction contains many isoprene, m-pentadiene and cyclopentadiene, which are important raw materials for synthetic rubber and fine chemicals. The hydrogenation of diolefin from C_5 fraction to olefin is a good component of gasoline. Further hydrogenation of C_5 fraction olefin to C_5 alkanes can be used as raw materials for hydrocarbon cracking.

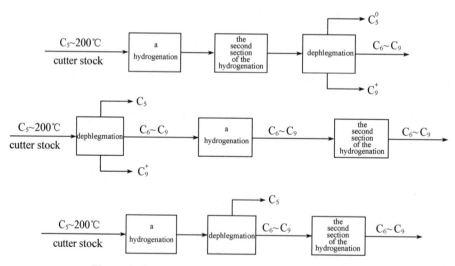

Figure 4 - 3   Fractionation diagram of cracking gasoline

② Cracking gasoline hydrogenation

Cracking gasoline hydrogenation is a widely used refining method. Due to the different process conditions for removing diolefin, single olefin and organic compounds from pyrolysis gasoline, two-stage hydrotreating process is generally adopted. The purpose of the first hydrogenation is to convert the diolefin into single olefin and the alkenyl aromatics into aromatic hydrocarbons. The purpose of the second hydrogenation is to saturate the single olefin and remove sulfur, oxygen, nitrogen and other organic compounds.

(3) Aromatization of light hydrocarbons and lightening of heavy aromatics

Naphtha is the main raw material for catalytic reforming, pyrolysis, and gasoline production. Due to the increasing demand for gasoline, people have to find other raw materials to produce aromatics besides naphtha. One is aromatization using LPG and other light hydrocarbons, the other is lightening of heavy aromatics. Important progress has been made in industrialization and basic research of these two raw material routes. Aromatization of liquefied petroleum gas to produce aromatic hydrocarbons has been preliminarily industrialized.

4.1.2　Separation of aromatic distillates

The aromatic fractions obtained from catalytic reforming and hydrorefining pyrolysis gasoline are mixtures of aromatic and non-aromatic hydrocarbons. Because the boiling points of aromatic hydrocarbons are very close to those non-aromatic hydrocarbons with the same carbon number, and azeotropes are sometimes formed. It is difficult to separate them by ordinary distillation methods. In order to meet the requirements of aromatic purity, solvent extraction and extractive distillation are mainly used in industry at present. The former is suitable for the separation of benzene, toluene, xylene and so on from the wide fraction. Extractive distillation is suitable for separating single aromatic hydrocarbon with high purity from narrow fraction with high aromatic content.

1. Solvent extraction

Solvent extraction is using the selective dissolution of aromatic hydrocarbons and non-aromatic hydrocarbons by one or more solvents (extractants) to separate aromatic hydrocarbons. The performance of solvent is directly related to aromatic yield, aromatic quality, utilities consumption and equipment investment. The basic requirements for solvent performance are: good dissolution selectivity for aromatic hydrocarbon, high solubility, large density difference with extraction raw material, small latent heat and heat capacity of evaporation, small steam pressure, good chemical and thermal stability, small corrosion, etc.

2. Extraction distillation

Extraction distillation is a process using the principle that when polar solvents and hydrocarbon are mixing, the steam pressure of hydrocarbon can reduce and the mixture of the initial boiling point can improve. Because this effect has the greatest impact on aromatics, followed by cycloalkanes, and the least on alkanes, so it is conducive to the separation of aromatics and non-aromatics. The solvent extraction and distillation processes are combined in an extractive distillation column. The material to be separated enters the middle part of the extraction distillation column after preheating, and the solvent enters the top of the column for counter-current mass transfer. The non-aromatic hydrocarbons containing trace aromatic hydrocarbons are steam out from the top of the tower. After condensation, part of them flow back and the rest are out of the device. The solvent and aromatics are discharged from the bottom of the column and enter the stripper. After the aromatics are removed by the steam, the solvent is recycled. Extractive distillation is a process of extracting some high purity aromatic hydrocarbons directly from aromatic hydrocarbon rich fractions. The raw material needs to be prefractionated first to remove the light and heavy fractions and the central fraction sent for extraction distillation. Extractive distillation can be used to extract benzene, toluene, or xylene from reformed oil or cracked gasoline (after hydrogenation), as well as to extract styrene directly from unhydrogenated cracked gasoline.

4.1.3 Transformation of aromatics

The compositions of aromatic fraction from different sources are different, and the production of aromatics obtained are also different. Benzene and p-xylene, followed by o-xylene, are the most widely used and the most demanding aromatic components. Toluene, m-xylene and C_9 arenes have not achieved significant chemical utilization, so there is a surplus so far.

It is worth noting that, due to the toxicity of benzene, the restrictions on benzene content in gasoline are more and more strict, and other uses, such as chemical raw materials and solvents, also try to use substitutes, so the demand for benzene is decreasing. In addition, due to the recent development of the petrochemical industry, o-xylene and m-xylene has been found with unique advantages of new uses in synthetic resins, dyes, drugs, plasticizers and a variety of intermediates, prompting the demand for o-xylene and m-xylene.

1. Chemical process of aromatic hydrocarbon conversion reaction

Aromatic hydrocarbon conversion reactions mainly include isomerization, disproportionation and alkyl transfer, alkylation and dealkylation. The main transformation reactions and their reaction mechanisms are as follows.

Isomerization reaction

$$(4-1)$$

$$\underset{\text{CH}_3}{\bigcirc}\text{—CH}_3 \xrightleftharpoons{\text{acid catalyst}} \underset{\text{CH}_3}{\bigcirc}\text{—CH}_3 \tag{4-2}$$

Disproportionation reaction

$$2\ \underset{\text{CH}_3}{\bigcirc} \rightleftharpoons \bigcirc + \underset{\text{CH}_3}{\bigcirc}\text{—CH}_3 \tag{4-3}$$

Alkylation reaction

$$\bigcirc (g) + H_2C{=}CH_2 \rightleftharpoons \underset{}{\overset{C_2H_5}{\bigcirc}} (g) \tag{4-4}$$

$$\bigcirc (g) + CH_3CH{=}CH_2 \rightleftharpoons \bigcirc\text{—CH(CH}_3)_2\ (g) \tag{4-5}$$

Alkyl transfer reaction

$$\bigcirc (l) + H_2C{=}CH_2 \rightleftharpoons \bigcirc\text{—C}_2H_5\ (l) \tag{4-6}$$

$$\bigcirc + 2\ \underset{}{\overset{C_2H_5}{\bigcirc}}\text{—C}_2H_5 \xrightleftharpoons{\text{acid catalyst}} 2\ \underset{}{\overset{C_2H_5}{\bigcirc}} \tag{4-7}$$

Dealkylation reaction

$$\underset{}{\overset{CH_3}{\bigcirc}} + H_2 \longrightarrow \bigcirc + CH_4 \tag{4-8}$$

$$\underset{}{\overset{CH_3}{\bigcirc\bigcirc}} + H_2 \longrightarrow \bigcirc\bigcirc + CH_4 \tag{4-9}$$

Aromatic hydrocarbon conversion reactions (except dealkylation reactions) are all carried out in the presence of acidic catalysts, with the same ionic reaction mechanism (but under special conditions, such as free radical initiation or high temperature conditions can also occur free radical reactions), the reaction process includes the generation of positive hydrocarbon ions and further reactions of positive hydrocarbon ions. Aromatic isomerization, disproportionation, alkyl transfer and alkylation are all based on ionic reaction mechanism. Positive hydrocarbon ions are very active and can participate in multiple reactions within their life, thus resulting in the complexity of various aromatic conversion reaction products. Competition between different conversion reactions is mainly determined by the life time of the ion and its activity in the reaction concerned.

2. Catalyst

Aromatic conversion reaction is acid-base catalytic reaction. The reaction rate is not only related to the alkalinity of the aromatics (and alkenes), but also to the activity of the acid catalyst, which is related to the acid concentration, acid strength and the state of the acid.

There are two main types of catalysts used in aromatic conversion reaction.

(1) Acid halides

Acid halide molecules such as $AlBr_3$, $AlCl_3$, BF_3 all have the ability to accept a pair of electron and they are all Lewis acids.

(2) Solid acid

① Protic acid impregnated on an appropriate carrier

Such as H_2SO_4, H_3PO_4, HF supported on a carrier.

② Acid halides impregnated on an appropriate carrier

Such as $AlCl_3$, $AlBr_3$, BF_3, $FeCl_3$, $ZnCl_2$ and $TiCl_4$ supported on a carrier.

③ Mixed oxide catalyst

$SiO_2 - Al_2O_3$ catalyst is commonly used, also known as aluminum silicate catalyst, mainly used in isomerization and alkylation reaction.

④ Noble metal-monox-alumina catalysts

It is mainly Pt/ $SiO_2 - Al_2O_3$ catalyst, which not only has acid function, but also has hydrogenation-dehydrogenation function.

⑤ Molecular sieve catalyst

Modified Y zeolite, mordenite (also known as M zeolite) and ZSM zeolite are widely used as catalysts for aromatics disproportionation, alkyl transfer, isomerization and alkylation.

4.2 Aromatic conversion

Benzene and p-xylene, followed by o-xylene, are the most widely used and greatest demanded aromatic components. Toluene, m-xylene and C_9 arenes have not been used in chemical industry, so are in excess. In order to meet the urgent demand for benzene and p-xylene, the process of dealkylation of benzene production was developed in the early 1960s. In the late 1960s, toluene disproportionation, toluene or C_9 aromatics alkyl transfer and xylene isomerization were developed. These processes are effective means to increase benzene and p-xylene production, so as to achieve rapid development.

4.2.1　Dealkylation of aromatics

Alkyl groups in aromatic molecules directly linked to benzene rings can be removed under certain conditions, such reactions are called aromatics dealkylation. In industry, it is mainly used in toluene demethylation to benzene and methyl naphthalene demethylation to naphthalene.

1. Chemical process of dealkylation reaction—take toluene hydrodealkylation to benzene as an example

Due to the rapid increase of benzene demand, hydrodealkylation of toluene to benzene is one of the ways to increase benzene production, which is developed to adjust the balance of benzene supply and demand since 1960s.

The principal reaction is shown in formula (4-8).

Side reactions

$$\bigcirc + 3H_2 \longrightarrow \bigcirc \qquad (4-10)$$

$$\bigcirc + 6H_2 \longrightarrow 6CH_4 \qquad (4-11)$$

$$CH_4 \longrightarrow C + 2H_2 \qquad (4-12)$$

$$\bigcirc + \bigcirc{-}CH_3 \longrightarrow \bigcirc{-}\bigcirc{-}CH_3 + H_2 \qquad (4-13)$$

$$\longrightarrow \bigcirc{-}\bigcirc + H_2$$

2. Dealkylation method

(1) Catalytic dealkylation of alkyl aromatics

Alkyl benzene can be dealkylated to benzene and olefin in catalytic cracking. This reaction is the reverse reaction of benzene alkylation and is a strong endothermic reaction. For example, isopropyl benzene catalytically dealkylated to benzene and propylene under the action of aluminum silicate catalyst at 350℃~550℃.

$$C_6H_5CH(CH_3)_2 \longrightarrow C_6H_6 + CH_3CHCH_2 \qquad (4-14)$$

(2) Catalytic oxidative dealkylation of alkyl aromatics

Alkyl aromatics can be oxidized by air with some oxidation catalysts to form maternal aromatics, carbon dioxide and water.

$$\bigcirc{-}C_nH_{2n+1} + \frac{3}{2}nO_2 \longrightarrow \bigcirc + nCO_2 + nH_2O \qquad (4-15)$$

(3) hydrodealkylation of alkyl aromatics

In the presence of a large amount of hydrogen and under pressure, the alkyl aromatics undergo hydrogenolysis reaction to remove the alkyl to form maternal aromatics and alkanes.

$$\text{(structure with R)} + H_2 \longrightarrow \text{(benzene ring)} + RH \qquad (4-16)$$

(4) Steam decalkyl method of alkyl benzene

Under the same reaction condition of hydrodealkylation, water vapor is used instead of hydrogen for dealkylation.

$$\text{(toluene)} + H_2O \longrightarrow \text{(benzene)} + CO + 2H_2 \qquad (4-17)$$

$$\text{(toluene)} + 2H_2O \longrightarrow \text{(benzene)} + CO_2 + 3H_2 \qquad (4-18)$$

3. Industrial production methods

Dealkylation to benzene is the largest chemical utilization of toluene and one of the main sources of benzene. However, as the use of benzene is restricted, the development of such devices tends to be stagnant, and some have even been closed. Only a small amount of toluene is used for dealkylation to benzene in China. Dealkylation of benzene production process is divided into catalytic dealkylation and thermal dealkylation.

4.2.2 Aromatics disproportionation and alkyl transfer

Aromatic disproportionation refers to the reaction of two identical aromatic molecules under the action of acid catalyst in which the side chain alkyl of one aromatics molecule transfers to another aromatics molecule like equation (4 - 3). Alkyl transfer reaction refers to the process of alkyl transfer between two different aromatic molecules, as equation (4 - 7). It can be seen from the above two equations that disproportionation and alkyl transfer reaction are reversible reactions. Disproportionation of toluene is the most widely used in industry. Through the toluene disproportionation reaction, less used and surplus toluene can be converted into benzene and xylene, which are two important aromatic raw materials. If the alkyl transfer reaction of C_9 aromatic is carried out at the same time, the production of xylene can also be increased.

1. Main and side reactions of toluene disproportionation

The main reaction of toluene disproportionation is shown in equation (4 - 3), which is a reversible endothermic reaction, but the reaction thermal effect is very small.

The side reactions of toluene disproportionation are as follows.

(1) Secondary disproportionation of the xylene

$$2\ \text{(toluene)} \rightleftharpoons \text{(benzene)} + \text{(xylene)}\!-\!(CH_3)_2 \qquad (4-19)$$

$$2\ \text{(toluene)}\!-\!(CH_3)_2 \rightleftharpoons \text{(toluene)}\!-\!CH_3 + \text{(toluene)}\!-\!(CH_3)_3 \qquad (4-20)$$

(2) Alkyl transfer reaction between product (xylene) and raw material (toluene) or by-product (polytoluene)

$$\text{(toluene)}\!-\!CH_3 + \text{(toluene)} \rightleftharpoons \text{(benzene)} + \text{(toluene)}\!-\!(CH_3)_2 \qquad (4-21)$$

$$\text{(toluene)}\!-\!CH_3 + \text{(toluene)}\!-\!(CH_3)_2 \rightleftharpoons \text{(toluene)} + \text{(toluene)}\!-\!(CH_3)_3 \qquad (4-22)$$

$$\text{(toluene)} + \text{(toluene)}\!-\!(CH_3)_2 \rightleftharpoons 2\ \text{(toluene)}\!-\!CH_3 \qquad (4-23)$$

(3) Dealkylation of toluene

$$\text{(toluene)} \longrightarrow \text{(benzene)} + C + H_2 \qquad (4-24)$$

(4) Dehydrogenation and condensation of aromatic hydrocarbons to form polycyclic aromatic hydrocarbons and coke

The side reaction will cause rapid coking on the surface of the catalyst and decrease its activity. In order to inhibit the formation of coke and prolong the life of catalyst, hydrogen disproportionation method is used in industrial production. Toluene disproportionation in the presence of hydrogen can not only inhibit the generation of coke, but also inhibit the side reaction (4-24) and avoid the deposition of carbon. However, in the presence of hydrogen, side reactions such as hydrodemethylation of toluene into benzene and methane and hydrogenolysis of benzene ring into alkane are also added. The latter one will reduce the yield of aromatic hydrocarbons and should be minimized.

2. Equilibrium composition of toluene disproportionation products

Toluene disproportionation is a reversible reaction with small thermal effect and little effect of temperature on equilibrium constant.

The disproportionation reaction of toluene is complicated. In addition to the isomerization of the generated xylene, a series of disproportionation and alkyl transfer reactions also occur. Therefore, the disproportionated product is a balanced mixture of aromatic hydrocarbons.

4.2.3　Isomerization of C_8 aromatics

Industrial isomerization of C_8 aromatics is based on C_8 aromatics with little or no p-xylene as raw materials, and through the action of catalyst, the concentration of C_8 aromatics is close to the equilibrium concentration, so as to achieve the purpose of increasing p-xylene production.

1. Principal and side reactions and thermodynamic analysis

In the isomerization of C_8 aromatics, the possible main reactions are the interconversion of three xylene isomers and the conversion between ethylbenzene and xylene. The side reactions are disproportionation and aromatics hydrogenation.

2. Dynamic analysis

(1) Isomerization process of xylene

There are two views on the reaction schema of xylene isomerization.

One is the interconversion between the three isoforms:

$$\text{m-xylene}$$

$$\Updownarrow \qquad \Updownarrow \qquad (4-25)$$

$$\text{o-xylene} \Longleftrightarrow \text{p-xylene}$$

The other is serial isomerization:

$$\text{o-xylene} \Longleftrightarrow \text{m-xylene} \Longleftrightarrow \text{p-xylene} \qquad (4-26)$$

(2) Isomerization process of ethylbenzene

According to the study of the gas phase hydroisomerization of ethylbenzene on Pt/Al_2O_3 catalyst, the whole isomerization process includes hydrogenation, isomerization and dehydrogenation. Low temperature is conducive to hydrogenation, high temperature is conducive to isomerization and dehydrogenation, so only by coordinating various conditions can ethylbenzene isomerization get better results.

$$(4-27)$$

3. Catalyst

The main catalysts are amorphous $SiO_2 - Al_2O_3$ catalysts, supported platinum catalysts, ZSM zeolite catalysts and $HF - BF_3$ catalysts, etc.

4.2.4　Alkylation of aromatics

Alkylation of aromatics is the reaction in which one or more hydrogens of benzene rings in aromatic molecules are replaced by alkyl groups to form alkyl aromatics. The alkylation of benzene is the most important in the alkylation of aromatic hydrocarbons. This kind of reaction is mainly used to produce ethylbenzene, isopropylbenzene and dodecylbenzene in industry. Substances that provide alkyl for alkylation of hydrocarbons are called alkylation agents. There are many alkylation agents available, olefins and halogenated alkanes are commonly used in industry. Alkenes such as ethylene, propylene, dodecane, not only have good reactivity, but also relatively easy to obtain. Since the normal hydrocarbon ions formed in the alkylation process of olefin will undergo skeleton rearrangement to obtain the most stable structure, only isomeric alkyl benzene can be obtained when the olefins above ethylene are alkylated with benzene, but not normal alkyl benzene. The active order of olefin from high to low is isobutene, n-butene, ethylene. Haloalkanes are mainly chlorinated alkanes, such as chloroethane, dodecyl chloride and so on. In addition, alcohols, esters and ethers can also be used as alkylation agents.

4.3　Separation of C_8 aromatics

4.3.1　Composition and properties of C_8 aromatics

C_8 aromatics from various sources are mixtures of three xylene isomers and ethylbenzene. The difference of boiling point between o-xylene and m-xylene is 5.3℃, which can be separated by distillation in industry. The difference of boiling point between ethylbenzene and paraxylene is 2.2℃, which can be separated by distillation tower with $300 \sim 400$ plates in industry. However, most processes do not use the energy-intensive distillation to recover ethylbenzene, but convert it in the isomerization device. The boiling point of m-xylene and p-xylene is close, so it is very difficult to separate p-xylene by means of ordinary distillation. So the technical difficulty of C_8 aromatic separation mainly lies in the separation of m-xylene and p-xylene. Originally the industry mainly uses freezing point difference and adopts cryogenic crystallization separation method, then adsorption separation and complexation separation technology are developed, especially adsorption separation occupies more and more important position.

4.3.2 Separation of C₈ aromatic monomers

1. Separation of o-xylene and ethylbenzene

（1）Separation of o-xylene

The boiling point of o-xylene in C_8 aromatics is the highest, 5.3℃ higher than that of the key component m-xylene. It can be separated by distillation. The distillation tower needs 150~200 plates, the two towers are in series, the reflux ratio is 7~10, and the product purity is 98%~99.6%.

（2）Separation of ethylbenzene

The boiling point of ethylbenzene in C_8 aromatics is the lowest, only 2.2℃ different from that of the key component paraxylene, The ethylbenzene can be separated by distillation, but it is difficult. The number of actual plates in the distillation column for industrial separation of ethylbenzene is 300~400(equivalent to the number of theoretical plates 200~250), the three towers are connected in series, tower reactor pressure is 0.35 MPa~0.4 MPa, and the reflux ratio is 50~100. The purity of ethylbenzene above 99.6% can be obtained. Other methods include complex extraction（CEA）methods such as Pomex method of Mitsubishi Gas Chemical Company, and absorption methods such as Ebex method of UOP Company.

2. Separation of p-xylene and m-xylene

Because the boiling point difference between p-xylene and m-xylene is only 0.75℃, it is difficult to separate p-xylene by distillation. At present, there are three main industrial separation methods for p-xylene: cryogenic crystallization separation, complexation separation and simulated moving bed adsorption separation.

（1）Cryogenic crystallization separation method

When C_8 arene was deeply cooled to $-60℃\sim-75℃$, p-xylene with the highest melting point was crystallized first. In the crystallization process of p-xylene, the crystal inevitably contains a part of C_8 aromatic mixture, which affects the purity of p-xylene. In order to improve the purity of p-xylene, two-stage crystallization process is often used in industry. The purity of p-xylene in the first crystallization is about 85%~90%. The purity of p-xylene in the second crystallization is 99.2%~99.5%. Due to the limit of eutectic temperature, if the temperature is further reduced, o-xylene and m-xylene will be crystallized at the same time, and the uncrystallized C_8 aromatic liquid still contains about 6.2%~6.9% p-xylene. Therefore, the one-way rate of crystallization separation is low, only about 60%.

Amoco crystallization and separation process of p-xylene is shown in Figure 4-4.

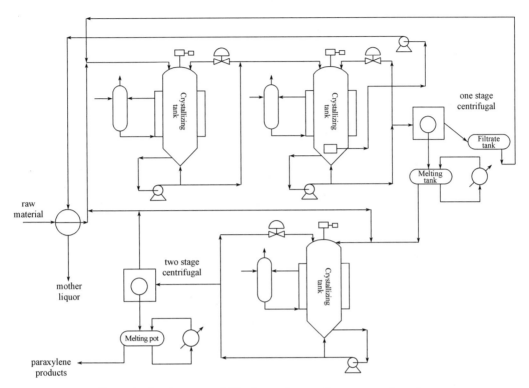

Figure 4 - 4　Amoco crystallization and separation process of p-xylene

(2) Complexation extraction separation method

According to the properties of coordination compounds formed between some compounds and xylene isomers, the separation of different isomers can be achieved. The most successful industrial example of complexation separation is the MGCC method developed by Mitsubishi Gas Chemical Company in Japan. This is the only industrial method for the effective separation of m-xylene, and it also simplifies the separation of other C_8 aromatics. The separation process of xylene by MGCC complexation extraction separation is shown in Figure 4 - 5.

When the four isomers of C_8 arene coexist with HF in one system, two separated liquid layers are formed: the upper layer is hydrocarbon layer and the lower layer is HF layer. When BF_3 is added, the following reactions occur to form coordination compounds with high solubility in HF.

$$X+HF+BF_3 \longrightarrow XHBF_4 \qquad (4-28)$$

$$MX+PXHBF_4 \longrightarrow PX+MXHBF_4 \qquad (4-29)$$

$$MX+OXHBF_4 \longrightarrow OX+MXHBF_4 \qquad (4-30)$$

(3) Adsorption separation method

Adsorption separation is a separation method which uses solid adsorbent to adsorb xylene isomers in different capacities. Adsorption separation was industrialized

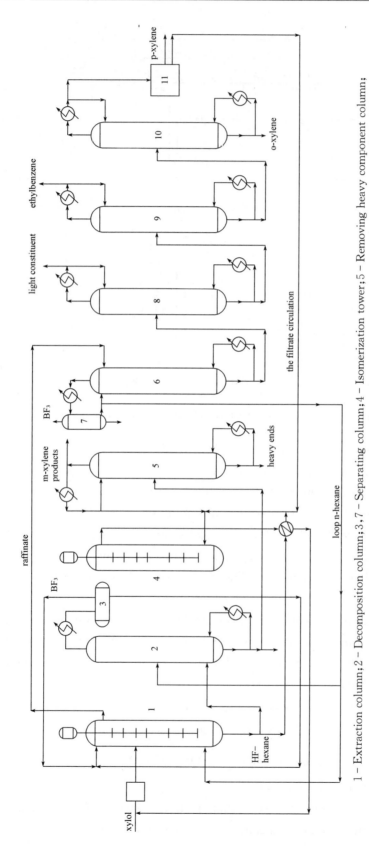

1 – Extraction column；2 – Decomposition column；3，7 – Separating column；4 – Isomerization tower；5 – Removing heavy component column；
6 – Extracting raffinate column；8 – Removing light component column；9 – Removing light component column；10 – O-xylene separation tower；
11 – Crystallization tank of p-xylene.

Figure 4－5 Schematic process of xylene separation by MGCC complexation extraction separation

by UOP company in the United States, which solved three key problems. One is the successful development of a solid adsorbent with high selective adsorption of various xylene isomers. The other is the successful development of a simulated moving bed technology with a 24 - channel rotary valve for switching operation. The third is to choose a reagent with the same adsorption affinity with p-xylene. Adsorption separation has more advantages than crystallization separation, such as simple process, one-way recovery of 98%, low production cost, and has replaced cryogenic crystallization, becoming a widely used xylene separation technology.

4.4   Development direction of aromatics production technology

Because of the important position of aromatic hydrocarbons in chemical raw materials, the development of their production technology has been widely paid attention to. In recent years, great technological progress has been made in opening up new sources of raw materials, developing higher levels of catalyst and process, improving the aromatics yield and selectivity, reducing energy consumption and operation cost, technical modification of existing equipment and improving the flexibility of production plan, greatly improving the aromatics production technology level. Some have already played an important role in industrial production, some have shown great potential. The development of aromatics production technology mainly includes the following aspects.

(1) Expand the source of aromatic hydrocarbon raw materials.

(2) Process innovation, improve the technical level.

(3) The new uses of products promote the adjustment of product structure.

(4) New technology of chemical engineering plays an important role.

(5) The development of old and new technologies.

Questions

4 - 1   Briefly describe the main sources and production process of aromatic hydrocarbons.

4 - 2   What are the main products of aromatics? What are their uses?

4 - 3   Try to discuss the necessity and significance of aromatics conversion. What are the main aromatics conversion reactions?

4 - 4   Try to analyze the characteristics and causes of aromatic hydrocarbon production in China, the United States and Japan.

4 - 5   Briefly introduce the main production process of benzene, toluene and various xylene monomers, and explain their characteristics.

4 - 6　Briefly introduce the new progress and main characteristics of aromatics production technology.

4 - 7　How to understand the relationship between aromatic production, conversion and separation processes? Please organize two different aromatics production schemes.

第5章 合成气生产

5.1 概述

合成气是指一氧化碳和氢气的混合气,英文缩写为 syngas。合成气中 H_2 与 CO 的比值随原料和生产方法的不同而异,其 H_2/CO(摩尔比)的值为 $\frac{1}{2} \sim 3$。合成气是有机合成原料之一,也是氢气和一氧化碳的来源,在化学工业中有着重要作用。制造合成气的原料是多种多样的,许多含碳资源如煤、天然气、石油馏分、农林废料、城市垃圾等均可用来制造合成气,其中"废料"的利用具有巨大的经济效益和社会效益,大大拓宽了化工原料的来源,所以发展合成气有利于资源优化利用,有利于化学工业原料路线和产品结构多元化发展。合成气可以转化成液体和气体燃料、大宗化学品和高附加值的精细有机合成产品,实现这种转化的重要技术是 C1 化工技术。凡包含一个碳原子的化合物,如 CH_4、CO、CO_2、HCN、CH_3OH 等参与反应的化学,称为 C1 化学,涉及 C1 化学反应的工艺过程和技术称为 C1 化工。自 20 世纪 70 年代后期,C1 化工得到世界各国极大重视,以天然气和煤炭为基础的合成气转化制备化工产品的研究广泛开展,并已实现部分 C1 化工过程工业化,未来将有更多 C1 化工过程实现工业化。今后,合成气的应用前景将越来越宽广。

5.1.1 合成气生产方法

按原料不同,合成气生产方法主要为三种:以煤为原料、以天然气为原料和以重油或渣油为原料。以天然气为原料制合成气的成本最低;重油与煤炭制造合成气的成本相仿,但重油和渣油制合成气可以使石油资源得到充分的利用。

其他含碳原料(包括各种含碳废料)制合成气在工业上尚未形成大规模生产,随着再生资源的开发、二次资源的广泛利用,今后会迅速发展。

1. 以煤为原料的生产方法

以煤为原料的生产方法有间歇和连续操作两种方式。连续式生产效率高、技术较先进。以煤制合成气是在高温下以水蒸气和氧气为气化剂,与煤反应生成 CO 和 H_2 等气体,这样的过程称为煤的气化。因为煤中氢的含量相当低,所以煤制合成气中 H_2/CO 的比值较低,适于合成有机化合物。因此在煤储量丰富的国家和地区,除了用煤发电外,应该大力发展以煤制合成气路线为基础的煤化工,更应该重视发展热电站与煤化工的联合生产企业,使煤资源得到充分的利用。

2. 以天然气为原料的生产方法

以天然气为原料制合成气主要有转化法和部分氧化法。目前工业上多采用水蒸气转化法(steam reforming),该方法制得的合成气中 H_2/CO 比值理论上为3,有利于制造合成氨或氢气,用来制造其他有机化合物(如甲醇、醋酸、乙烯、乙二醇等)时此比值需要再加调整。近年来,部分氧化法的工艺因其热效率较高、H_2/CO 比值易于调节而逐渐受到重视和应用,但需要有廉价的氧源,才能有满意的经济性。

3. 以重油或渣油为原料的生产方法

以重油或渣油为原料制合成气主要采用部分氧化法,即在反应器中通入适量的氧和水蒸气,使氧与原料油中的部分烃类燃烧,放出热量并产生高温,另一部分烃类则与水蒸气发生吸热反应而生成 CO 和 H_2,调节原料中油、H_2O 与 O_2 的比例,可达到自热平衡而不需要外供热。

5.1.2 合成气应用实例

1. 工业化主要产品

(1) 合成氨

$$N_2 + 3H_2 \Longrightarrow 2NH_3 \tag{5-1}$$

(2) 合成甲醇

$$CO + 2H_2 \Longrightarrow CH_3OH \tag{5-2}$$

(3) 合成醋酸

$$CH_3OH + CO \Longrightarrow CH_3COOH \tag{5-3}$$

(4) 烯烃的氢甲酰化产品

$$CH_3CHCH_2 + CO + H_2 \longrightarrow CH_3CH_2CH_2CHO + (CH_3)_2CHCHO \tag{5-4}$$

(5) 合成天然气、汽油和柴油

$$nCO + (2n+1)H_2 \Longrightarrow C_nH_{2n+2} + nH_2O \tag{5-5}$$

2. 合成气应用新途径

在合成气基础上制备化工产品的新途径有三种,即将合成气转化为乙烯或其他烃类,然后再进一步加工成化工产品;先合成甲醇,然后再将其转化为其他产品;直接将合成气转化为化工产品。这些新应用中,有的正在研究,有的已进入工业开发阶段,有的已具有一定生产规模。

(1) 直接合成乙烯等低碳烯烃

$$2CO + 4H_2 \Longrightarrow C_2H_4 + 2H_2O \tag{5-6}$$

(2) 合成气经甲醇再转化为烃类

$$2nCH_3OH \xrightarrow{-H_2O} nCH_3OCH_3 \xrightarrow{-H_2O} C_2 \sim C_4 \tag{5-7}$$

(3) 甲醇同系化制乙烯

$$CH_3OH + CO + 2H_2 \Longrightarrow CH_3CH_2OH + H_2O \tag{5-8}$$

$$CH_3CH_2OH \Longrightarrow C_2H_4 + H_2O \tag{5-9}$$

（4）合成低碳醇

（5）合成乙二醇

$$4CH_3OH + 2CO + O_2 \Longrightarrow 2CO(OCH_3)_2 + 2H_2O \qquad (5-10)$$

$$(COOCH_3)_2 + 4H_2 \Longrightarrow (CH_2OH)_2 + 2CH_3OH \qquad (5-11)$$

（6）合成气与烯烃衍生物羰基化产物

5.2 由煤制造合成气

煤的气化过程是热化学过程。它是以煤或焦炭为原料，以氧气（空气、富氧或纯氧）、水蒸气等为气化剂，在高温条件下，通过化学反应把煤或焦炭中的可燃部分转化为气体的过程。气化时所得的气体也称为煤气，其有效成分包括一氧化碳、氢气和甲烷等。气化煤气可用作城市煤气、工业燃气、合成气和工业还原气。在各种煤转化技术中，特别是开发洁净煤技术中，煤的气化是最有应用前景的技术之一。

5.2.1 煤气化过程工艺原理

1. 煤气化基本反应

煤气化过程的反应主要有

$$2C + O_2 \Longrightarrow 2CO \qquad (5-12)$$

$$C + O_2 \Longrightarrow CO_2 \qquad (5-13)$$

$$C + H_2O \Longrightarrow CO + H_2 \qquad (5-14)$$

$$C + 2H_2O \Longrightarrow CO_2 + 2H_2 \qquad (5-15)$$

$$C + CO_2 \Longrightarrow 2CO \qquad (5-16)$$

$$C + 2H_2 \Longrightarrow CH_4 \qquad (5-17)$$

气化生成的混合气称为水煤气。以上均为可逆反应，总过程是强吸热的。

这些反应中，碳与水蒸气反应的意义最大。它参与各种煤气化过程，为强吸热过程。碳与二氧化碳的还原反应也是重要的气化反应。碳燃烧反应放出的热量与上述的吸热反应相匹配，对自热式气化过程有重要的作用。加氢气化反应对于制取合成天然气很重要。

2. 煤气化反应条件

（1）温度

温度对煤气化影响最大，至少要在900℃以上才有满意的气化速率，一般操作温度在1100℃以上。近年来新工艺采用1500℃～1600℃进行气化，使生产强度大大提高。

（2）压力

降低压力有利于提高CO和H_2的平衡浓度，但加压有利于提高反应速率并减小反应体积，目前气化压力一般为2.5MPa～3.2MPa，因而CH_4含量比常压法高些。

（3）水蒸气和氧气的比例

氧的作用是与煤燃烧放热,此热供给水蒸气与煤的气化反应。H_2O/O_2 的比值对温度和煤气组成有影响。具体的 H_2O/O_2 的比值要视采用的煤气化生产方法来定。

5.2.2 煤气化生产方法

煤气化过程需要高温,工业上采用燃烧煤来实现。气化过程按操作方式来分,有间歇式和连续式,前者的工艺较后者落后,现在逐渐被淘汰。目前最通用的是按反应器分类,分为固定床(移动床)、流化床、气流床和熔融床。至今熔融床还处于中试阶段,而固定床(移动床)、流化床和气流床是工业化或建立示范装置的方法。此外,不同生产方法对煤质要求不同。

5.3 由天然气制造合成气

5.3.1 天然气制合成气工艺技术

天然气中甲烷含量一般大于 90%,其余为少量的乙烷、丙烷等气态烷烃,有些还含有少量氯和硫化物。其他含甲烷等气态烃的气体,如炼厂气、焦炉气、油田气和煤层气等均可用来制造合成气。

目前工业上由天然气制合成气的技术主要有蒸汽转化法和部分氧化法。蒸汽转化法是在催化剂存在及高温条件下,使甲烷等烃类与水蒸气反应,生成 H_2、CO 等混合气,其主反应为

$$CH_4 + H_2O \Longrightarrow CO + 3H_2 \tag{5-18}$$

该反应是强吸热的,需要外界供热。此技术成熟,目前广泛应用于生产合成气、纯氢气和合成氨原料气。

部分氧化法是由甲烷等烃类与氧气进行不完全氧化生成合成气。

$$2CH_4 + O_2 \Longrightarrow 2CO + 4H_2 \tag{5-19}$$

5.3.2 天然气蒸汽转化过程的工艺原理

因为天然气中甲烷含量在 90% 以上,而甲烷在烷烃中是热力学最稳定的(其他烃类较易反应),因此在讨论天然气转化过程时,只需考虑甲烷与水蒸气的反应。

1. 甲烷水蒸气转化反应和化学平衡

甲烷水蒸气转化过程的主要反应有

$$CH_4 + H_2O \Longrightarrow CO + 3H_2 \tag{5-20}$$

$$CH_4 + 2H_2O \Longrightarrow CO_2 + 4H_2 \tag{5-21}$$

$$CO + H_2O \Longrightarrow CO_2 + H_2 \tag{5-22}$$

可能发生的副反应主要是析碳反应,它们是

$$CH_4 \Longrightarrow C + 2H_2 \qquad\qquad (5-23)$$

$$2CO \Longrightarrow C + CO_2 \qquad\qquad (5-24)$$

$$CO + H_2 \Longrightarrow C + H_2O \qquad\qquad (5-25)$$

甲烷水蒸气转化反应只有在催化剂存在下才有足够的反应速率。倘若操作条件不适当,析碳反应严重,生成的碳会覆盖在催化剂内外表面,致使催化活性降低,反应速率下降。析碳更严重时,会导致床层堵塞,阻力增加,催化剂毛细孔内的碳遇水蒸气会剧烈气化,致使催化剂崩裂或粉化,迫使停工,经济损失巨大。所以,对于烃类蒸汽转化过程要特别注意防止析碳。

影响甲烷水蒸气转化反应平衡的主要因素有温度、水碳比和压力。

(1) 温度的影响

甲烷与水蒸气反应生成 CO 和 H_2 是吸热的可逆反应。高温对平衡有利,即 H_2 及 CO 的平衡产率高,CH_4 的平衡含量低。一般情况下,当温度提高 10℃,甲烷的平衡含量可降低 1%~1.3%。高温对一氧化碳变换反应的平衡不利,可以少生成二氧化碳,而且高温也会抑制一氧化碳歧化和还原析碳的副反应。但是,温度过高,会有利于甲烷裂解,当温度高于 700℃时,甲烷均相裂解速率很快,会大量析出碳,并沉积在催化剂和器壁上。

(2) 水碳比的影响

水碳比对于甲烷转化影响重大,高的水碳比有利于甲烷的蒸气重整反应,在 800℃、2 MPa 条件下,水碳比由 3 提高到 4 时,甲烷平衡含量由 8% 降至 5%,可见水碳比对甲烷平衡含量影响是很大的。同时,高水碳比也有利于抑制析碳副反应。

(3) 压力的影响

甲烷蒸汽转化反应是体积增大的反应,低压有利平衡。当温度为 800℃、水碳比为 4,压力由 2 MPa 降低到 1 MPa 时,甲烷平衡含量由 5% 降至 2.5%。低压也可抑制一氧化碳的两个析碳反应,但是低压对甲烷裂解析碳反应平衡有利,适当加压可抑制甲烷裂解。压力对一氧化碳变换反应平衡无影响。

总之,从反应平衡考虑,甲烷水蒸气转化过程应该采用适当的高温、稍低的压力和高水碳比。

2. 甲烷水蒸气转化催化剂

甲烷水蒸气转化反应在无催化剂时的反应速率很慢,在 1 300℃以上才有满意的速率,然而在此高温下大量甲烷裂解,没有工业生产价值,所以必须采用催化剂。催化剂的组成和结构决定了其催化性能,而催化剂使用是否得当会影响其性能的发挥。生产中催化剂因其老化、中毒和积碳而失去活性。

(1) 转化催化剂的组成和外形

研究表明,一些贵金属和镍均具有对甲烷蒸汽转化的催化活性,其中镍最便宜,又具有足够高的活性,所以工业上一直采用镍催化剂,并添加一些助催化剂以提高活性或改善诸如机械强度、活性组分分散、抗结碳、抗烧结、抗水合等性能。转化催化剂的促进剂有铝、镁、钾、钙、钛、镧、铈等金属氧化物。甲烷与水分子的反应是在固体催化剂活性表面上进行的,所以催化剂应该具有较大的镍表面。提高镍表面的最有效的方法是采用大比表面的载体,来支承、分散活性组分,并通过载体与活性组分间的强相互作用而使镍晶粒不易烧结。载体还应具有足够的机械强度,使催化剂在使用中不易破碎。为了抑制烃类在催化剂表面酸性中心上裂解析碳,往往在载体中添加碱性物质中和表面酸性。

(2)转化催化剂的使用和失活

转化催化剂在使用前是氧化态,装入反应器后应先进行严格的还原操作,使氧化镍还原成金属镍才有活性。还原气可以是氢气、甲烷或一氧化碳。纯氢还原可得到很高的镍表面积,但镍表面积不稳定,在反应时遇水蒸气会减少,故工业上是采用通入水蒸气并升温到500℃以上,然后添加一定量的天然气和少量氢气来进行还原。

转化催化剂在使用中出现活性下降现象的原因主要有老化、中毒、积碳等。

5.3.3 天然气蒸汽转化过程工艺条件

在选择工艺条件时,理论依据包括热力学、动力学分析和化学工程原理,此外,还需要结合技术经济、生产安全等进行综合优化。转化过程主要工艺条件有压力、温度、水碳比和空速,这几个条件之间互有关系,要恰当匹配。

(1)压力

从热力学特征看,低压有利于转化反应。从动力学看,在反应初期,增加系统压力,相当于增加了反应物分压,反应速率加快;但到反应后期,反应接近平衡,反应物浓度高,加压反而会降低反应速率。所以,从化学角度看,压力不宜过高。但从工程角度考虑,适当提高压力对传热有利,因为甲烷转化过程需要外部供热,大的给热系数是强化传热的前提。

(2)温度

从热力学角度看,高温下甲烷平衡浓度低;从动力学看,高温使反应速率加快。所以出口残余甲烷含量低。因加压对平衡的不利影响,更要提高温度来弥补。

(3)水碳比

水碳比是诸操作变量中最便于调节的一个条件,又对一段转化过程影响较大。水碳比高,有利于防止积碳,残余甲烷含量也低。实验指出,当原料气中无不饱和烃时,水碳比若小于2,温度到400℃时会析碳,而当水碳比大于2时,温度要高达1 000℃才有碳析出;但若原料气有较多不饱和烃存在时,即使水碳比大于2,温度达到400℃时就会析碳。为了防止积碳,操作中一般控制水碳比在3.5左右。

(4)气流速度

反应炉管内气体流速高有利于传热,降低炉管外壁温度,延长炉管寿命。当催化剂活性足够时,高流速也能强化生产,提高生产能力。但流速不宜过高,否则床层阻力过大,能耗增加。

5.3.4 天然气蒸汽转化流程和主要设备

天然气蒸汽转化制合成气的基本步骤如图5-1所示。

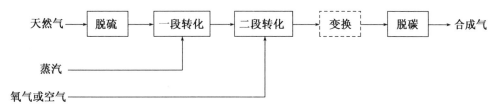

图 5-1　天然气蒸汽转化制合成气过程

图 5-2 是以天然气为原料日产千吨氨的大型合成氨厂转化工段流程图。

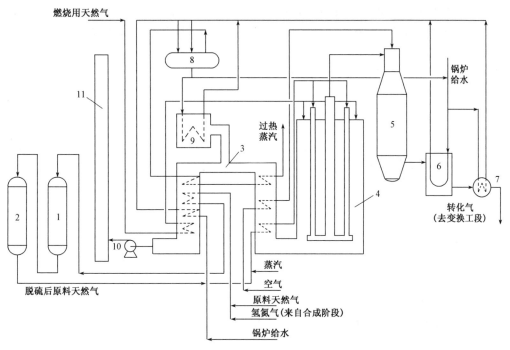

1—钴钼加氢脱硫器；2—氧化锌脱硫罐；3——段炉对流段；4——段炉辐射段；5—二段转化炉；
6—第一废热锅炉；7—第二废热锅炉；8—汽包；9—辅助锅炉；10—排风机；11—烟囱。

图 5-2　天然气蒸汽转化流程

5.4　由渣油制造合成气

　　制造合成气用的渣油是石油减压蒸馏塔底的残余油，亦称减压渣油。由渣油转化为 CO、H_2 等气体的过程称为渣油气化。气化技术有部分氧化法和蓄热炉深度裂解法，目前常用的是部分氧化法，由美国德士古(Texaco)公司和荷兰谢尔(Shell)公司在 20 世纪 50 年代开发成功，分别称为德士古法和谢尔法，当时用于重油。在 20 世纪 80 年代经改进，得到德士古新工艺，可用于渣油气化。

　　渣油制合成气的加工步骤主要如图 5-3 所示。

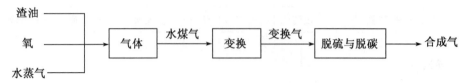

图 5-3 渣油制合成气的加工步骤

渣油是许多大分子烃类的混合物,沸点很高,所含元素的重量组成包括 C 为 $84\%\sim87\%$,H 为 $11\%\sim12.5\%$,其余有 S、N、O,以及微量元素 Ni、V 等。

氧化剂是氧气。当氧气充分时,渣油会完全燃烧生成 CO_2 和 H_2O,只有当氧量低于完全氧化理论值时,才发生部分氧化,生成以 CO 和 H_2 为主的气体。

渣油在常温时是黏稠的、黑色半固体状物,要将渣油预热变成易流动的液态,才能进入反应器。渣油在反应器中经历的变化如下:

先是渣油分子(C_mH_n)吸热升温、气化,气态渣油与氧气混合均匀,然后与氧气反应,如果氧气量充足,则会发生完全燃烧反应,即

$$C_mH_n + \left(m+\frac{n}{4}\right)O_2 \rightarrow mCO_2 + \frac{n}{2}H_2O(放热) \tag{5-26}$$

如果氧气量低于完全氧化理论量,则发生部分氧化,放热量少于完全燃烧,反应式为

$$C_mH_n + \left(\frac{m}{2}+\frac{n}{4}\right)O_2 \rightarrow mCO + \frac{n}{2}H_2O(放热) \tag{5-27}$$

$$C_mH_n + \frac{m}{2}O_2 \rightarrow mCO + \frac{n}{2}H_2(放热) \tag{5-28}$$

5.5 一氧化碳变换过程

一氧化碳与水蒸气反应生成氢气和二氧化碳的过程,称为 CO-水煤气变换(water gas shift)。通过变换反应可产生更多氢气,同时降低 CO 含量,可用于调节 H_2/CO 的比例,满足不同生产需要。

5.5.1 一氧化碳变换反应化学平衡

一氧化碳变换的反应式为

$$CO + H_2O(气) \Longrightarrow CO_2 + H_2 \tag{5-29}$$

变换反应的平衡受温度、水碳比(即原料气中 H_2O/CO 的摩尔比)、原料气中 CO_2 含量等因素影响,低温和高水碳比有利于平衡右移,压力对平衡无影响。

变换反应可能发生的副反应主要有

$$2CO \Longrightarrow C + CO_2 \tag{5-30}$$

$$CO + 3H_2 \Longrightarrow CH_4 + H_2O \tag{5-31}$$

$$CO_2 + 4H_2 \Longrightarrow CH_4 + 2H_2O \tag{5-32}$$

当水碳比低时,更有利于这些副反应。CO 歧化会使催化剂积碳,后两反应是甲烷化,消耗氢气,所以都要抑制它们。

5.5.2　一氧化碳变换催化剂

无催化剂存在时,变换反应的速率极慢,即使温度升至 700℃ 以上,反应仍不明显;因此必须采用催化剂,使反应在不太高的温度下有足够高的反应速率,以达到较高的转化率。目前工业上采用的变换催化剂有以下三大类。

(1) 铁铬系变换催化剂

该类催化剂的化学组成以 Fe_2O_3 为主,反应前还原成 Fe_3O_4 才有活性,促进剂有 Cr_2O_3 和 K_2CO_3。适用温度范围为 300℃～530℃。该类催化剂称为中温或高温变换催化剂,因为温度较高,反应后气体中残余 CO 含量最低为 3%～4%。

(2) 铜基变换催化剂

该类催化剂的化学组成以 CuO 为主,ZnO 和 Al_2O_3 为促进剂和稳定剂,反应前也要将其还原成具有活性的细小铜晶粒。若还原操作中或正常运转中超温,均会造成铜晶粒烧结而失活。该类催化剂另一弱点是易中毒,所以原料气中硫化物的体积分数不得超过 $0.1×10^{-6}$。铜基催化剂适用温度范围为 180℃～260℃,称为低温变换催化剂,反应后残余 CO 可降至 0.2%～0.3%。铜基催化剂活性高,若原料气中 CO 含量高时,应先经高温变换,将 CO 降至 3% 左右,再进行低温变换,以防剧烈放热而烧坏催化剂。

(3) 钴钼系耐硫催化剂

该类催化剂为钴、钼氧化物并负载在氧化铝上,反应前将钴、钼氧化物转变为硫化物(预硫化)才有活性,反应中原料气必须含硫化物,其适用温度范围为 160℃～500℃,属宽温变换催化剂。该类催化剂的特点是耐硫抗毒,使用寿命长。

5.5.3　一氧化碳变换反应动力学

1. 反应机理和动力学方程

目前提出的 CO 变换反应机理很多,流行的有两种:一种观点是 CO 和 H_2O 分子先吸附到催化剂表面,两者在表面进行反应,然后生成物脱附;另一观点是被催化剂活性位吸附的 CO 与晶格氧结合形成 CO_2 并脱附,被吸附的 H_2O 解离脱附出 H_2,而氧则补充到晶格中,这就是有晶格氧转移的氧化还原机理。由不同机理可推导出不同的动力学方程;不同催化剂,其动力学方程亦不同。

2. 反应条件对变换反应速率的影响

(1) 压力影响

加压可提高反应物分压。在 3.0 MPa 以下,反应速率与压力的平方根成正比;压力超过 3.0 Mpa 时,影响就不明显了。

(2) 水蒸气影响

水蒸气用量决定了 H_2O 与 CO 的比值,该水碳比对反应速率的影响规律与其对平衡转化率的影响相似。在水碳比低于 4 时,提高水碳比,可使反应速率增长较快;但当水碳比大于 4 后,反应速率增长就不明显了,故一般选用水碳比为 4 左右。

（3）温度影响

CO 变换是一个放热可逆反应，此类反应存在最佳反应温度（T_{op}）。

5.5.4 一氧化碳变换操作条件

1. 压力

压力虽对平衡无影响，但加压对反应速率有利。反应速率不宜过高，一般中、小型厂采用常压或 2 MPa，大型厂多用 3 MPa，有些用 8 MPa。

2. 水碳比

高水碳比对反应平衡和反应速率均有利，但太高时效果不明显，反而能耗过高，现常用水碳比为 4（水蒸气/水煤气为 1.1～1.2）。近年来节能工艺很受重视，希望水碳比能降到 3 以下，其中，关键是变换催化剂的选择性要提高，有效抑制 CO 加 H_2 副反应。

3. 温度

变换反应的温度最好沿最佳反应温度曲线变化。反应初期，转化率低，最佳温度高；反应后期，转化率高，最佳温度低，但是 CO 变换反应是放热的，需要不断地将此热量排出体系才可能使温度下降。在工程实际中，降温措施不可能完全符合最佳温度曲线，因此变换过程是采用分段冷却来降温，即反应一段时间后进行冷却，然后再反应，如此分段越多，操作温度越接近最佳温度曲线。应特别注意的是，操作温度必须控制在催化剂活性温度范围内，低于此范围，催化剂活性太低，反应速率太慢；高于此范围，催化剂易过热而受损，失去活性。各类催化剂均有各自的活性温度范围，只能在其范围内使操作温度尽可能地接近最佳反应温度曲线。

5.5.5 变换过程工艺流程

一氧化碳变换流程有许多种，包括常压、加压；两段中温变换（简称高变）、三段中温变换（简称高变）、高-低变串联等。主要根据制造合成气的生产方法、水煤气中 CO 含量、对残余 CO 含量的要求等因素来选择。一氧化碳高-低温变换串联流程如图 5-4，一氧化碳三段中温变换流程示意图如图 5-5。

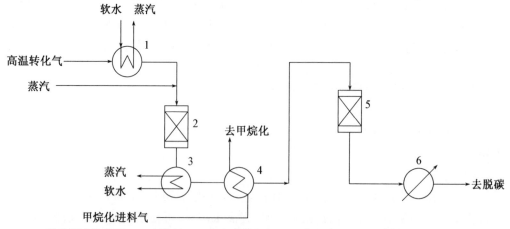

1—转化器废热锅炉；2—高变炉；3—高变废热锅炉；4—热交换器；5—低变炉；6—热交换器。

图 5-4　一氧化碳高-低温变换串联流程

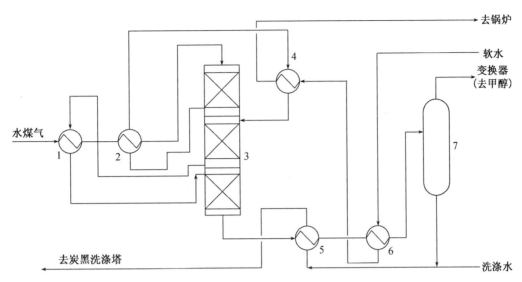

1,2,4,5,6—换热器；3—变换反应器；7—冷凝液分离器。

图 5-5　一氧化碳三段中温变换流程示意图

5.6　气体中硫化物和二氧化碳脱除

在制造合成气时，所用的气、液、固三类原料均含有硫化物。石油馏分中含有硫醇（RSH）、硫醚（RSR）、二硫化碳（CS_2）、噻吩（C_4H_4S）等，它们多集中于重质油馏分，尤其是渣油中，煤中常含有羰基硫（COS）和硫铁矿。这些原料制造合成气时，其中的硫化物转化成硫化氢和有机硫气体，会使催化剂中毒，腐蚀金属管道和设备，危害很大，必须脱除，并回收利用这些硫资源。

粗合成气中所含硫化物种类和含量与所用原料的种类及加工方法有关。用天然气或轻油制造合成气时，为避免蒸汽转化催化剂中毒，已预先将原料彻底脱硫，转化生成的气体中无硫化物；用煤或重质油制合成气时，气化过程不用催化剂，故不需对原料预先脱硫，因此产生的气体中含有硫化氢和有机硫化物，在后续加工之前，必须进行脱硫。含硫量高的无烟煤气化生成的气体（标准状态）中，硫化氢可达 $4\sim6$ g/m^3，有机硫总量为 $0.5\sim0.8$ g/m^3。重油中若含硫 $0.3\%\sim1.5\%$ 时，气化后的气体（标准状态）中含硫化氢为 $1.1\sim2.0$ g/m^3，有机硫为 $0.03\sim0.4$ g/m^3。一般情况下，气体中硫化氢的含量为有机硫总量的 $10\sim20$ 倍。

不同用途或不同加工过程对气体脱硫净化度要求不同。例如，天然气转化过程对原料气的脱硫要求是总含硫量（体积分数）小于 0.1×10^{-6}，最高不能超过 0.5×10^{-6}；一氧化碳高温变换要求原料气中硫化氢体积分数小于 5.0×10^{-4}，有机硫体积含量小于 1.5×10^{-4}；合成甲醇时用的铜基催化剂则要求总硫体积分数小于 0.5×10^{-6}；合成氨的铁催化剂则要求原料气不含硫。

气、液、固原料经转化或气化制造合成气的过程中会生成一定量的 CO_2，尤其当有一氧化碳变换过程时，生成更多的 CO_2，其含量可高达 $28\%\sim30\%$，因此也需要脱除 CO_2，回收的 CO_2 可加以利用。例如，CO_2 可供给天然气转化以降低合成气的 H_2/CO 比例；可供合成氨厂合成尿素；可供给制碱厂生产纯碱（Na_2CO_3）；还可加工一些有机化学品。CO_2 的回收利用不仅增加了经济效益，还减少了造成温室效应的危害。脱除二氧化碳的过程通常简称为脱碳。

5.6.1 脱硫方法及工艺

脱硫方法要根据硫化物的含量、种类和要求的净化度来选定，还要考虑具体的技术条件和经济性，有时可用多种脱硫方法组合来达到对脱硫净化度的要求。按脱硫剂的状态来分，脱硫有干法和湿法两大类。

1. 干法脱硫

干法脱硫又分为吸附法和催化转化法。

吸附法是采用对硫化物有强吸附能力的固体来脱硫，吸附剂主要有氧化锌、活性炭、氧化铁、分子筛等。

催化转化法是使用加氢脱硫催化剂，将烃类原料中所含的有机硫化物氢解，转化成易于脱除的硫化氢，再用其他方法除之。加氢脱硫催化剂是以 Al_2O_3 为载体，负载 CoO 和 MoO_3，亦称钴钼加氢脱硫剂。使用时需预先用 H_2S 或 CS_2 将其硫化变成 Co_9S_8 和 MoS_2 才有活性。

有机硫的氢解反应举例如下：

$$COS+H_2 \rightleftharpoons CO+H_2S \tag{5-33}$$

$$C_2H_5SH+H_2 \rightleftharpoons C_2H_6+H_2S \tag{5-34}$$

$$CH_3SC_2H_5+2H_2 \rightleftharpoons CH_4+C_2H_6+H_2S \tag{5-35}$$

$$C_2H_5SC_2H_5+2H_2 \rightleftharpoons 2C_2H_6+H_2S \tag{5-36}$$

$$C_4H_4S+4H_2 \rightleftharpoons C_4H_{10}+H_2S \tag{5-37}$$

钴钼加氢转化后可用氧化锌脱除生成的 H_2S。因此，用氧化锌—钴钼加氢转化—氧化锌组合，可达到精脱硫的目的。

2. 湿法脱硫

湿法脱硫剂为液体，一般用于含硫高、处理量大的气体的脱硫。按其脱硫机理的不同又分为化学吸收法、物理吸收法、物理-化学吸收法和湿式氧化法。

化学吸收法是常用的湿式脱硫工艺。有一乙醇胺法（MEA）、二乙醇胺法（DEA）、二甘醇胺法（DGA）、二异丙醇胺法（DIPA）以及近年来发展很快的改良甲基二乙醇胺法（MDEA）（MDEA 添加有促进剂，净化度很高）。以上几种统称为烷醇胺法或醇胺法。

物理吸收法是利用有机溶剂在一定压力下进行物理吸收脱硫，然后减压而释放出硫化物气体，溶剂得以再生。主要有冷甲醇法（Rectisol），还有碳酸丙烯酯法（Fluar）和 N-甲基吡啶烷酮法（Purisol）等。

物理-化学吸收法是将具有物理吸收性能和化学吸收性能的两类溶液混合在一起，

脱硫效率较高。常用的吸收剂为环丁砜-烷基醇胺(如甲基二乙醇胺)混合液,前者对硫化物是物理吸收,后者是化学吸收。

湿式氧化法,其脱硫的基本原理是利用含催化剂的碱性溶液吸收 H_2S,以催化剂作为载氧体,使 H_2S 氧化成单质硫,催化剂本身被还原。再生时通入空气将还原态的催化剂氧化复原,如此循环使用。

3. 硫化氢的回收

湿法脱硫后,吸收剂再生时释放的气体含有大量硫化氢,为了保护环境和充分利用硫资源,应予以回收。工业上成熟的技术是克劳斯工艺,克劳斯法的基本原理是首先在燃烧炉内使 1/3 的 H_2S 和 O_2 反应,生成 SO_2,剩余 2/3 的 H_2S 与此 SO_2 在催化剂作用下发生克劳斯反应,生成单质硫。反应式为

$$H_2S + \frac{3}{2}O_2 \Longrightarrow H_2O + SO_2 + Q_1 \tag{5-38}$$

$$2H_2S + SO_2 \Longrightarrow 2H_2O + 3S + Q_2 \tag{5-39}$$

5.6.2　脱除二氧化碳的方法和工艺

脱除 CO_2 要根据不同的情况来选择适宜的方法。目前国内外多采用溶液吸收剂来吸收 CO_2,根据吸收机理可分为化学吸收和物理吸收两大类。近年来出现了变压吸附法、膜分离等固体脱除二氧化碳法。

1. 化学吸收法

化学吸收方法在早期曾有过一乙醇胺法(MEA)和氨水法,现已少用。目前常用的化学吸收法是改良的热钾碱法,即在碳酸钾溶液中添加少量活化剂,以加快吸收 CO_2 的速率和解吸速率,活化剂作用类似于催化剂。在吸收阶段,碳酸钾与 CO_2 生成碳酸氢钾;在再生阶段,碳酸氢钾受热分解,析出 CO_2,溶液复原,循环使用。

2. 物理吸收法

目前国内外使用的物理吸收法主要有冷甲醇法、聚乙二醇二甲醚法和碳酸丙烯酯法。物理吸收法在加压(2 MPa～5 MPa)和较低温度条件下吸收 CO_2,溶液的再生靠减压解吸,而不是加热分解,属于冷法,能耗较低。

3. 物理-化学吸收法

物理-化学吸收法是将物理吸收剂与化学吸收剂结合起来的气体净化法,如MDEA 法中用甲基二乙醇胺-环丁砜混合液作吸收剂,能同时脱硫和脱碱,可与改良热钾碱法相竞争,但溶剂较贵。

4. 变压吸附法(PSA)

变压吸附技术是利用固体吸附剂在加压下吸附 CO_2,使气体得到净化。吸附剂再生时减压脱附析出 CO_2。该法一般在常温下进行,能耗小、操作简便、无环境污染,PSA法还可用于分离提纯 H_2、N_2、CH_4、CO、C_2H_4 等气体。我国已有国产化的 PSA 装置,规模和技术均达到国际先进水平。

思考题

5-1 有哪些原料可生产合成气? 合成气的生产方法有哪些, 近年来出现哪些生产合成气的新方法? 它们与原有生产方法相比有什么优点?

5-2 合成气可用来制造什么化工产品? 为什么近年来合成气的生产和应用受到重视?

5-3 以天然气为原料生产合成气的过程有哪些主要反应? 从热力学角度考虑, 对反应条件有哪些要求? 从动力学角度考虑又有哪些要求?

5-4 如何根据化学热力学、化学动力学原理和工程实际来优化天然气-水蒸气转化制合成气的工艺条件?

5-5 天然气-水蒸气转化法制合成气过程有哪些步骤? 为什么天然气要预先脱硫才能进行转化? 用哪些脱硫方法较好?

5-6 为什么天然气-水蒸气转化过程需要供热? 供热形式是什么? 一段转化炉有哪些型式?

5-7 为什么说一段转化管属于变温反应器? 为什么天然气-水蒸气转化要用变温反应器?

5-8 为什么转化炉的对流室内要设置许多热交换器? 转化气的显热是如何回收利用的?

5-9 由煤制合成气有哪些生产方法? 这些方法相比较各有什么优点? 较先进的方法是什么?

5-10 一氧化碳变换的反应是什么? 影响该反应的平衡和速度的因素有哪些? 如何影响? 为什么该反应存在最佳反应温度? 最佳反应温度与哪些常数有关?

5-11 为什么一氧化碳变换过程要分段进行, 要用多段反应器? 段数的选定依据是什么? 有哪些形式的反应器?

5-12 一氧化碳变换催化剂有哪些类型? 各适用于什么场合? 使用中注意哪些事项?

5-13 由渣油制合成气的过程包括哪几个步骤? 渣油气化的主要设备是什么, 有何结构特点?

5-14 在合成气制造过程中, 为什么要有脱碳(CO_2)步骤? 通常有哪些脱碳方法, 各适用于什么场合?

5-15 工业上气体脱硫有哪些方法, 各适用于什么场合?

Chapter 5   Production of syngas

5.1   Summary

Syngas is a mixture of carbon monoxide and hydrogen. The ratio of H_2 to CO in syngas varies with different raw materials and production methods, and the H_2/CO (molar ratio) varies from $\frac{1}{2}$ to 3. Syngas is one of the raw materials for organic synthesis and the source of hydrogen and carbon monoxide, which plays an important role in chemical industry. Syngas can be made from a variety of raw materials, including coal, natural gas, petroleum distillate, agricultural and forestry wastes, municipal waste. The utilization of "waste" has great economic and social benefits, and greatly expands the source of chemical raw materials. Therefore, the development of syngas is conducive to the optimal utilization of resources and the diversification of raw material route and product structure in the chemical industry. Syngas can be converted into liquid and gas fuels, bulk chemicals and high value added fine organic synthetic products. The important technology to achieve this transformation is the C1 chemical technology. Chemical reactions involving compounds containing one carbon atom, such as CH_4, CO, CO_2, HCN, CH_3OH, are called C1 chemistry. Processes and technologies involving C1 chemical reactions are called C1 chemistry. Since the late 1970s, C1 chemical industry has received great attention by all countries in the world. The research of chemical products prepared from syngas based on natural gas and coal has been widely carried out. More and more C1 chemical processes have been and will be industrialized. In the future, the application prospect of syngas will be wider and wider.

5.1.1   Production method of syngas

According to the different raw materials, the production methods of syngas are mainly divided into three types: using coal as raw material, using natural gas as raw material and using heavy oil or residual oil as raw material. The cost of making syngas from natural gas is the lowest. The cost of producing syngas from heavy oil is similar to that from coal, but it can make full use of oil resources.

The production of syngas from other carbon-containing raw materials (including various carbon-containing wastes) has not yet been produced on a large scale. With the development of renewable resources and the wide utilization of secondary resources, it will develop rapidly in the future.

1. Production method using coal as raw material

There are two modes of operation: intermittent and continuous. Continuous production is more efficient, the technology is more advanced. The process is using water vapor and oxygen as the gasification agent at high temperature, and reacting with coal to generate CO and H_2 gases, such a process is called coal gasification. Because the hydrogen content in coal is relatively low, so the H_2/CO ratio of coal-formed syngas is low, suitable for synthesis of organic compounds. Therefore, in countries and regions with rich coal reserves, in addition to using coal to produce electricity, we should vigorously develop coal chemical industry on the basis of coal to syngas route, and should attach great importance to the development of joint production enterprises of thermal power plants and coal chemical industry to make full use of the coal resources.

2. Production method using natural gas as raw material

The methods mainly include conversion and partial oxidation. At present, steam reforming method is commonly used in industry. Theoretically, the H_2/CO ratio of syngas prepared by this method is 3, which is good for the production of synthetic ammonia or hydrogen. When it is used to produce other organic compounds such as methanol, acetic acid, ethylene, ethylene glycol and so on, the ratio needs to be adjusted. In recent years, due to its high thermal efficiency and the easily adjustable H_2/CO ratio, the partial oxidation process gradually attracted attention and application, but it needs cheap oxygen source to be satisfactorily economical.

3. Production methods using heavy oil or residual oil as raw material

Partial oxidation is mainly used, that is, an appropriate amount of oxygen and water vapor is injected into the reactor to make oxygen and part of the hydrocarbons in the raw oil burn, release heat, generate high temperature, and the other part of the hydrocarbons react with water vapor to generate CO and H_2, so as to adjust the mutual proportion of oil, H_2O and O_2 in the raw material, which can achieve self-heating balance without the need for external heating.

5.1.2 Application examples of syngas

1. Main products of industrialization

(1) Synthetic ammonia

$$N_2 + 3H_2 \rightleftharpoons 2NH_3 \qquad (5-1)$$

(2) Synthetic methanol

$$CO + 2H_2 = CH_3OH \qquad (5-2)$$

(3) Synthetic acetic acid

$$CH_3OH + CO = CH_3COOH \qquad (5-3)$$

(4) Hydroformylation products of olefins

$$CH_3CHCH_2 + CO + H_2 \longrightarrow CH_3CH_2CH_2CHO + (CH_3)_2CHCHO \qquad (5-4)$$

(5) Synthetic natural gas, gasoline and diesel

$$nCO + (2n+1)H_2 = C_nH_{2n+2} + nH_2O \qquad (5-5)$$

2. New way of syngas application

There are three new ways of making chemicals from syngas: converting syngas to ethylene or other hydrocarbons, then processing it further into chemicals; converting syngas first into methanol, then into other products; directly converting syngas into chemicals. Some of these new applications are under study, some have entered the stage of industrial development, and some have a certain scale of production.

(1) Direct synthesis of low carbon olefins such as ethylene

$$2CO + 4H_2 = C_2H_4 + 2H_2O \qquad (5-6)$$

(2) Syngas is converted into hydrocarbons by methanol

$$2nCH_3OH \xrightarrow{-H_2O} nCH_3OCH_3 \xrightarrow{-H_2O} C_2 \sim C_4 \qquad (5-7)$$

(3) Methanol is homogenized to produce ethylene

$$CH_3OH + CO + 2H_2 = CH_3CH_2OH + H_2O \qquad (5-8)$$

$$CH_3CH_2OH = C_2H_4 + H_2O \qquad (5-9)$$

(4) Synthesis of low carbon alcohol

(5) Synthetic ethylene glycol

$$4CH_3OH + 2CO + O_2 = 2CO(OCH_3)_2 + 2H_2O \qquad (5-10)$$

$$(COOCH_3)_2 + 4H_2 = (CH_2OH)_2 + 2CH_3OH \qquad (5-11)$$

(6) Carbonylation products of syngas and olefin derivatives

5.2 Synthesis gas from coal

Coal gasification is a thermochemical process. It uses coal or coke as raw material, oxygen (air, rich oxygen or pure oxygen) and water vapor as gasification agent. The gasification process is converting the combustible part of coal or coke into gas through chemical reaction under high temperature conditions. The gas produced by gasification is also called coal gas, and its active components include carbon monoxide, hydrogen and methane. Gasification gas can be used as city gas, industrial gas, syngas and industrial reduction gas. Coal gasification is one of the most promising technologies in coal conversion, especially in developing clean coal technology.

5.2.1 Principle of coal gasification process

1. Basic reaction of coal gasification

The main reactions in the coal gasification process are

$$2C+O_2 \Longleftrightarrow 2CO \qquad\qquad (5-12)$$

$$C+O_2 \Longleftrightarrow CO_2 \qquad\qquad (5-13)$$

$$C+H_2O \Longleftrightarrow CO+H_2 \qquad\qquad (5-14)$$

$$C+2H_2O \Longleftrightarrow CO_2+2H_2 \qquad\qquad (5-15)$$

$$C+CO_2 \Longleftrightarrow 2CO \qquad\qquad (5-16)$$

$$C+2H_2 \Longleftrightarrow CH_4 \qquad\qquad (5-17)$$

The gas mixture produced by gasification is called water gas. All the above reactions are reversible and the overall process is strongly endothermic.

In these reactions, the reaction of carbon with water vapor makes the most sense. It participates in various coal gasification processes, and this reaction is a strong endothermic process. Reduction of carbon and carbon dioxide is also an important gasification reaction. The heat released from the carbon combustion reaction matches the above endothermic reaction and plays an important role in the autothermal gasification process. Hydrogasification is very important to produce synthetic natural gas.

2. Reaction conditions of coal gasification

(1) Temperature

Temperature has the greatest influence on coal gasification. A satisfactory gasification rate can be achieved only above 900℃, and the general operating temperature is above 1 100℃. In recent years, the new process uses 1 500℃~1 600℃ for gasification, which greatly improves the production intensity.

(2) Pressure

Reducing the pressure is beneficial to improve the equilibrium concentration of CO and H_2, but increasing the pressure is beneficial to improve the reaction rate and reduce the reaction volume. At present, the gasification pressure is generally 2. 5 MPa~ 3. 2 MPa, so the content of CH_4 is higher than that of atmospheric pressure method.

(3) The ratio of water vapor to oxygen

The function of oxygen is to release heat from coal combustion, and this heat supplies the gasification reaction of water vapor and coal. The ratio of H_2O/O_2 has an effect on temperature and gas composition. The exact ratio of H_2O/O_2 depends on the coal gasification process used.

5.2.2　Production method of coal gasification

The process of coal gasification requires high temperature, which is achieved by burning coal in industry. Gasification process can be divided into intermittent type and continuous type by operation mode, the former technology is backward compared with the latter, so it is gradually eliminated now. At present, the most common classification method is by reactor, which divides the process into fixed bed (moving bed), fluidized bed, airflow bed and molten bed. The molten bed is still in the pilot stage, while the fixed bed (moving bed), fluidized bed and airflow bed are the methods for industrialization or the establishment of demonstration units. In addition, different production methods have different requirements on coal quality.

5.3　Synthesis gas from natural gas

5.3.1　Technology of natural gas to syngas

The content of methane in natural gas is generally more than 90%, and the rest is a small amount of gaseous alkanes such as ethane and propane, some of which also contain a small amount of chlorine and sulfide. Other gases containing gaseous hydrocarbons, such as refinery gas, coke oven gas, oilfield gas and coalbed methane, can be used to make syngas.

At present, steam conversion and partial oxidation are the main technologies to produce syngas from natural gas in industry. Steam conversion method is in the presence of catalyst and high temperature, so that methane and other hydrocarbons can react with water vapor to form H_2, CO and other mixtures, the main reaction is

$$CH_4 + H_2O \Longleftrightarrow CO + 3H_2 \qquad (5-18)$$

The reaction is strongly endothermic and requires external heating. This method is mature and widely used in the production of syngas, pure hydrogen and synthetic ammonia raw gas.

Partial oxidation is the incomplete oxidation of hydrocarbons such as methane and oxygen to produce syngas.

$$2CH_4 + O_2 \Longrightarrow 2CO + 4H_2 \qquad (5-19)$$

5.3.2　Principle of natural gas and steam conversion process

Because the content of methane in natural gas is more than 90%, and methane is the most thermodynamically stable among alkanes (other hydrocarbons are easier to react), so the reaction between methane and vapor is the only one needed to be considered when discussing the natural gas conversion process.

1. Methane vapor conversion reaction and chemical equilibrium

The main reactions in the methane vapor conversion process are

$$CH_4 + H_2O \Longrightarrow CO + 3H_2 \qquad (5-20)$$

$$CH_4 + 2H_2O \Longrightarrow CO_2 + 4H_2 \qquad (5-21)$$

$$CO + H_2O \Longrightarrow CO_2 + H_2 \qquad (5-22)$$

The main side reactions that can occur are carbon-out reactions, which are

$$CH_4 \Longrightarrow C + 2H_2 \qquad (5-23)$$

$$2CO \Longrightarrow C + CO_2 \qquad (5-24)$$

$$CO + H_2 \Longrightarrow C + H_2O \qquad (5-25)$$

Methane vapor conversion reaction must be in the presence of catalyst to have sufficient reaction rate. If the operating conditions are not appropriate, the carbon evolution reaction will be serious, the generated carbon will cover the inner and outer surface of the catalyst, resulting in the reduction of catalytic activity and reaction rate. When the carbon evolution reaction is more serious, the bed will block up, the resistance will increase, the carbon in the catalyst pores will be violently vaporized when meeting water vapor, causing the catalyst to collapse or pulverize, forcing shutdown and huge economic loss. Therefore, special attention should be paid to the prevention of carbon evolution in the process of hydrocarbon vapor transformation.

The main factors affecting the equilibrium of methane vapor conversion reaction are temperature, water-carbon ratio and pressure.

(1) Influence of temperature

The reaction of methane with water vapor to form CO and H_2 is endothermic and reversible. High temperature is favorable to equilibrium, that is, the equilibrium yield of H_2 and CO is high, and the equilibrium content of CH_4 is low. Under normal circumstances, when the temperature increases by $10°C$, the equilibrium content of methane can be reduced by $1\% \sim 1.3\%$. High temperature is unfavorable to the balance of carbon monoxide transformation reaction and can produce less carbon dioxide. High temperature will also inhibit carbon monoxide disproportionation and carbon reduction side reaction. However, when the temperature is too high, it is conducive to methane cracking. When the temperature is higher than $700°C$, the homogeneous cracking rate of methane is very fast, and a large amount of carbon will be precipitated and deposited on the catalyst and the container wall.

(2) Influence of water-carbon ratio

The high water-carbon ratio is favorable to the steam reforming reaction of methane. When the water-carbon ratio increases from 3 to 4 at 800℃ and 2 MPa, the methanol equilibrium content decreases from 8% to 5%, indicating that the water-carbon ratio has a great influence on the methane equilibrium content. At the same time, the high water-carbon ratio is also beneficial to inhibit the carbon evolution side reaction.

(3) Influence of pressure

Methane vapor conversion reaction is a reaction of increasing volume, which favorable balance is at low pressure. When the temperature is 800℃, the ratio of water to carbon is 4, and the pressure drops from 2 MPa to 1 MPa, the equilibrium content of methane drops from 5% to 2.5%. Low pressure can also inhibit the two carbon evolution reactions of carbon monoxide, but low pressure is favorable to the carbon evolution reaction balance of methane cracking, and appropriate pressure can inhibit methane cracking. Pressure has no effect on the equilibrium of carbon monoxide transformation reaction.

In conclusion, considering the balance of the reaction, the methane vapor conversion process should use appropriate high temperature, slightly lower pressure and high water-carbon ratio.

2. Methane vapor conversion catalyst

In the absence of catalyst, methane water vapor conversion reaction rate is very slow, only above 1 300℃ has a satisfactory rate. But at this high temperature, a large amount of methane cracking, there is no industrial production value, so it must use catalyst. The composition and structure of catalyst determine its catalytic performance, and whether it is used properly will affect its performance. The catalyst becomes inactive due to its aging, poisoning and carbon deposition.

(1) Composition and shape of conversion catalyst

Studies have shown that some precious metals and nickel have catalytic activity of methane steam reforming, in which nickel is the most cheap and has a high enough activity, so the industry has been using nickel catalyst, and adding some cocatalysts in order to improve the activity of catalysts or improve other properties such as mechanical strength, the active component dispersed, anti-carbon, anti-sintering, anti-hydration. The promoter of the transformation catalyst includes metal oxides such as aluminum, magnesium, potassium, calcium, titanium, lanthanum and cerium. The reaction of methane and water molecules is carried out on the active surface of solid catalyst, so the catalyst should have a large nickel surface. The most effective way to improve the surface of nickel is to use a carrier with a large surface to support and disperse the active component, and make the nickel grains difficult to sinter through the strong interaction between the carrier and the active component. The carrier should also have sufficient mechanical strength to make the catalyst not easy to break in use. In order to prevent hydrocarbons from splitting in the surface acidic center of the catalyst, alkaline substances are often added to the support to neutralize the surface acidity.

(2) Use and inactivation of conversion catalyst

The conversion catalyst is an oxidation state before use, and after being loaded into the reactor, strict reduction operation should be carried out to reduce nickel oxide to metal nickel before it becomes active. The reducing gas may be hydrogen, methane or carbon monoxide. Pure hydrogen reduction can obtain a high nickel surface area, but the nickel surface area is unstable, in the reaction will be reduced by water vapor, so the water vapor is introduced and heated above 500℃, and then a certain amount of natural gas and a small amount of hydrogen are added for reduction in industry.

The main reasons for the decrease of conversion catalyst activity are aging, poisoning and carbon deposition.

5.3.3　Conditions of natural gas steam conversion process

In the selection of process conditions, the theoretical bases include thermodynamics, dynamics analysis and chemical engineering principles. In addition, it also needs to combine technical economy, production safety and other comprehensive optimization. The main technological conditions of the conversion process are pressure, temperature, water-carbon ratio and space velocity, which are interrelated and should be properly matched.

(1) Pressure

From thermodynamic characteristics, low pressure is favorable for conversion reaction. From the perspective of kinetics, at the initial stage of the reaction, increasing the system pressure is equivalent to increasing the partial pressure of the reactants and speeding up the reaction rate. However, in the later stage of the reaction, the reaction is close to equilibrium, and the concentration of reactants is high. Pressurization will reduce the reaction rate, so from the chemical point of view, the pressure should not be too high. However, from the engineering point of view, appropriately increasing the pressure is beneficial to heat transfer, because the methane conversion process needs external heating, and a large heat transfer coefficient is the premise of enhanced heat transfer.

(2) Temperature

From the thermodynamic point of view, the equilibrium concentration of methane at high temperature is low, and from the kinetic point of view, high temperature accelerates the reaction rate. So the residual methane at the outlet is low. Because of the adverse effect of pressure on balance, it is necessary to raise temperature to compensate.

(3) Water-carbon ratio

The ratio of water to carbon is the most easy one to adjust among the operational variables, and it has a great influence on the transformation process. High water-carbon ratio is conducive to preventing carbon accumulation and the residual methane content is also low. The experiment shows that when there is no saturated hydrocarbon in the raw gas, the water-carbon ratio is less than 2, carbon will be precipitated at 400℃, and when the water-carbon ratio is greater than 2, carbon will be precipitated at 1 000℃. However, when there are more unsaturated hydrocarbons in the raw gas, even if the water-carbon ratio is greater than 2, carbon will be precipitated at 400℃. In order to prevent carbon accumulation, the water-carbon ratio is generally controlled at about 3.5 during operation.

(4) Airflow speed

High gas flow rate in the reactor tube is beneficial to heat transfer, it can reduce the temperature of the outer wall of the furnace tube, and prolong the life of the furnace tube. When catalyst activity is sufficient, high flow rates can also enhance production and increase productivity. But the flow rate should not be too high, otherwise the bed resistance is too large, energy consumption increases.

5.3.4　Principle and main equipment of natural gas steam conversion process

The basic steps of converting natural gas steam to syngas is shown in Figure 5 - 1.

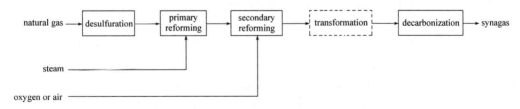

Figure 5 - 1　Conversion of natural gas vapor to syngas

Flow chart of conversion section in a large-scale ammonia plant with natural gas as raw material is shown in Figure 5 - 2.

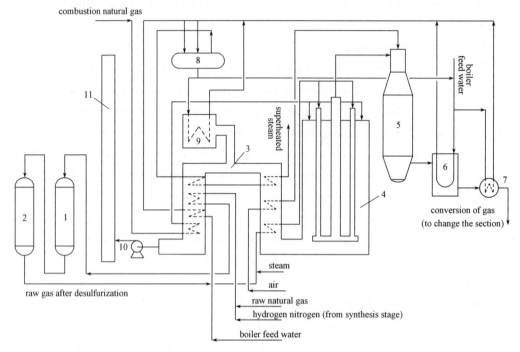

1 – Cobalt molybdenum hydrodesulfurizer; 2 – Zinc oxide desulfurization tank;
3 – A convection section of furnace; 4 – A furnace radiation section; 5 – Two stage reformer;
6 – First waste heat boiler; 7 – Second waste heat boiler; 8 – Drum; 9 – Auxiliary boiler;
10 – Exhaust fan; 11 – Chimney.

Figure 5 – 2   Flow chart of conversion section in a large-scale ammonia plant
with natural gas as raw material

5.4   Synthesis gas from residue oil

The residue used to produce syngas is the residual oil at the bottom of vacuum distillation tower, also known as vacuum residue. The conversion of residual oil into CO, H_2 and other gases is called residue gasification. Gasification technology includes partial oxidation method and regenerative furnace deep cracking method. Currently, the commonly used technology is partial oxidation method, which was successfully developed by Texaco and Shell in the 1950s, respectively called Texaco method and Shell method, and was used for heavy oil at that time. In the 1980s, it was modified and become Texaco's new process, which can be used for residue gasification.

The processing steps of residuum to syngas are as follows.

Figure 5 – 3   Processing steps of residue oil to syngas

Residual oil is a mixture of many macromolecular hydrocarbons with high boiling point. The weight composition of elements contained in residual oil is 84%~87% C, 11%~12.5% H, and the rest are S, N, O, and trace elements such as Ni, V, etc.

The oxidant is oxygen. When oxygen is sufficient, residual oil will completely burn and generate CO_2 and H_2O. Only when oxygen is lower than the theoretical value of complete oxidation, partial oxidation will occur and gas dominated by CO and H_2 will be generated.

The residue, which is a thick, black semi-solid at room temperature, must be preheated to a fluid state before entering the reactor. The residue undergoes the following changes in the reactor.

The residue molecules ($C_m H_n$) are thermally heated and vaporized first. The gaseous residue is evenly mixed with oxygen and then reacts with oxygen. If oxygen is sufficient, a complete combustion reaction will occur, that is

$$C_m H_n + \left(m+\frac{n}{4}\right)O_2 \longrightarrow mCO_2 + \frac{n}{2}H_2O(\text{exothermic}) \qquad (5-26)$$

If the oxygen is lower than the theoretical amount of complete oxidation, partial oxidation occurs and the heat release is less than that of complete combustion.

The equations are

$$C_m H_n + \left(\frac{m}{2}+\frac{n}{4}\right)O_2 \longrightarrow mCO + \frac{n}{2}H_2O(\text{exothermic}) \qquad (5-27)$$

$$C_m H_n + \frac{m}{2}O_2 \longrightarrow mCO + \frac{n}{2}H_2(\text{exothermic}) \qquad (5-28)$$

5.5 Carbon monoxide conversion process

The process by which carbon monoxide reacts with water vapor to form hydrogen and carbon dioxide is called the CO to water gas shift. The transformation reaction can produce more hydrogen and reduce CO content at the same time, which can be used to adjust the ratio of H_2 to CO to meet different production needs.

5.5.1 Chemical equilibrium of carbon monoxide shift reaction

The equation for the carbon monoxide transformation is

$$CO + H_2O(\text{vapour}) \Longrightarrow CO_2 + H_2 \qquad\qquad (5-29)$$

The equilibrium of the transformation reaction is affected by temperature, water-carbon ratio (the molar ratio of H_2O to CO in the feed gas), CO_2 content in the feed gas and other factors. Low temperature and high water-carbon ratio are favorable to the right equilibrium shift, while pressure has no effect on the equilibrium.

The possible side reactions of the transformation reaction are

$$2CO \Longrightarrow C + O_2 \qquad\qquad (5-30)$$

$$CO + 3H_2 \Longrightarrow CH_4 + H_2O \qquad\qquad (5-31)$$

$$CO_2 + 4H_2 \Longrightarrow CH_4 + 2H_2O \qquad\qquad (5-32)$$

When the water-carbon ratio is low, it is more favorable for these side reactions. CO disproportionation causes the catalyst to accumulate carbon, and the latter two reactions are methanation and hydrogen consumption, so they should be suppressed.

5.5.2　Carbon monoxide conversion catalyst

When there is no catalyst, the rate of transformation reaction is very slow, even when the temperature rises to 700℃, the reaction is still not obvious. Therefore, a catalyst must be used to ensure that the reaction has a high enough reaction rate at a not too high temperature to achieve a high conversion rate. There are three main types of conversion catalysts used in industry at present.

(1) Fe-Cr conversion catalyst

Its chemical composition is mainly Fe_2O_3, the promoter is Cr_2O_3 and K_2CO_3. Before the reaction, it needs to be reduced to Fe_3O_4 to have activity. Its applicable temperature range is 300℃ ～ 530℃. This type of catalyst is called medium temperature or high temperature transformation catalyst. Because of the high temperature, the residual CO content in the gas after the reaction is at least 3% ～ 4%.

(2) Copper-based transformation catalyst

Its chemical composition is mainly CuO. ZnO and Al_2O_3 are promoter and stabilizer. The catalyst should also be reduced to active fine copper grains before the reaction. If the temperature is exceeded during the reduction operation or normal operation, it will lead to sintering and inactivation of copper grains. Another weakness of the catalyst is toxicity, so the volume fraction of sulfide in the gas should not exceed 0.1×10^{-6}. Copper-based catalyst is suitable for the temperature range of 180℃～260℃, known as low temperature transformation catalyst. The residual CO can be reduced to 0.2% ～0.3% after the reaction. Copper-based catalyst has high activity. If the CO content in the raw gas is high, the CO should be reduced to about 3% by high temperature transformation first, and then undergoes low temperature transformation to prevent severe heat release and burn out the low variation catalyst.

(3) Cobalt-molybdenum sulphur-resistant catalysts

Its chemical composition is cobalt and molybdenum oxide, which are loaded on alumina. The cobalt and molybdenum oxide should be converted into sulfide (pre-sulfide) before the reaction, and the reaction gas must contain sulfide. Applicable temperature range is $160\,^\circ\text{C} \sim 500\,^\circ\text{C}$, so it is a wide temperature conversion catalyst. It is characterized by sulfur resistance, toxicity resistance and long service life.

5.5.3   Carbon monoxide transformation kinetics

1. Reaction mechanism and kinetic equation

At present, a lot of CO shift reaction mechanism is put forward, there are two popular ones: a view is that CO and H_2O molecules are adsorbed on the catalyst surface to react, then the products are attached; the other view is that CO absorbed on the active site of the catalyst combines with lattice oxygen to form CO_2 and then desorbs, the absorbed H_2O dissociated to attach H_2, and oxygen is added to the lattice, so that's a REDOX mechanism with lattice oxygen transfer. Different dynamic equations can be derived from different mechanisms. Different catalysts have different kinetic equations.

2. Influence of reaction conditions on transformation reaction rate

(1) Influence of pressure

Pressure can increase the partial pressure of the reactants. Below 3.0 MPa, the reaction rate is proportional to the square root of the pressure, and the effect is not obvious at higher pressures.

(2) Influence of water vapor

Water-carbon ratio is determined by the amount of water vapor, and the influence of the water-carbon ratio on the reaction rate is similar to its influence on the equilibrium conversion rate. When the water-carbon ratio is lower than 4, the increase of the water-carbon ratio can make the reaction rate grow faster, but when the water-carbon ratio is greater than 4, the increase of the reaction rate is not obvious, so the water-carbon ratio is generally selected as about 4.

(3) Temperature influence

CO transformation is an exothermic reversible reaction, which has an optimal reaction temperature (T_{op}).

5.5.4   Operating conditions of carbon monoxide conversion

1. Pressure

Although the pressure has no effect on the balance, and the pressure is beneficial to the reaction rate, it should not be too high. Generally, medium and small plants use normal pressure or 2 MPa, large plants use 3 MPa, some use 8 MPa.

2. Water-carbon ratio

A high water-carbon ratio is beneficial to reaction balance and reaction rate, but when it is too high, the effect is not obvious, and the energy consumption is too high. Currently, the commonly used ratio of H_2O to CO is 4 (the ratio of water vapor to water gas is 1.1~1.2). In recent years, energy saving technology has been attached great importance, and it is hoped that the water-carbon ratio can be reduced to below 3. The key is to improve the selectivity of transformation catalyst, and effectively inhibit the side reaction of CO and H_2.

(3) Temperature

The temperature of the transformation reaction is best along the curve of the optimal reaction temperature. The initial reaction, the conversion rate is low, the optimal temperature is high; the later reaction, the conversion rate is high, the optimal temperature is low, but the CO transformation reaction is exothermic, the heat needs to be constantly removed from the system to make the temperature drop. In engineering practice, the cooling can't completely conform to the best temperature curve, the transformation process is using subsection cooling, that is, cooling after a period of reaction, then reaction, so the more segments, the operating temperature is closer to the best temperature curve. It should be particularly noted that the operating temperature must be controlled in catalyst activity temperature range. Lower than this range, the catalyst activity is too low and the reaction rate is too slow. Above this range, the catalyst is vulnerable to overheating and damage, loss of activity. All kinds of catalysts have their own active temperature range, within which the operating temperature can be as close to the optimum reaction temperature curve as possible.

5.5.5 Technological flow of transformation process

There are many kinds of carbon monoxide transformation processes, including atmospheric pressure and pressure; two-stage medium temperature transformation (referred to as high variation), three-stage medium temperature transformation (referred to as high variation), high-low variation series, etc. The selection is mainly based on the production method of syngas, CO content in water gas and the requirement of residual CO content.

CO high-low temperature transformation series process is shown in Figure 5 - 4. Middle temperature transformation process diagram of carbon monoxide three sections is shown in Figure 5 - 5.

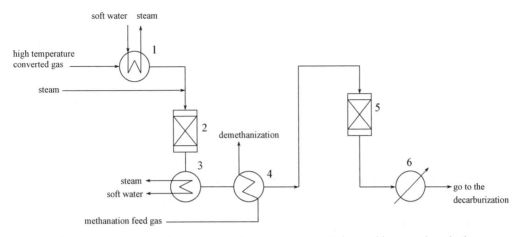

1 – Converter waste heat boiler; 2 – High converter; 3 – High variable waste heat boiler;
4 – Heat exchanger; 5 – Low converter; 6 – Heat exchanger.

Figure 5 – 4　CO high-low temperature transformation series process

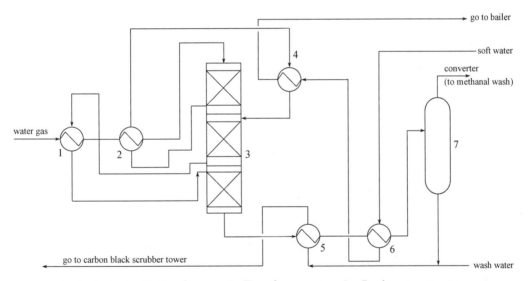

1, 2, 4, 5, 6 – Heat exchanger; 3 – Transform reactor; 7 – Condensate separator.

Figure 5 – 5　Middle temperature transformation flow diagram of carbon monoxide three stages

5.6　Removal of sulfide and carbon dioxide from gas

In the manufacture of syngas, the gas, liquid and solid materials used contain sulfide. Petroleum distillates contain mercaptan (RSH), thioether (RSR), carbon disulfide (CS_2) and thiophene (C_4H_4S), which are mostly concentrated in heavy oil distillates, especially in residual oil. Coal often contains carbonyl sulfur (COS) and pyrite. When using these raw materials to make syngas, the sulfide in them is converted into hydrogen sulfide and organic sulfur gas, which can poison the catalyst and corrode metal pipes and equipment. It is very harmful and must be removed and recycled.

The type and content of sulfide in crude syngas are related to the type of raw materials used and processing method. When making synthesis gas with natural gas or light oil, the raw materials have been thoroughly desulfurized in advance to avoid steam reforming catalyst poisoning, and there are no sulfur in gas; when using coal or heavy oil to make syngas, gasification process does not use catalyst, therefore, it does not need for raw materials desulfurization in advance, and the produced gas contains hydrogen sulfide and organic sulfide. So desulfurization must be carried out before subsequent processing. In the gas (standard state) generated from anthracite gasification with high sulfur content, hydrogen sulfide can reach $4 \sim 6 \text{ g/m}^3$, and total organic sulfur can reach $0.5 \sim 0.8 \text{ g/m}^3$. If the heavy oil contains $0.3\% \sim 1.5\%$ sulfur, the gas after gasification (standard state) contains hydrogen sulfide $1.1 \sim 2.0 \text{ g/m}^3$, organic sulfur $0.03 \sim 0.4 \text{ g/m}^3$. In general, the content of hydrogen sulfide in the gas is $10 \sim 20$ times of the total organic sulfur.

Different use or processing process have different requirements of gas desulfurization purification degree. For example, in the process of natural gas conversion, the total sulfur content (volume fraction) of the raw gas is less than 0.1×10^{-6} and the maximum is 0.5×10^{-6}. The hydrogen sulfide content of the raw gas is less than 5.0×10^{-4} for carbon monoxide high-temperature transformation. The organic sulfur content is less than 1.5×10^{-4}, the copper based catalyst for methanol synthesis requires the total sulfur content to be less than 0.5×10^{-6}, and the iron catalyst for ammonia synthesis requires the raw gas to be sulfur-free.

During the conversion or gasification of gas, liquid and solid raw materials to produce syngas, a certain amount of CO_2 will be generated, especially when there is a carbon monoxide transformation process, more CO_2 will be generated, the content of which can be up to $28\% \sim 30\%$. Therefore, CO_2 also needs to be removed, and the recovered CO_2 can be used. For example, it can supply natural gas converting to reduce the H_2/CO ratio of syngas; it can be used to synthesize urea in ammonia plant; it can be supplied to alkali plant to produce soda (Na_2CO_3); some organic chemicals can also be processed with CO_2. The recycling of CO_2 not only increases the economic benefits, but also reduces the harm caused by the greenhouse effect. The process of removing carbon dioxide is often referred to simply as decarbonization.

5.6.1 Desulfurization method and process

The desulfurization method should be selected according to the content, type of sulfide, the required purification degree, and the specific technical conditions and economy. Sometimes a variety of desulfurization methods can be combined to achieve the requirements of desulfurization purification. According to the state of desulfurizer, there are two categories of desulfurization, dry desulfurization and wet desulfurization.

1. Dry desulfurization

Dry desulfurization is divided into adsorption and catalytic conversion method.

Adsorption method is using solid with strong adsorption capacity to desulfurization, and the adsorbent mainly includes zinc oxide, activated carbon, iron oxide, molecular sieve and so on.

Catalytic conversion method is to use hydrodesulfurization catalyst to hydrolyse the organic sulfur compounds contained in hydrocarbon raw materials into hydrogen solution, which is easy to be removed, and then remove it by other methods. Hydrodesulfurization catalyst is supported by Al_2O_3 and loaded with CoO and MoO_3, also known as cobalt molybdenum hydrodesulfurizer. When used, it is necessary to be vulcanized into Co_9S_8 and MoS_2 by H_2S or CS_2 in advance.

Hydrolytic reactions of organic sulfur are as follows.

$$COS + H_2 \Longrightarrow CO + H_2S \qquad (5-33)$$

$$C_2H_5SH + H_2 \Longrightarrow C_2H_6 + H_2S \qquad (5-34)$$

$$CH_3SC_2H_5 + 2H_2 \Longrightarrow CH_4 + C_2H_6 + H_2S \qquad (5-35)$$

$$C_2H_5SC_2H_5 + 2H_2 \Longrightarrow 2C_2H_6 + H_2S \qquad (5-36)$$

$$C_4H_4S + 4H_2 \Longrightarrow C_4H_{10} + H_2S \qquad (5-37)$$

After hydrogenation of cobalt and molybdenum, H_2S is removed by zinc oxide. Therefore, the combination of zinc oxide, cobalt molybdenum hydrogenation and zinc oxide can achieve the purpose of fine desulfurization.

2. Wet desulfurization

Wet desulfurizer is liquid, generally used for desulfurization of gas with high sulfur content and large processing capacity. According to the different desulfurization mechanism, it is divided into chemical absorption method, physical absorption method, physical-chemical absorption method and wet oxidation method.

Chemical absorption is a common wet desulfurization process. There are monoethanolamine method (MEA), diethanolamine method (DEA), diethylene glycol amine method (DGA), diisopropanolamine method (DIPA) and the modified methyl diethanolamine method (MDEA) developed rapidly in recent years. MDEA is added with accelerant, high purification. These are collectively referred to as the alkanolamine method or alcohol amine method.

Physical absorption method is the use of organic solvents under a certain pressure for physical absorption desulfurization, and then decompression and release of sulfide gas, solvent regeneration. It is mainly composed of cold methanol (Rectisol), propylene carbonate (Fluar) and N-methylpyridine-anone (Purisol).

Physical-chemical absorption method is a mixture of two kinds of solutions with physical absorption properties and chemical absorption properties, and the desulfurization efficiency is higher. A commonly used absorbent is a mixture of sulfolane and alkyl alcoholamine (e. g. methyldiethanolamine), the former is a physical absorption of the sulfide and the latter is a chemical absorption.

In wet oxidation method, the basic principle of desulfurization is to use the alkaline solution containing catalyst to absorb H_2S, and take the catalyst as the oxygen carrier, so that H_2S is oxidized into elemental sulfur, the catalyst itself is reduced. During regeneration, air is passed through to oxidize and restore the reduced catalyst, so that it can be recycled.

3. Recovery of hydrogen sulfide

After wet desulfurization, the gas released during the regeneration of absorbent contains a lot of hydrogen sulfide, which should be recovered in order to protect the environment and make full use of sulfur resources. The mature technology in industry is Krauss process. The basic principle of Krauss process is to react 1/3 of H_2S and O_2 in the combustion furnace to generate SO_2, and the remaining 2/3 of H_2S and SO_2 will undergo Krauss reaction under the action of catalyst to generate elemental sulfur. The equations are

$$H_2S + \frac{3}{2}O_2 \Longrightarrow H_2O + SO_2 + Q_1 \tag{5-38}$$

$$2H_2S + SO_2 \Longrightarrow 2H_2O + 3S + Q_2 \tag{5-39}$$

5.6.2 Method and process for removing carbon dioxide

To remove CO_2, appropriate methods should be selected according to different specific conditions. At present, various decarbonization methods at home and abroad mostly use solution absorbent to absorb CO_2, which can be divided into chemical absorption and physical absorption according to the absorption mechanism. In recent years, pressure swing adsorption, membrane separation and other solid removal of carbon dioxide have emerged.

1. Chemical absorption method

Chemical absorption methods in the early stage include monoethanolamine method (MEA) and ammonia method, which have been used less. At present, the commonly used chemical absorption method is the modified hot potassium alkali method, that is, a small amount of activator is added to the potassium carbonate solution to accelerate the absorption rate of CO_2 and desorption rate, the function of activator is similar to that of catalyst. In the absorption stage, potassium carbonate and CO_2 produce potassium bicarbonate. In the regeneration stage, potassium bicarbonate is decomposed by heat, CO_2 is precipitated, and the solution is restored and recycled.

2. Physical absorption method

At present, the physical absorption methods used at home and abroad are cold methanol method, polyethylene glycol dimethyl ether method and propylene carbonate method. The physical absorption method absorbs CO_2 under pressure (2 MPa $\sim$ 5 MPa) and lower temperature. The regeneration of the solution depends on decompression desorption rather than heating decomposition, which belongs to the cold method and has lower energy consumption.

3. Physical-chemical absorption method

Physical-chemical absorption method is the gas purification method that combines physical absorption agent and chemical absorption agent. For example, MDEA method uses mixed liquid of methyldiethanolamine-sulfolane as absorben, which can desulphurize and take off alkali at the same time, and can compete with improved hot potassium alkali method, but the solvent is more expensive.

4. Pressure swing adsorption method (PSA)

Pressure swing adsorption technology is the use of solid adsorbent under pressure to adsorb CO_2, so that the gas is purified. CO_2 was precipitated by decompression desorption during adsorbent regeneration. PSA method can also be used to separate and purify H_2, N_2, CH_4, CO, C_2H_4 and other gases. Our country has made PSA device, the scale and technology have reached the international advanced level.

Questions

5 - 1   What raw materials are available to produce syngas? What are the production methods of syngas and what new methods have emerged in recent years? What are their advantages over the original production methods?

5 - 2   What chemical products can syngas be used to make? Why are syngas production and application valued in recent years?

5 - 3   What are the main reactions in the process of producing syngas from natural gas? From a thermodynamic point of view, what are the requirements for reaction conditions? What are the dynamics requirements?

5 - 4   How to optimize the process conditions of natural gas-water vapor conversion according to chemical thermodynamics, chemical dynamics and engineering practice?

5 - 5   What are the steps in the syngas process of natural gas-water vapor conversion?

Why does natural gas need to be desulfurized before it can be converted? Which desulphurization method is better?

5 – 6 Why does the gas-water vapor conversion process require heating? What is the form of heating? What are the types of primary reformer?

5 – 7 Why is primary reformer tube a variable temperature reactor? Why is a variable temperature reactor used for gas-water vapor conversion?

5 – 8 Why are there many heat exchangers in the convection chamber of the reformer? How is the sensible heat recycled?

5 – 9 What are the production methods of syngas from coal? What are the advantages of these methods compared to each other? What are the more advanced methods?

5 – 10 What is the reaction of the carbon monoxide transformation? What factors affect the balance and speed of the reaction? How? Why is there an optimal reaction temperature for this reaction? What constants are associated with the optimal reaction temperature?

5 – 11 Why is the carbon monoxide conversion process staged and multi-stage reactor used? What is the basis for selecting the number of segments? What types of reactors are there?

5 – 12 What are the types of carbon monoxide conversion catalysts? For what occasions? What matters should be paid attention to in use?

5 – 13 What are the steps in the process of making syngas from residue? What is the main equipment of residue gasification, and what are its structural characteristics?

5 – 14 Why is there a decarbonization (CO_2) step in syngas manufacturing? What are the common decarbonization methods and where are they applicable?

5 – 15 What are the methods of industrial gas desulfurization, and for what occasions?

第6章 加氢与脱氢过程

6.1 概 述

通常,催化加氢是指有机化合物中一个或几个不饱和的官能团在催化剂作用下与氢气加成。H_2 和 N_2 反应生成合成氨以及 CO 和 H_2 反应合成甲醇及烃类亦为加氢反应。而在催化剂作用下,烃类脱氢生成两种或两种以上的新物质称为催化脱氢。催化加氢和催化脱氢在有机化工生产中得到广泛应用,如合成氨、甲醇、丁二烯、苯乙烯的制取等都是极为重要的化工过程。催化加氢反应分为多相催化加氢和均相催化加氢两种,相比之下,多相催化加氢的选择性较低,反应方向不易控制,而均相催化加氢采用可溶性催化剂,选择性较高,反应条件较温和。

加氢反应又可细分为加氢和氢解两大类。前者氢分子进入化合物内,使化合物还原,或提高不饱和化合物的饱和度,如烯烃加氢合成烷烃,棉籽油经加氢可变为饱和的硬化油;后者又称为破坏加氢,在加氢的同时有机化合物分子发生分解,此时氢分子一部分进入生成物大分子中,另一部分进入裂解所得的小分子中,如重质石油馏分经氢解变为轻质油料,含硫、氧、氮的有机化合物变为烃类、硫化物、水和氨等。

催化加氢除了合成有机产品外,还用于许多化工产品的精制过程,如烃类裂解制得的乙烯和丙烯产物中,含有少量乙炔、丙炔和丙二烯等杂质,可采用催化加氢的方法进行选择加氢,将炔类和二烯烃类转化为相应的烯烃而除去;再如氢气的精制,氢气中含有极少量的一氧化碳、二氧化碳,这些杂质对后续工序的催化剂有中毒作用,可通过催化加氢生成甲烷而得到精制。此外,还有苯的精制、裂解汽油的加氢精制等。

利用催化脱氢反应,可将低级烷烃、烯烃及烷基芳烃转化为相应的烯烃、二烯烃及烯基芳烃,这些都是高分子材料的重要单体,其中苯乙烯和丁二烯是最重要的两个产量大、用途广的化工产品。

加氢与脱氢过程在反应机理、所用催化剂等方面有很多共同之处。

6.1.1 加氢反应类型

(1) 不饱和炔烃、烯烃双键的加氢

$$-C\equiv C- \ +H_2 \longrightarrow \ \diagup\!\!C=C\!\!\diagdown \tag{6-1}$$

$$\diagup\!\!C=C\!\!\diagdown \ +H_2 \longrightarrow \ -\overset{|}{\underset{|}{C}}-\overset{|}{\underset{|}{C}}- \tag{6-2}$$

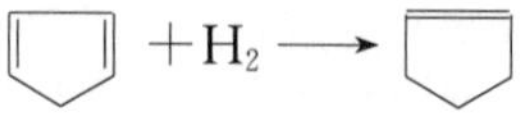

$$\square + H_2 \longrightarrow \square \qquad (6-3)$$

又如乙炔催化加氢生成乙烯,乙烯加氢生成乙烷等。

$$HC\equiv CH + H_2 \longrightarrow H_2C\equiv CH_2 \qquad (6-4)$$

$$H_2C\equiv CH_2 + H_2 \longrightarrow H_3C-CH_3 \qquad (6-5)$$

（2）芳烃加氢

可以对苯环直接加氢,也可以对苯环外的双键加氢,或两者兼有。不同的催化剂有不同的选择,如苯乙烯在 Ni 催化剂下生成乙基环己烷,而在 Cu 催化剂下则生成乙苯。

$$C_6H_6 + 3H_2 \longrightarrow C_6H_{12} \qquad (6-6)$$

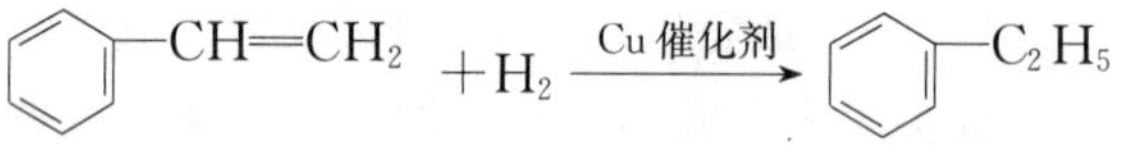

$$(6-7)$$

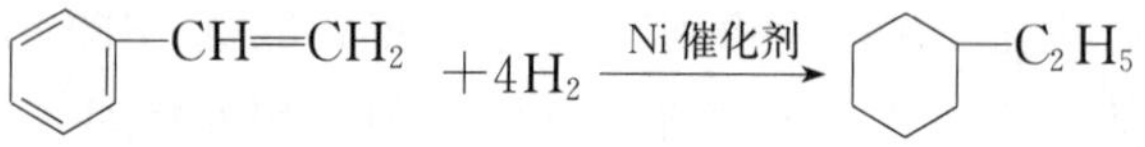

$$(6-8)$$

（3）含氧化合物加氢

带有 $\diagdown C\equiv O$ 基的化合物经催化加氢后可转化为相应的醇类。如一氧化碳加氢在铜催化剂作用下生成甲醇,丙酮在铜催化剂作用下生成异丙醇,羧酸加氢生成相应的醇。

$$CO + 2H_2 \longrightarrow CH_3OH \qquad (6-9)$$

$$(CH_3)_2CO + H_2 \longrightarrow (CH_3)_2CHOH \qquad (6-10)$$

$$RCOOH + 2H_2 \longrightarrow RCH_2OH + H_2O \qquad (6-11)$$

（4）含氮化合物加氢

N_2 和 H_2 合成的氨是目前产量最大的化工产品之一。对于含有—CN、—NO_2 等官能团的化合物,加氢后可得到相应的胺类。如己二腈在 Ni 催化剂作用下加氢可合成己二胺,硝基苯催化加氢可合成苯胺等。

$$N_2 + 3H_2 \rightleftharpoons 2NH_3 \qquad (6-12)$$

$$N\equiv C(CH_2)_4C\equiv N + 4H_2 \longrightarrow H_2N(CH_2)_6NH_2 \qquad (6-13)$$

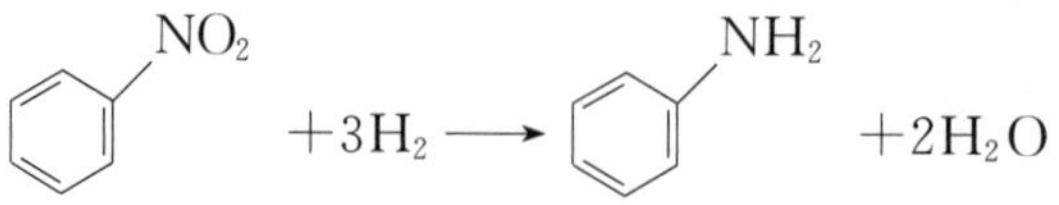

$$(6-14)$$

（5）氢解

在加氢反应过程中,有些原子或官能团被氢气所置换,生成相对分子量较小的一种或两种产物。如甲苯加氢脱烷基生成苯和甲烷,硫醇氢解生成烷烃和硫化氢气体,吡啶氢解生成烷烃和氨。

$$C_6H_5CH_3 + H_2 \longrightarrow C_6H_6 + CH_4 \qquad (6-15)$$

$$C_2H_5SH + H_2 \longrightarrow C_2H_6 + H_2S \qquad (6-16)$$

$$C_5H_5N + 5H_2 \longrightarrow C_5H_{12} + NH_3 \qquad (6-17)$$

6.1.2 脱氢反应类型

(1) 烷烃脱氢生成烯烃、二烯烃及芳烃

$$n\text{-}C_4H_{10} \longrightarrow n\text{-}C_4H_8 \longrightarrow H_2C{=}CH{-}CH{=}CH_2 \qquad (6-18)$$

$$C_{12}H_{26} \longrightarrow n\text{-}C_{12}H_{24} + H_2 \qquad (6-19)$$

$$n\text{-}C_6H_{14} \longrightarrow C_6H_6 + 4H_2 \qquad (6-20)$$

(2) 烯烃脱氢生成二烯烃

$$i\text{-}C_5H_{10} \longrightarrow H_2C{=}CH{-}C(CH_3){=}CH_2 + H_2 \qquad (6-21)$$

(3) 烷基芳烃脱氢生成烯基芳烃

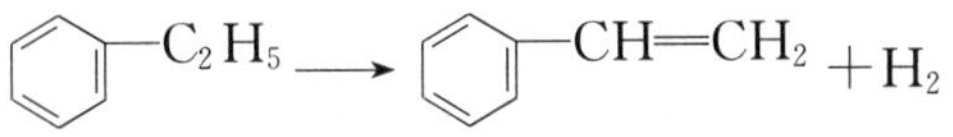

$$\text{〔苯〕}{-}C_2H_5 \longrightarrow \text{〔苯〕}{-}CH{=}CH_2 + H_2 \qquad (6-22)$$

$$C_2H_5C_6H_4C_2H_5 \longrightarrow H_2C{=}CH{-}C_6H_4{-}CH{=}CH_2 + 2H_2 \qquad (6-23)$$

(4) 醇类脱氢可制得醛和酮类

$$CH_3CH_2OH \longrightarrow CH_3CHO + H_2 \qquad (6-24)$$

$$CH_3CHOHCH_3 \longrightarrow CH_3COCH_3 + H_2 \qquad (6-25)$$

6.2 加氢、脱氢反应的一般规律

6.2.1 加氢反应的一般规律

1. 热力学分析

催化加氢反应是放热反应过程,由于有机化合物的官能团结构不同,加氢时放出的热量也不尽相同。

影响加氢反应的因素有温度、压力及反应物中氢的用量。

(1) 温度影响

当加氢反应的温度低于 $100℃$ 时,绝大多数的加氢反应平衡常数都非常大,可看作不可逆反应。由于加氢反应是放热反应,其热效应 $\Delta H^{\ominus} < 0$,所以加氢反应的平衡常数 K 随温度的升高而减小。

(2) 压力影响

加氢是分子数减少的反应,因此,增大反应压力,可以提高 K_p 值,从而提高加氢反应的平衡产率,如提高反应压力,可以提高氨合成产率、甲醇合成产率等。

(3) 氢用量比

从化学平衡分析,提高反应物 H_2 的用量,有利于反应向右进行,可以提高其平衡转化率,同时氢作为良好的载热体,可及时移走反应热,有利于反应的进行。但氢用量比也不能过大,以免造成产物浓度降低,而且大量氢气的循环,既消耗了动力,又增加了产物分离的困难。

2. 动力学分析

有关加氢反应的机理,许多研究者提出不同的看法,如氢气是否发生化学吸附,催化剂表面活性中心是单位吸附还是多位吸附,吸附在催化剂活性表面的分子是如何反应生成产物的,是否有中间产物的生成等。一般认为催化剂的活性中心对氢分子进行化学吸附,并使其解离为氢原子,同时催化剂又使不饱和的双键或三键的 π 键打开,形成了活泼的吸附化合物,活性氢原子与不饱和化合物中双键碳原子结合,形成加氢产物。

影响反映速率的因素有温度、压力、氢的用量比及加氢物质的结构。

(1) 反应温度的影响

热力学上十分有利的加氢反应,可视为不可逆反应,温度升高反应速率常数 k 也升高,反应速率加快。但温度升高会影响加氢反应的选择性,增加副产物的生成,加重产物分离的难度,甚至使催化剂表面积碳,活性下降。对于可逆加氢反应,反应速率常数 k 随温度升高而升高,但平衡常数则随温度的升高而下降,其反应速率与温度的变化是:当温度较低时,反应速率随温度的升高而加快;而在较高的温度下,平衡常数变得很小,反应速率随温度的升高而下降,故有一个最适宜的温度,在该温度下反应速率最大。

(2) 反应压力的影响

一般而言,提高氢分压和被加氢物质的分压均有利于反应速率的增加。但当被加氢物质的级数是负值时,反应速率反而下降。若产物在催化剂上是强吸附的就会占据一部分催化剂的活性中心,抑制加氢反应的进行,此时产物分压越高,加氢反应速率就越慢。

对于液相加氢反应,一般情况下,其反应速率与液相中氢的浓度成正比,故增加氢的分压,有利于增大氢气的溶解度,提高加氢反应速率。

(3) 氢用量比的影响

一般采用过量的氢。氢过量不仅可以提高被加氢物质平衡转化率和加快反应速率,而且可提高传热系数,有利于导出反应热和延长催化剂的寿命。但氢过量太多,会导致产物浓度下降,增加分离难度。

(4) 加氢物质结构的影响

加氢物质在催化剂表面的吸附能力不同、难易程度不同,加氢时受到空间障碍的影响以及催化剂活性组分的差异等都会影响加氢反应速率。

3. 加氢催化剂

为了提高加氢反应速率和选择性,一般都要使用催化剂。当然不同类型的加氢反应所选用的催化剂不同,即使是同一类型的反应因选用不同的催化剂,其反应条件也不尽相同。

加氢催化剂种类很多,其活性组分主要是第Ⅵ族和第Ⅷ族的过渡元素,这些元素对氢有较强的亲和力。最常采用的元素有铁、钴、镍、铂、钯和铑,其次是铜、钼、锌、铬、钨等,其氧化物或硫化物也可作加氢催化剂。Pt‐Rh、Pt‐Pd、Pd‐Ag、Ni‐Cu 等都是很有前途的新型加氢催化剂。

（1）金属催化剂

金属催化剂是把活性组分如 Ni、Pd、Pt 等载于载体上,以提高活性组分的分散性和均匀性,增加催化剂的强度和耐热性。载体是多孔性的惰性物质,常用的载体有氧化铝、硅胶和硅藻土等。在这类催化剂中 Ni 催化剂最常使用,其价格相对较便宜。

金属催化剂的优点是活性高,在低温下也可以进行加氢反应,适用于大多数官能团的加氢反应。其缺点是容易中毒,如 S、As、Cl 等化合物都是催化剂的毒物,故对原料中的杂质含量要求严格,一般在体积分数 10^{-6} 以下。

（2）骨架催化剂

将金属活性组分和载体铝或硅制成合金形式,然后将制成的催化剂用氢氧化钠溶液熔解合金中的硅或铝,得到由活性组分构成的骨架状物质,称为骨架催化剂。最常用的催化剂有骨架镍催化剂,其中镍含量占合金的 $40\% \sim 50\%$,该催化剂的特点是具有很高的活性、足够的机械强度。此外骨架铜催化剂、骨架钴催化剂等,在加氢反应中也得到应用。

（3）金属氧化物催化剂

用于加氢反应的金属氧化物催化剂主要有 MoO_3、Cr_2O_3、ZnO、CuO、NiO 等,这些氧化物既可以单独使用,也可以混合使用,如 $ZnO\text{‐}Cr_2O_3$、$CuO\text{‐}ZnO\text{‐}Al_2O_3$、$Co\text{‐}Mo\text{‐}O_n$。这类加氢催化剂的活性较低,需要有较高的反应温度与压力,以弥补活性差的缺陷。为此在催化剂中常加入 Cr_2O_3、MoO_3 等高熔点的组分,以提高其耐热性能。

（4）金属硫化物催化剂

金属硫化物催化剂主要用于含硫化合物的氢解反应,也可用于加氢精制过程,被加氢原料气中不必预先进行脱硫处理。这类金属硫化物主要有 MoS_2、NiS_2、$Co\text{‐}Mo\text{‐}S$ 等。该类催化剂活性较低,需要较高的反应温度。

（5）金属配位催化剂

用于加氢反应的配位催化剂除了采用贵金属 Ru、Rh、Pd 之外,还有 Ni、Co、Fe、Cu 等为中心原子。该类催化剂的优点是活性较高,选择性好,反应条件温和;其不足是催化剂和产物同一相,分离困难,特别是采用贵金属时,催化剂回收显得非常重要。

6.2.2　脱氢反应一般规律

1. 热力学分析

（1）温度影响

与烃类加氢反应相反,烃类脱氢反应是吸热反应,$\Delta H > 0$,其吸热量与烃类的结构有关。大多数脱氢反应在低温下平衡常数很小,由于 $\Delta H > 0$,随着反应温度升高而平衡常数增大,平衡转化率也升高。

（2）压力影响

脱氢反应是分子数增加的反应,从热力学分析可知,降低总压力,可使产物的平衡浓度增大。

2. 脱氢催化剂

由于脱氢反应是吸热反应,需要在较高的温度条件下进行,同时,反应伴随的副反应较多,要求脱氢催化剂有较好的选择性和耐热性,而金属氧化物催化剂的耐热性好于金属催化剂,所以该催化剂在脱氢反应中受到重视。

脱氢催化剂应满足下列要求:第一是具有良好的活性和选择性,能够尽量在较低的温度条件下进行反应;第二是催化剂的热稳定性要好,能耐较高的操作温度而不失活;第三是化学稳定性好,金属氧化物在氢气的存在下不被还原成金属态,同时在大量的水蒸气下催化剂颗粒能长期运转而不粉碎,保持足够的机械强度;第四是有良好的抗结焦性和易再生性能。

工业生产中常用的脱氢催化剂有 $Cr_2O_3 - Al_2O_3$ 系列,活性组分是氧化铬,氧化铝作载体,助催化剂是少量的碱金属或碱土金属。其中,Cr_2O_3 为 $18\% \sim 20\%$。水蒸气对此类催化剂有中毒作用,故不能采用水蒸气稀释法,应直接用减压法,该催化剂易结焦,再生频繁。

氧化铁系列催化剂,其活性组分是氧化铁(Fe_2O_3),助催化剂是 Cr_2O_3 和 K_2O。氧化铬可以提高催化剂的热稳定性,还可以起着稳定铁的价态的作用;氧化钾可以改变催化剂表面的酸度,以减少裂解反应的发生,同时提高催化剂的抗结焦性。据研究,脱氢反应中起催化作用的可能是 Fe_3O_4,这类催化剂有较高的活性和选择性。但在氢的还原气氛中,其选择性很快下降,这可能是二价铁、三价铁和四价铁之间相互转化而引起的,为此需在大量 Cr_2O_3 和水蒸气存在下,阻止氧化铁被过度还原。所以氧化铁系列脱氢催化剂必须用水蒸气作稀释剂。由于 Cr_2O_3 毒性较大,已采用 Mo 和 Ce 来代替制成无铬的氧化铁系列催化剂。

6.3 氮加氢制合成氨

合成氨用氢气约占氢气生产量的 50%,主要用来制氮肥,还可用来制造硝酸、纯碱、氨基塑料、聚酰胺纤维、丁腈橡胶、磺胺类药物及其他含氮的无机和有机化合物。在国防部门,用氨制备的硝酸是制造硝化甘油、硝化纤维、三硝基甲苯(TNT)、三硝基苯酚等炸药及导弹火箭推进剂的重要原料。氨还是常用的冷冻剂之一。由于氨在农业生产中的特殊地位,其在国民经济中占有重要地位。

1. 氨合成反应热力学

氢与氮合成氨的化学反应式为

$$\frac{1}{2}N_2 + \frac{3}{2}H_2 \Longrightarrow NH_3 \tag{6-26}$$

式(6-26)是可逆放热反应,反应热与温度、压力有关。

2. 氨合成反应动力学

从动力学角度分析,提高温度可以加快反应速率,但温度升高平衡常数 K_p 则下降,影响平衡氨浓度,实际氨浓度也下降,故不能采用过高的反应温度。实现氨产量的增加最有效的方法是改进催化剂性能,提高反应速率。

大型合成氨厂流程见图 6-1。

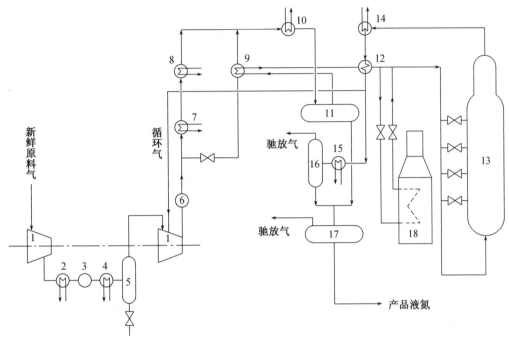

1—离心式合成气压缩机；2,9,12—换热器；3—中间水冷却器；4,7,8,10,15—氨冷器；
5—水分离器；11—高压氨分离器；13—氨合成塔；14—锅炉给水的加热器；16—氨分离器；
17—低压氨分离器；18—开工炉。

图 6-1　大型合成氨厂流程

6.4　甲醇合成

甲醇作为化工原料，主要用于制备甲醛、对苯二甲酸二甲酯、卤甲烷、炸药、医药、染料、农药及其他有机化工产品。随着世界能源的消耗日益增加，天然气和石油资源日趋紧张，因此，在甲醇的应用方面开发了许多新的领域，如甲醇作为非石油基燃料迅速进入燃料市场，成为汽油的代用燃料，汽油掺烧得到迅速发展；甲醇可直接合成汽油，也可合成甲基叔丁基醚（MTBE），并作为无铅汽油的优质添加剂，具有重要的经济效益和社会效益；此外，甲醇还可作为合成蛋白质的碳来源，具有广阔的前途。

甲醇合成的原料气为 CO 和 H_2，可由煤、天然气、轻油、重油、裂解气及焦炉气来制取。近年的发展可以看出，从天然气出发生产甲醇的原料路线备受重视，其投资费用和消耗指标都低于煤和石油，且天然气储量较石油丰富。目前世界上以天然气作原料合成甲醇的能力占总能力的 80% 以上。

6.4.1 合成甲醇的基本原理

1. 合成甲醇的反应热力学

一氧化碳加氢合成甲醇的反应式为

$$CO+2H_2 \Longrightarrow CH_3OH(g) \qquad (6-27)$$

当合成气中有 CO_2 时,也可合成甲醇。

$$CO_2+3H_2 \Longrightarrow CH_3OH(g)+H_2O \qquad (6-28)$$

CO 加氢反应除了生成甲醇外,还有许多副反应发生,例如

$$2CO+4H_2 \Longrightarrow (CH_3)_2O+H_2O \qquad (6-29)$$

$$CO+3H_2 \Longrightarrow CH_4+H_2O \qquad (6-30)$$

$$4CO+8H_2 \Longrightarrow C_4H_9OH+3H_2O \qquad (6-31)$$

$$CO_2+H_2 \Longrightarrow CO+H_2O \qquad (6-32)$$

生成的副产物主要是二甲醚、异丁醇及甲烷气体,此外还有少量的乙醇及微量的醛、酮、醚及酯等。因此,冷凝得到的产物是含有杂质的粗甲醇,需有精制过程。

2. 合成甲醇反应动力学

许多学者对合成甲醇的反应机理进行了研究,也有很多报道,归纳起来有三种假定。

第一种假定认为,甲醇是由 CO 直接加氢生成的,CO_2 通过逆变换生成 CO 后再合成甲醇。

第二种假定认为,甲醇是由 CO_2 直接合成的,而 CO 通过变换反应后合成甲醇。

第三种假定认为,甲醇是由 CO 和 CO_2 同时直接生成。

第一、二种假定认为合成甲醇是连串反应,第三种假定则认为是平行反应。各种假定都有一定的实验数据作为依据,有待进一步研究和探索。至于反应的活性中心和吸附类型,仍然是众说纷纭,如对于铜基催化剂而言,有几种看法:表面零价金属铜 Cu 是活性中心;溶解在 ZnO 中的 Cu^+ 是活性中心;$Cu-Cu^+$ 构成活性中心,CO、CO_2 的吸附中心与 Cu 有关,而 H_2 和 H_2O 的吸附中心与 ZnO 无关。

6.4.2 合成甲醇的催化剂

目前工业生产上采用的催化剂大致可分为锌-铬系和铜-锌(或铝)系两大类。不同类型的催化剂其性能不同,要求的反应条件也不同。

1. 锌-铬系催化剂

锌-铬系催化剂是早期的合成甲醇催化剂,该催化剂活性较低,需要较高的反应温度(380℃~400℃)。由于高温下受平衡转化率的限制,必须提高压力(30 MPa)才能满足,故该催化剂要求高温高压。此外,该催化剂的机械强度和耐热性能较好,使用寿命长,一般为2~3年。

2. 铜基催化剂

铜基催化剂的活性组分是 Cu 和 ZnO,还需添加一些助催化剂,促进该催化剂的活性,如加入铝和铬时活性较高。Cr_2O_3 的添加可以提高铜在催化剂中的分散度,同时又能阻止分散的铜晶粒在受热时被烧结、长大,可延长催化剂的寿命。由于添加 Al_2O_3 助催化剂可使催化剂活性更高,而且 Al_2O_3 价廉、无毒,用 Al_2O_3 代替 Cr_2O_3 的铜基催化剂更好。

6.4.3　合成甲醇的工艺条件

合成甲醇反应是多个反应同时进行的,除主反应之外,还有生成二甲醚、异丁醇、甲烷等副反应。因此,如何提高合成甲醇反应的选择性,从而提高甲醇收率是核心问题,它涉及催化剂的选择以及操作条件的控制,诸如反应温度、压力、空速及原料气组成等。

（1）反应温度和压力

合成甲醇是个可逆的放热反应,平衡产率与温度、压力有关。温度升高,反应速率增加,而平衡常数下降,因此,与氨的合成一样,存在一个最适宜温度。催化剂床层的温度分布要尽可能接近最适宜温度曲线,为此,反应器内部结构比较复杂,便于及时移出反应热。一般采用冷激式和间接式两种。从热力学分析,合成甲醇是体积缩小的反应,增加压力有利于甲醇平衡产率的提高。

（2）空速

从理论上讲,空速高,反应气体与催化剂接触的时间短,转化率降低;而空速低,转化率提高。对合成甲醇来说,由于副反应多,空速过低,会使副反应增加,降低合成甲醇的选择性和生产能力。当然,空速过高也是不利的,会使甲醇含量太低,增加产品的分离困难。因此,应选择适当的空速,以提高生产能力,减少副反应,提高甲醇产品的纯度。对 Zn-Cr 催化剂,空速以 $20\,000\sim40\,000\ h^{-1}$ 为宜,而对 Cu-Zn-Al 催化剂,空速以 $10\,000\ h^{-1}$ 为宜。

（3）合成甲醇原料气配比

H_2 与 CO 的化学计量比（摩尔比）为 2：1,而工业生产原料气除 H_2 和 CO 外,还有一定量的 CO_2,因此,常用 $H_2-CO_2/(CO+CO_2)=2.1\pm0.1$ 作为合成甲醇新鲜原料气组成。而实际进入合成塔的混合气中 $H_2/CO\gg2$,其原因是氢含量高可提高反应速率,降低副反应的发生,而且氢气的热导率（导热系数）大,有利于反应热的导出,易于反应温度的控制。

6.4.4　合成甲醇的工艺流程及反应器

合成甲醇的工艺流程与氨合成工艺流程相似,由于合成率低,未反应的合成气必须循环使用,并采用循环压缩机升压,产物甲醇必须从反应尾气中分离出来。由于甲醇的沸点较高,在合成压力下用水冷却即可将甲醇冷凝下来。为了使惰性气体含量保持在一定范围,循环气需要放空一部分。由于合成甲醇的副反应较多,粗甲醇必须经过精制,才能符合产品要求。

早期的高压合成甲醇工艺流程存在很多缺陷,且低压法技术经济指标先进,动力消耗仅为高压法的60%左右,故高压法逐渐被低压法所代替。

低压分离甲醇流程见图6-2。

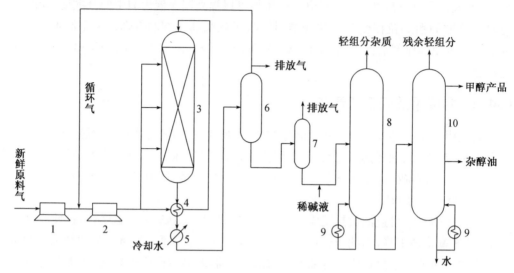

1—合成压缩机;2—循环压缩机;3—合成塔;4—换热器;5—冷凝器;6—分离器;
7—闪蒸罐;8—轻组分脱除塔;9—再沸器;10—精馏塔。

图6-2 低压分离甲醇流程简图

6.4.5 合成甲醇的技术进展

虽然开发了高活性的铜基催化剂,使合成甲醇从高压法转向低压法,完成了合成甲醇技术的一次重大飞跃,但仍存在许多问题,如反应器结构复杂;单程转化率低,气体压缩和循环的能耗大;反应温度不易控制,反应器热稳定性差。这些问题向人们揭示,在合成甲醇技术方面仍有很大的潜力,更新更高的技术等待我们去开发,下面介绍20世纪80年代来所取得的新成果。

(1)气液固三相合成甲醇工艺

该法首先由美国化学系统公司提出,采用三相流化床,液相是惰性介质,催化剂是ICI的Cu-Zn改进型催化剂。对液相介质的要求是在甲醇合成条件下有很好的热稳定性和化学稳定性。既是催化剂的流化介质,又是反应热吸收介质,甲醇在液相介质中的溶解度越小越好,产物甲醇以气相的形式离开反应器。

(2)液相法合成甲醇工艺

液相合成甲醇工艺的特点是采用活性更高的过渡金属配位催化剂。催化剂均匀分布在液相介质中,不存在催化剂表面不均一性和内扩散影响问题,反应温度低,一般不超过200℃。

(3)新型GSSTFR和RSIPR反应系统

该系统采用反应、吸附、和产物交换三步骤交替进行的一种新型反应装置。GSSTFR是指气-固-固滴流流动反应系统,CO和H_2在催化剂作用下,在此系统内进行反应合成甲醇,而甲醇马上被固态粉状吸附剂所吸附,并滴流带出反应系统。RSIPR是级间产品脱除反应系统,当已吸附气态甲醇的粉状吸附剂流入该系统时,与该系统内的液相四甘醇二甲醚进行交换,气态的甲醇被液相所吸附(气态的甲醇溶于液相四甘醇二甲醚中),然后再将四甘醇二甲醚中的甲醇分离出来。这样合成甲醇反应不断向右进行,CO 的单程转化率可达 100%,气相反应物不循环。这项新工艺仍处于研究之中,尚未投入工业生产,还有许多技术问题需要解决和完善。

6.5　乙苯脱氢制苯乙烯

苯乙烯是不饱和芳烃,无色液体,沸点为 145℃,难溶于水,能溶于甲醇、乙醇、四氯化碳及乙醚等溶剂中。

苯乙烯是高分子合成材料的一种重要单体,自身均聚可制得聚苯乙烯树脂,其用途十分广泛,与其他单体共聚可得到多种有价值的共聚物,如与丙烯腈共聚制得色泽光亮的 SAN 树脂,与丙烯腈、丁二烯共聚得 ABS 树脂,与丁二烯共聚可得丁苯橡胶及 SBS塑性橡胶等。此外,苯乙烯还广泛用于制药、涂料、纺织等工业。

苯乙烯聚合物于 1827 年被发现,1867 年 Berthelot 发现乙苯通过赤热瓷管时能生成苯乙烯。1930 年,美国道化学公司首创了乙苯热脱氢法生产苯乙烯过程,1945 年实现了苯乙烯工业化生产。现如今,苯乙烯生产技术不断进步,已趋于完善。2000 年世界苯乙烯生产能力超过 2 230 万吨。由于市场需求旺盛,苯乙烯生产将会高速发展。

6.5.1　制取苯乙烯方法简介

1. 乙苯脱氢法

乙苯脱氢法是目前生产苯乙烯的主要方法。工业上,乙苯主要由苯和乙烯在催化剂作用下采用烷基化法生成,其次是从炼油厂的重整油、烷烃裂解过程中的裂解汽油及炼焦厂的煤焦油中通过精馏分离出来。

$$\text{⟨苯⟩} + C_2H_4 \longrightarrow \text{⟨苯⟩}C_2H_5 \tag{6-33}$$

$$\text{⟨苯⟩}C_2H_5 \rightleftharpoons \text{⟨苯⟩}CH{=}CH_2 + H_2 \tag{6-34}$$

2. 乙苯共氧化法

该法分三步进行,首先是生成乙苯过氧化氢,其次将乙苯过氧化氢与丙烯反应,生成 α-甲基苯甲醇和环氧丙烷,最后 α-甲基苯甲醇脱水生成苯乙烯。

$$\text{⟨苯⟩}C_2H_5 + O_2 \longrightarrow \text{⟨苯⟩}\underset{O-OH}{CH-CH_3} \tag{6-35}$$

$$\text{⟨苯⟩}\underset{O-OH}{CH-CH_3} + CH_3CHCH_2 \longrightarrow \text{⟨苯⟩}\underset{OH}{CH-CH_3} + CH_3-\underset{O}{CH-CH_2}$$

$$\tag{6-36}$$

$$\text{（苯环）}CH{-}CH_3\ ({-}OH) \longrightarrow \text{（苯环）}CH{=}CH_2 + H_2O \qquad (6-37)$$

3. 甲苯为原料合成苯乙烯

第一步采用 $PbO \cdot MgO/Al_2O_3$ 催化剂，在水蒸气存在下使甲苯脱氢缩合生成苯乙烯基苯；第二步将苯乙烯基苯与乙烯在 $WO \cdot K_2O/SiO_2$ 催化剂作用下生成苯乙烯。

$$2\ \text{（苯环）}CH_3 \longrightarrow \text{（苯环）}CH{=}CH\text{（苯环）} + 2H_2 \qquad (6-38)$$

$$\text{（苯环）}CH{=}CH\text{（苯环）} + C_2H_4 \longrightarrow 2\ \text{（苯环）}CH{=}CH_2 \qquad (6-39)$$

另一种方法是甲苯与甲醇直接合成苯乙烯。

$$CH_3OH \longrightarrow HCHO + H_2 \qquad (6-40)$$

$$HCHO + \text{（苯环）}CH_3 \longrightarrow \text{（苯环）}CH_2CH_2OH \qquad (6-41)$$

$$\text{（苯环）}CH_2CH_2OH \longrightarrow \text{（苯环）}CH{=}CH_2 + H_2O \qquad (6-42)$$

$$\text{（苯环）}CH_2CH_2OH + H_2 \longrightarrow \text{（苯环）}C_2H_5 + H_2O \qquad (6-43)$$

4. 乙烯和苯直接合成苯乙烯

$$\text{（苯环）} + C_2H_4 + \frac{1}{2}O_2 \longrightarrow \text{（苯环）}CH{=}CH_2 + H_2O \qquad (6-44)$$

5. 乙苯氧化脱氢

$$\text{（苯环）}C_2H_5 + \frac{1}{2}O_2 \longrightarrow \text{（苯环）}CH{=}CH_2 + H_2O \qquad (6-45)$$

6.5.2 乙苯催化脱氢基本原理

1. 乙苯催化脱氢的主、副反应

主反应

$$\text{（苯环）}C_2H_5 \rightleftharpoons \text{（苯环）}CH{=}CH_2 + H_2 \qquad (6-46)$$

副反应主要有乙苯的裂解、加氢裂解、水蒸气裂解、聚合和缩合而形成焦油等。

$$\text{（苯环）}C_2H_5 \rightleftharpoons \text{（苯环）} + C_2H_4 \qquad \Delta H(873\,\text{K}) = -102\,\text{kJ/mol} \qquad (6-47)$$

$$\text{（苯环）}C_2H_5 + H_2 \rightleftharpoons \text{（苯环）}CH_3 + CH_4 \qquad \Delta H(873\,\text{K}) = 64.4\,\text{kJ/mol}$$

$$(6-48)$$

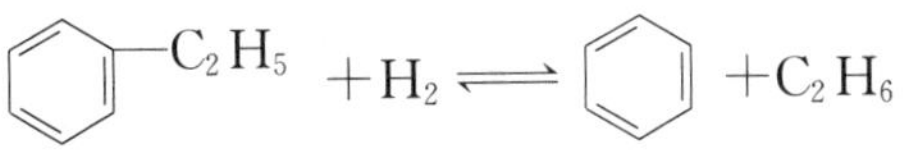

$$\text{C}_2\text{H}_5 + \text{H}_2 \Longrightarrow \text{C}_2\text{H}_6 \qquad \Delta H(298\,\text{K}) = 41.8\,\text{kJ/mol}$$

$$\tag{6-49}$$

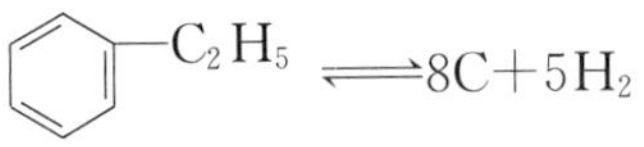

$$\text{C}_2\text{H}_5 \Longrightarrow 8\text{C} + 5\text{H}_2 \qquad \Delta H(873\,\text{K}) = 1.72\,\text{kJ/mol} \tag{6-50}$$

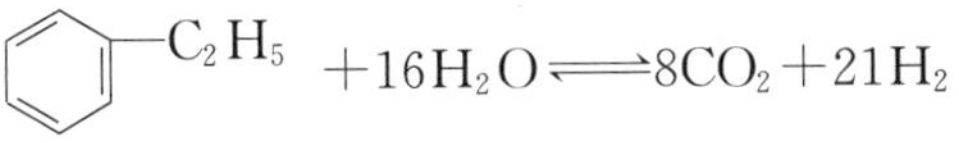

$$\text{C}_2\text{H}_5 + 16\text{H}_2\text{O} \Longrightarrow 8\text{CO}_2 + 21\text{H}_2 \qquad \Delta H(873\,\text{K}) = 793\,\text{kJ/mol}$$

$$\tag{6-51}$$

2. 乙苯脱氢催化剂

从热力学上讲,裂解反应比脱氢反应有利,加氢裂解反应也比脱氢反应有利,即使在700℃下,加氢裂解的平衡常数仍然很大。故乙苯在高温下进行脱氢时,主要产物是苯,要使主反应进行顺利,必须采用高活性、高选择性的催化剂。

除了活性和选择性外,脱氢反应是在高温、有氢和大量水蒸气存在下进行的,这要求催化剂具有良好的热稳定性和化学稳定性,此外催化剂还应能抗结焦和易于再生。

脱氢催化剂的活性组分是氧化铁,助催化剂有钾、钒、钼、钨、铈等氧化物,如按 Fe_2O_3：K_2O：$Cr_2O_3 = 87：10：3$ 组成的催化剂,可使乙苯的转化率达到60%,选择性为87%。

助催化剂 K_2O 能改变催化剂表面酸度,减少裂解反应的发生,并能提高催化剂的抗结焦性能和消炭作用,促进催化剂的再生能力,延长再生周期。

助催化剂 Cr_2O_3 是高熔点金属氧化物,可以提高催化剂的耐热性,稳定铁的价态。但氧化铬对人体及环境有毒害作用,应采用无铬催化剂,如 $Fe_2O_3 - Mo_2O_3 - CeO - K_2O$ 催化剂,以及国产 XH-02 和 335 型无铬催化剂等。

6.5.3　乙苯脱氢反应条件选择

1. 温度

乙苯脱氢反应是可逆吸热反应,温度升高有利于平衡转化率的提高,也有利于反应速率的提高。而温度升高同时有利于乙苯的裂解和加氢裂解,结果是随着温度的升高,乙苯的转化率增加,而苯乙烯的选择性下降。温度降低时,副反应虽然减少,有利于苯乙烯选择性的提高,但因反应速率下降,产率也不高。

2. 压力

对反应速度而言,增加压力,反应速率会加快,但对脱氢的平衡不利。工业上采用水蒸气来稀释原料气,以降低乙苯的分压,提高乙苯的平衡转化率。如果水蒸气对催化剂性能有影响,只有采取降压操作的方法。

3. 空速

乙苯脱氢反应是个复杂反应,空速低,接触时间增加,副反应加剧,选择性会下降,故需采用较高的空速,以提高选择性。虽然可以通过循环使用未反应的原料气解决转化率不是很高的问题,但必然造成耗能增加。因此需综合考虑,选择最佳空速。

4. 催化剂颗粒度的影响

催化剂颗粒的大小影响乙苯脱氢反应的反应速率,脱氢反应的选择性随粒度的增加而降低。可解释为主反应受内扩散影响大,而副反应受内扩散影响小。所以,工业上常用较小颗粒度的催化剂,以减少催化剂的内扩散阻力。同时还可以将催化剂进行高温焙烧改性,以减少催化剂的微孔结构。

5. 水蒸气的用量

用水蒸气作为脱氢反应的稀释剂具有下列优点:① 降低了乙苯的分压,利于提高乙苯脱氢的平衡转化率;② 可以抑制催化剂表面的结焦,同时有消炭作用;③ 提供反应所需的热量,且易于产物的分离。增加水蒸气用量对上述三点有利,但水蒸气用量不是越多越好,超过一定比值以后平衡转化率的提高就不明显了。

6.6 正丁烯氧化脱氢制丁二烯

丁二烯是最简单的具有共轭双键的二烯烃,易发生齐聚和聚合反应,也易与其他具有双键的不饱和化合物共聚,是最重要的聚合物单体,主要用来生产合成橡胶,也可用于合成塑料和树脂。

6.6.1 生产方法

1. 从烃类热裂解制低级烯烃的副产 C_4 馏分得到丁二烯

这是目前获取丁二烯的最经济和最主要的方法。C_4 馏分产量约为乙烯产量的 $30\%\sim50\%$,其中丁二烯的含量可高达 40% 左右,西欧和日本的全部丁二烯、美国 80% 的丁二烯均是通过这一途径得到的。由于 C_4 馏分各组分的沸点相近(正丁烯、异丁烯和丁二烯的沸点分别为 $-6.3℃$、$-6.9℃$ 和 $-4.4℃$),工业上通常采用萃取精馏法将它们分离,所用萃取剂有 N-甲基吡咯烷酮、二甲基甲酰胺和乙腈等。

2. 由正丁烷和正丁烯脱氢生产丁二烯

正丁烷脱氢是连串可逆反应

$$CH_3CH_2CH_2CH_3 \underset{+H_2}{\overset{-H_2}{\rightleftharpoons}} \begin{cases} H_2C\!=\!CHCH_2CH_3 \\ (反-)\ CH_3HC\!=\!CHCH_3 \underset{+H_2}{\overset{-H_2}{\rightleftharpoons}} HCH_2C\!=\!CHCH_2 \\ (顺+)\ CH_3HC\!=\!CHCH_3 \end{cases}$$

$$(6-52)$$

脱氢反应第一阶段得到三种正丁烯异构体,第二阶段三种丁烯异构体继续脱氢得到 1,3-丁二烯。两个阶段的热效应分别为 -126 kJ/mol 和 -113.7 kJ/mol。脱氢反应是吸热而且物质的量(mol)增加的反应,因而采用高温和低压(甚至负压)对脱氢反应是有利的。由于高温下副反应激烈,副产物增加,故要采用催化活性高、选择性好的催化剂。

3. 正丁烯氧化脱氢法制丁二烯

在脱氢反应气中加入适量的氧来迅速除去脱氢反应中产生的氢,这就是氧化脱氢法。它打破了化学平衡限制并提高了反应速率。氧化脱氢法有如下优点:

(1) 反应温度较低。正丁烯制丁二烯只需 $400℃\sim500℃$,比通常的脱氢反应低

$100℃\sim200℃$。

(2) 通常脱氢反应是吸热反应(约 126 kJ/mol),需补给热量,而氧化脱氢却是放热的,可省去原先的供热设备。

(3) 由于催化剂在较低温度和氧的气氛下工作,结焦极少,所以催化剂可以长期运转。

(4) 在通常的脱氢反应中,压力对平衡转化率有很大影响,但在氧化脱氢时,压力影响甚微,所以减压或用水蒸气稀释并非一定必要。

(5) 由于反应温度较低,升温有可能让转化率及选择性都获得提高。而在通常的脱氢法中,升高温度转化率可以提高,但选择性因副反应激烈往往是下降的。

因此,烃类的氧化脱氢工艺有着光明的前景。例如,正丁烯氧化脱氢已逐步取代脱氢法,成为制造丁二烯的重要方法。

6.6.2 工艺原理

1. 化学反应

正丁烯氧化脱氢生成丁二烯的主反应是一个放热反应。

$$C_4H_8+\frac{1}{2}O_2 \longrightarrow H_2C=CHCH=CH_2+H_2O(气) \qquad \Delta H(773\ K)=134.31\ kJ/mol$$

$$(6-53)$$

主要的副反应有:

(1) 正丁烯氧化降解生成饱和及不饱和的小分子醛、酮、酸等含氧化合物,如甲醛、乙醛、丙烯醛、丙酮、饱和及不饱和低级有机酸等。

(2) 正丁烯氧化生成呋喃、丁烯醛和丁酮等。

(3) 完全氧化生成一氧化碳、二氧化碳和水。

(4) 正丁烯氧化脱氢环化生成芳烃。

(5) 深度氧化脱氢生成乙烯基乙炔、甲基乙炔等。

(6) 产物和副产物的聚合结焦。

2. 催化剂和催化机理

工业应用的正丁烯氧化脱氢催化剂主要有两大系列。

(1) 钼酸铋系列催化剂

钼酸铋系列催化剂是以 Mo-Bi 氧化物为基础的二组分或多组分催化剂,初期主要为 Mo-Bi-O 二组分和 Mo-Bi-P-O 三组分催化剂,但其活性和选择性都较低,后经改进,发展为六组分、七组分或更多组分混合氧化物催化剂,如 Mo-Bi-P-Fe-Ni-K-O,Mo-Bi-P-Fe-Co-Ni-Ti-O 等,催化活性和选择性均有明显的提高。

(2) 铁酸盐尖晶石系列催化剂

$ZnFe_2O_4$、$MnFe_2O_4$、$MgFe_2O_4$、$ZnCrFeO_4$ 和 $Mg_{0.1}Zn_{0.9}Fe_2O_4$(原子比)等铁酸盐是具有尖晶石型($A^{2+}B_2^{3+}O_4$)结构的氧化物,是 20 世纪 60 年代后期开发的一类正丁烯氧化脱氢催化剂。

（3）其他类型

主要有以 Sb 或 Sn 氧化物为基础的混合氧化物催化剂。

6.6.3　工艺条件选择

工艺条件与采用的催化剂和反应器有关,现以铁酸盐尖晶石催化剂及绝热式反应器为例讨论正丁烯氧化脱氢制取丁二烯的工艺条件选择。

（1）原料纯度的要求

正丁烯的 3 个异构体在铁酸盐尖晶石催化剂上的脱氢反应速率和选择性虽有所差异,但差别不大。因此,原料中 3 个异构体的组成分布的影响对工艺条件的选择影响不大。

原料中异丁烯的量要严格控制,因为异丁烯易氧化,会使氧的消耗量增加,并影响反应温度的控制。

C_3 或 C_3 以下烷烃性质稳定,不会被氧化,但其含量太高会影响反应器的生产能力,在操作条件下也有可能少量被氧化生成 CO_2 和水。

（2）氧与正丁烯的用量比

一般采用空气为氧化剂,由于丁二烯的收率与所用氧量有直接关系,故氧与正丁烯的用量比要严格控制。

（3）水蒸气与正丁烯的用量比

水蒸气的存在可以提高丁二烯的选择性,其反应选择性随 $H_2O/n-C_2H_8$ 的增加而增加,直至达到最大值。

（4）反应温度

由于氧化脱氢是放热反应,因此出口温度会明显高于进口温度,两者温差可达 220℃或更大。适宜的反应温度一般为 327℃～547℃。

（5）正丁烯的空速

正丁烯的空速在一定范围内变化,对选择性影响甚微。一般空速增加,需相应提高进口温度,以保持一定的转化率。工业上正丁烯质量空速（GHSV）为 600 h^{-1} 或更高。

（6）压力

反应器的进口压力虽然对转化率影响甚微,但对选择性有影响。进口压力升高,选择性下降,收率也下降。因此希望在较低压下操作,并要求催化剂床层的阻力降应尽可能小,为此采用径向绝热床反应器将更适宜。选择性下降的原因可能是原料、中间产物和丁二烯等在催化剂表面滞留时间过长,发生降解或完全氧化之故。

6.6.4　工艺流程

正丁烯氧化脱氢生产丁二烯的工艺流程分为反应、丁二烯分离和精制,以及未转化的正丁烯回收三部分。正丁烯氧化脱氢制丁二烯反应部分流程见图 6 - 3。

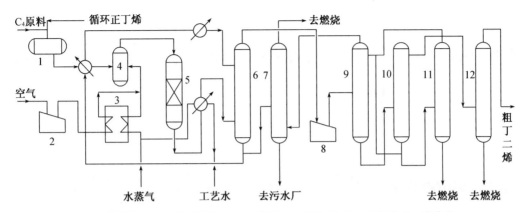

1—C₄原料罐；2—空气压缩机；3—加热炉；4—混合器；5—反应器；6—淬冷器；

7—吹脱塔；8—压缩机；9—吸收塔；10—解吸塔；11—油再生塔；12—脱重组分塔。

图6-3 正丁烯氧化脱氢制丁二烯反应部分流程

思考题

6-1 加氢反应和脱氢反应对催化剂有什么要求？

6-2 以煤为原料和以天然气为原料合成氨的生产过程有什么不同之处？

6-3 试分析比较合成气的三种精制方法。

6-4 影响氨平衡浓度的因素有哪些？

6-5 简述温度和压力对氨合成反应的平衡氨浓度及反应速率的影响。

6-6 简述惰性气体对氨合成反应的平衡浓度及反应速率的影响。

6-7 简述氨合成催化剂活性组分与助剂的作用。

6-8 在氨合成工艺流程中，排放气为什么在循环压缩机前，而氨冷则在循环压缩机之后？

6-9 合成氨与合成甲醇有哪些相似的地方？

6-10 高低压合成甲醇的比较。

6-11 简述乙苯脱氢制苯乙烯生产过程中温度和空速对选择性的影响。

Chapter 6　Hydrogenation and dehydrogenation process

6.1　Summary

In general, catalytic hydrogenation refers to the addition of one or more unsaturated functional groups in organic compounds with hydrogen under the action of catalysts. The reaction of H_2 and N_2 to synthetic ammonia and the reaction of CO and H_2 to methanol and hydrocarbons are also hydrogenation reactions. Under the action of catalyst, hydrocarbon dehydrogenation into two or more new substances called catalytic dehydrogenation. Catalytic hydrogenation and catalytic dehydrogenation are widely used in organic chemical production. Such as synthesis of ammonia, methanol, butadiene, styrene and so on are very important chemical processes. Catalytic hydrogenation can be divided into heterogeneous catalytic hydrogenation and homogeneous catalytic hydrogenation. In contrast, heterogeneous catalytic hydrogenation has low selectivity and the reaction direction is not easy to control, while homogeneous catalytic hydrogenation uses soluble catalyst with high selectivity and mild reaction conditions.

Hydrogenation reaction can be subdivided into hydrogenation and hydrogenolysis. Hydrogen molecules enter the compound to reduce the compound or improve the saturation of unsaturated compounds in hydrogenation. For example, alkenes are hydrogenated to produce alkenes, and cottonseed oil can become saturated hardened oil by hydrogenation. In hydrogenolysis, at the same time of hydrogenation, organic compound molecules decompose, part of the hydrogen molecules becomes the product macromolecules, the other part becomes the small molecules obtained by pyrolysis, such as heavy petroleum fractions convert into light oil by hydrogenation, organic compounds containing sulfur, oxygen, nitrogen convert into hydrocarbons, sulfide, water and ammonia, etc.

In addition to the synthetic organic products, catalytic hydrogenation is also used in many refining process of chemical products, such as ethylene and propylene produced by hydrocarbon pyrolysis, containing a small amount of acetylene, allene and propiolic impurities, can choose catalytic hydrogenation method, alkadiene and acetylene can be converted to the corresponding olefin. Besides, in the refining of hydrogen, hydrogen contains a very small amount of carbon monoxide, carbon dioxide, these impurities have toxic effect on the catalyst in the subsequent processes, it can be refined by catalytic hydrogenation to generate methane. In addition, there are benzene refining, cracking gasoline hydrorefining and so on.

By catalytic dehydrogenation, lower alkanes, olefin and alkyl aromatics can be converted into corresponding olefin, diolefin and alkenyl aromatics, which are important monomers of polymer materials. Among them, styrene and butadiene are the two most important chemical products with large yield and wide application.

Hydrogenation and dehydrogenation have much in common in reaction mechanism and catalyst used.

6.1.1   Type of hydrogenation reaction

(1) Hydrogenation of unsaturated alkynes and olefin

$$-C\equiv C- \; + H_2 \longrightarrow \; \overset{\diagdown}{\underset{\diagup}{C}}=\overset{\diagup}{\underset{\diagdown}{C}} \tag{6-1}$$

$$\overset{\diagdown}{\underset{\diagup}{C}}=\overset{\diagup}{\underset{\diagdown}{C}} \; + H_2 \longrightarrow \; -\overset{|}{\underset{|}{C}}-\overset{|}{\underset{|}{C}}- \tag{6-2}$$

$$\square\!\!\!| \; + H_2 \longrightarrow \; \square \tag{6-3}$$

Such as acetylene catalytic hydrogenation to produce ethylene, ethylene hydrogenation to produce ethane, etc.

$$HC\equiv CH \; + H_2 \longrightarrow H_2C=CH_2 \tag{6-4}$$

$$H_2C=CH_2 + H_2 \longrightarrow H_3C-CH_3 \tag{6-5}$$

(2) Aromatics hydrogenation

The hydrogenation of styrene to ethyl cyclohexane over Ni, and styrene to ethylbenzene over Cu can be done, which means it can directly hydrogenated with the benzene core or with the double bond outside the benzene core, or both.

$$C_6H_6 + 3H_2 \longrightarrow C_6H_{12} \tag{6-6}$$

$$\text{C}_6\text{H}_5-CH=CH_2 \; + H_2 \xrightarrow{\ Cu\ } \text{C}_6\text{H}_5-C_2H_5 \tag{6-7}$$

$$\text{C}_6\text{H}_5-CH=CH_2 \; + 4H_2 \xrightarrow{\ Ni\ } \text{C}_6\text{H}_{11}-C_2H_5 \tag{6-8}$$

（3）Hydrogenation of oxygen-containing compounds

The compounds with $\overset{\diagdown}{\underset{\diagup}{C}}=O$ group can be converted into corresponding alcohols by catalytic hydrogenation. For example, methanol is generated by hydrogenation of carbon monoxide under the action of copper catalyst, isopropanol is generated by acetone under the action of copper catalyst, and corresponding alcohols are generated by hydrogenation of carboxylic acid.

$$CO+2H_2 \longrightarrow CH_3OH \tag{6-9}$$

$$(CH_3)_2CO+H_2 \longrightarrow (CH_3)_2CHOH \tag{6-10}$$

$$RCOOH+2H_2 \longrightarrow RCH_2OH+H_2O \tag{6-11}$$

（4）Hydrogenation of nitrogen-containing compounds

Ammonia synthesized by N_2 and H_2 is one of the largest chemical products at present. For compounds containing —CN, —NO_2 and other functional groups, corresponding amines can be obtained by hydrogenation. For example, adiponitrile can be hydrogenated to hexanediamine under the action of Ni catalyst, nitrobenzene can catalytic hydrogenation to aniline.

$$N_2+3H_2 \Longleftrightarrow 2NH_3 \tag{6-12}$$

$$N\equiv C(CH_2)_4C\equiv N+4H_2 \longrightarrow H_2N(CH_2)_6NH_2 \tag{6-13}$$

$$\tag{6-14}$$

（5）Hydrogenolysis

In the hydrogenation reaction process, some atoms or functional groups are replaced by hydrogen, resulting in one or two products with relatively small molecular weight. Such as toluene hydrodealkylation to benzene and methane, mercaptan hydrogenolysis to produce alkanes and hydrogen sulfide gas, pyridine hydrogenolysis to form alkanes and ammonia.

$$C_6H_5CH_3+H_2 \longrightarrow C_6H_6+CH_4 \tag{6-15}$$

$$C_2H_5SH+H_2 \longrightarrow C_2H_6+H_2S \tag{6-16}$$

$$C_5H_5N+5H_2 \longrightarrow C_5H_{12}+NH_3 \tag{6-17}$$

6.1.2　Type of dehydrogenation reaction

（1）Dehydrogenation of alkanes to monoenes, dienes and aromatics

$$n\text{-}C_4H_{10} \longrightarrow n\text{-}C_4H_8 \longrightarrow H_2C=CH-CH=CH_2 \tag{6-18}$$

$$C_{12}H_{26} \longrightarrow n\text{-}C_{12}H_{24}+H_2 \tag{6-19}$$

$$n\text{-}C_6H_{14} \longrightarrow C_6H_6 + 4H_2 \qquad (6-20)$$

(2) Dehydrogenation of olefins to dienes

$$i\text{-}C_5H_{10} \longrightarrow H_2C\!=\!CH\!-\!C(CH_3)\!=\!CH_2 + H_2 \qquad (6-21)$$

(3) Dehydrogenation of alkyl aromatics to alkenyl aromatics

$$(6-22)$$

$$C_2H_5C_6H_4C_2H_5 \longrightarrow H_2C\!=\!CH\!-\!C_6H_4\!-\!CH_2\!=\!CH_2 + 2H_2 \qquad (6-23)$$

(4) Dehydrogenation of alcohols can produce aldehydes and ketones

$$CH_3CH_2OH \longrightarrow CH_3CHO + H_2 \qquad (6-24)$$

$$CH_3CHOHCH_3 \longrightarrow CH_3COCH_3 + H_2 \qquad (6-25)$$

6.2　General rules of hydrogenation and dehydrogenation reaction

6.2.1　General rules of hydrogenation reaction

1. Thermodynamic analysis

Catalytic hydrogenation reaction is an exothermic reaction process, because of the different functional group structure of organic compounds, the heat released during hydrogenation is not the same.

The factors affecting the hydrogenation reaction are temperature, pressure and the amount of hydrogen in the reactants.

(1) Influence of temperature

When the hydrogenation reaction temperature is below 100℃, most of the hydrogenation reaction equilibrium constant values are very large and can be regarded as irreversible reaction. Because hydrogenation reaction is exothermic, the value of the thermal effect is below zero, the equilibrium constant K for hydrogenation decreases with increasing temperature.

(2) Influence of pressure

Hydrogenation is a reaction in which the number of molecules is reduced. Therefore, increasing the reaction pressure can improve K_p value, thus improving the equilibrium yield of hydrogenation reaction. For example, increasing the reaction pressure can improve the yield of ammonia synthesis and methanol synthesis, etc.

(3) Hydrogen dosage ratio

From chemical equilibrium analysis, increasing the amount of H_2 can facilitate the reaction to move to the right, which improve its equilibrium conversion rate. At the same time, hydrogen, as a good heat carrier, can remove the reaction heat in time, which is conducive to the reaction. However, the ratio of hydrogen consumption should not be too large, so as to avoid the reduction of product concentration. The circulation of a large amount of hydrogen not only consumes power, but also increases the difficulty of product separation.

2. Kinetic analysis

Regarding the mechanism of hydrogenation reaction, many researchers have proposed different views, such as whether hydrogen chemical adsorption occurs, whether the active center of catalyst surface is unit adsorption or multi-adsorption, and how the molecules adsorbed on the active surface of the catalyst react to form products, whether there is the formation of intermediate products, etc. It is generally believed that the active center of the catalyst carries out chemisorption of hydrogen molecules and dissociates them into hydrogen atoms. At the same time, the catalyst also opens the π bond of the unsaturated double bond or triple bond, forming active adsorption compounds. The active hydrogen atoms combine with double bond carbon atoms of unsaturated compounds to form hydrogenation products.

The factors affecting the reaction rate are temperature, pressure, hydrogen dosage ratio and the structure of hydrogenated substance.

(1) Influence of reaction temperature

Hydrogenation reaction, which is very favorable in thermodynamics, can be regarded as an irreversible reaction. The reaction rate constant k increases with the increase of temperature, and the reaction rate accelerates as well. However, the increase of temperature will affect the selectivity of hydrogenation reaction, increase the generation of by-products, increase the difficulty of product separation, and even decrease the activity of catalyst owing to carbon deposit. For reversible hydrogenation, the reaction rate constant k increases with the increase of temperature, but the equilibrium constant decreases with the increase of temperature. The change of reaction rate and temperature is as follows: When the temperature is low, the reaction rate increases with the increase of temperature, while at a higher temperature, the equilibrium constant becomes very small, the reaction rate decreases with the increase of temperature, so there is an optimal temperature, at which the reaction rate is the maximum.

(2) Influence of reaction pressure

Generally speaking, increasing the partial pressure of hydrogen and hydrotreated substance is beneficial to increasing the reaction rate. However, when the order of the hydrogenated material is negative, the reaction rate decreases. If the product is strongly adsorbed on the catalyst, it will occupy part of the active center of the catalyst and inhibit the hydrogenation reaction. At this time, the higher the partial pressure of the product, the slower the hydrogenation reaction rate.

For liquid phase hydrogenation reaction, generally speaking, the reaction rate is proportional to the concentration of hydrogen in the liquid phase, so increasing the partial pressure of hydrogen is beneficial to increasing the solubility of hydrogen and can improve the hydrogenation reaction rate.

(3) Influence of hydrogen dosage ratio

Excessive amount of hydrogen is commonly used. The excessive hydrogen can not only improve the equilibrium conversion of the hydrogenated substance and accelerate the reaction rate, but also improve the heat transfer coefficient, which is beneficial to deriving the reaction heat and prolonging the life of the catalyst. But too much hydrogen will result in a decline in product concentration and separation difficulty.

(4) Influence of hydrogenation material structure

Because of the different adsorption capacity of hydrogenation materials on the surface of catalyst and the different degree of difficulty, the hydrogenation reaction rate is affected by the space barrier and the active components of catalyst.

3. Hydrogenation catalyst

In order to improve the hydrogenation reaction rate and selectivity, catalysts are usually used. Different types of hydrogenation reaction using different catalysts, even the same type of reaction due to the selection of different catalysts, the reaction conditions are not the same.

There are many kinds of hydrogenation catalysts, and the active components are mainly the transition elements of group Ⅵ and group Ⅷ, which have strong affinity for hydrogen. The most commonly used elements are iron, cobalt, nickel, platinum, palladium and rhodium, followed by copper, molybdenum, zinc, chromium, tungsten, etc. Their oxides or sulfides can also be used as hydrogenation catalysts. Pt-Rh, Pt-Pd, Pd-Ag and Ni-Cu are promising new hydrogenation catalysts.

(1) Metal catalyst

Metal catalyst loads active component such as Ni, Pd, Pt on the carrier to improve the dispersity and uniformity of the active component, as well as increase the strength and heat resistance of the catalyst. Carriers are inert porous materials. Alumina, silica gel and diatomite are commonly used. Ni catalyst is the most commonly used catalyst in this class and its price is relatively cheap.

Metal catalyst has the advantage of high activity, so it can be hydrogenated at low temperature, suitable for most of the hydrogenation of functional groups. Its disadvantage is easy to poison. S, As, Cl and other compounds are toxic to the metal catalyst, so the content of impurities in the raw material are strictly required, generally within the volume fraction of 10^{-6}.

(2) Skeleton catalyst

The metal active component and the carrier aluminum or silicon are made into alloy form, and then using sodium hydroxide solution to melt the silicon or aluminum in the alloy, the skeleton like material composed of active components is called skeleton catalyst. The most commonly used catalyst is skeleton nickel catalyst, nickel content accounted for $40\% \sim 50\%$ of the alloy. This catalyst is characterized by high activity and enough mechanical strength. In addition, there are skeleton copper catalyst, skeleton cobalt catalyst and so on, which are also used in hydrogenation reaction.

(3) Metal oxide catalyst

Metal oxide catalysts used for hydrogenation reaction mainly include MoO_3, Cr_2O_3, ZnO, CuO, NiO, etc. These oxides can be used alone or mixed, such as $ZnO - Cr_2O_3$, $CuO - ZnO - Al_2O_3$, $Co - Mo - O_n$. This kind of hydrogenation catalyst has low activity, so higher reaction temperature and pressure are needed to make up for the defect of poor activity. For this reason, Cr_2O_3, MoO_3 and other components with high melting point are often added to the catalyst to improve its heat resistance.

(4) Metal sulfide catalyst

Metal sulfide catalyst is mainly used for hydrolytic reaction of sulfur-containing compounds and hydrorefining process. There is no need to carry out desulfurization treatment in advance in the hydrotreated feed gas. These metal sulfides mainly include MoS_2, NiS_2, Co-Mo-S and so on. This kind of catalyst has low activity and needs high reaction temperature.

(5) Metal coordination catalyst

The coordination catalyst used for hydrogenation reaction not only adopts precious metals Ru, Rh, Pd, but also Ni, Co, Fe, Cu as the center atom. The advantages of this kind of catalyst are high activity, good selectivity, mild reaction conditions; the disadvantage is that the catalyst and the product are in the same phase, so it is difficult to separate, especially when using precious metals, catalyst recovery is very important.

6.2.2　General rules of dehydrogenation

1. Thermodynamic analysis

(1) Influence of temperature

In contrast to hydrocarbon hydrogenation, hydrocarbon dehydrogenation is an endothermic reaction, $\Delta H > 0$, its heat absorption is related to the structure of hydrocarbons. Most dehydrogenation reactions have very small equilibrium constants at low temperatures, the equilibrium constant increases with the increase of reaction temperature due to $\Delta H > 0$, and the equilibrium conversion rate also increases.

（2）Influence of pressure

Dehydrogenation is a reaction in which the number of molecules increases. It can be seen from thermodynamic analysis that the equilibrium concentration of the product can be increased by decreasing the total pressure.

2. Dehydrogenation catalyst

Because dehydrogenation is an endothermic reaction, it is required to be carried out at a higher temperature, accompanied by more side reactions, requiring dehydrogenation catalyst with better selectivity and heat resistance. As metal oxide catalyst has better heat resistance than metal catalyst, so it is paid attention to in dehydrogenation.

Dehydrogenation catalyst should meet the following requirements. Firstly, it should have good activity and selectivity, and can react at the temperature as low as possible. Secondly, the catalyst should have good thermal stability and can withstand higher operating temperature without losing activity. The third requirement is good chemical stability, metal oxide in the presence of hydrogen should not be reduced to metal state, at the same time in a large amount of water vapor catalyst particles, it should run for a long time without crushing, maintain sufficient mechanical strength. The fourth requirement is to have good anti-coking and easy regeneration performance.

The dehydrogenation catalyst commonly used in industrial production is $Cr_2O_3 - Al_2O_3$ series, the active component is chromium oxide, alumina as a carrier, and the cocatalyst is a small amount of alkali metal or alkaline earth metal. General composition: Cr_2O_3 accounts for $18\% \sim 20\%$. Water vapor has toxic effect on such catalyst, so direct decompression method should be used instead of water vapor dilution method, the catalyst is easily coking. Regeneration is frequent.

Iron oxide series catalyst. The active component is ferric oxide (Fe_2O_3), and the cocatalyst is Cr_2O_3 and K_2O. Chromium oxide can improve the thermal stability of the catalyst and can also play a role in stabilizing the valence state of iron. Potassium oxide can change the acidity of the catalyst surface to reduce the cracking reaction and improve the coking resistance of the catalyst. According to studies, the catalyst for dehydrogenation may be Fe_3O_4, which has high activity and selectivity. However, in the reducing atmosphere of hydrogen, its selectivity decreases rapidly, which may be caused by the mutual transformation between divalent iron, trivalent iron and tetravalent iron. Therefore, in the presence of a large amount of Cr_2O_3 and water vapor, to prevent the excessive reduction of iron oxide. So iron oxide series dehydrogenation catalyst must use water vapor as diluent. Because of the toxicity of Cr_2O_3, Mo and Ce have been used to replace the chromium free iron oxide series catalyst.

6.3 Nitrogen hydrogenation to synthetic ammonia

Hydrogen for synthetic ammonia accounts for about 50% of the hydrogen production. Synthetic ammonia mainly used for nitrogen fertilizer, which can also be used to manufacture nitric acid, soda, amino plastics, polyamide fiber, nitrile rubber, sulfonamide drugs and other inorganic and organic compounds containing nitrogen. In the national defense department, nitric acid prepared with ammonia is an important raw material for the manufacture of nitroglycerin, nitrocellulose fiber, trinitrotoluene (TNT) and other explosives, missile rocket propellant. Ammonia is also a common refrigerant. Ammonia plays an important role in the national economy because of its special position in agricultural production.

1. Thermodynamics of ammonia synthesis reaction

The chemical reaction formula for hydrogen and nitrogen to synthesize ammonia is

$$\frac{1}{2}N_2 + \frac{3}{2}H_2 \Longrightarrow NH_3 \qquad (6-26)$$

Equation (6 – 26) is a reversible exothermic reaction, and the heat of the reaction is related to temperature and pressure.

2. Kinetics of ammonia synthesis reaction

From the perspective of kinetics, the reaction rate can be accelerated by increasing the temperature, but the equilibrium constant K_p decreases when the temperature increases, which affects the equilibrium ammonia concentration and the actual ammonia concentration. Therefore, too high reaction temperature cannot be used. The most effective method to increase ammonia production is to improve the performance of catalyst and increase the reaction rate.

The process of a large ammonia plant is shown in Figure 6 – 1.

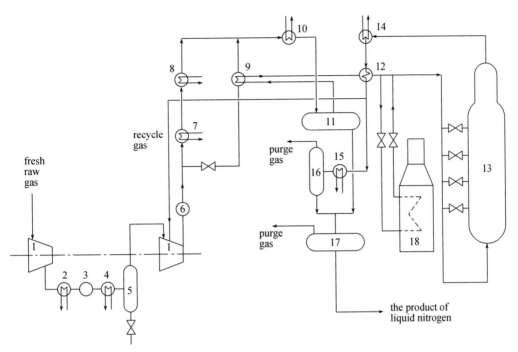

1 - Centrifugal syngas compressor; 2, 9, 12 - Heat exchanger; 3 - Intermediate water cooler; 4, 7, 8, 10, 15 - Ammonia coolers; 5 - Water separator; 11 - High pressure ammonia separator; 13 - Ammonia synthesis tower; 14 - Heater for boiler feed water; 16 - Ammonia separator; 17 - Low pressure ammonia separator; 18 - Start-up furnace.

Figure 6 - 1　Flow of large ammonia plant

6.4　Synthesis of methanol

As a chemical raw material, methanol is mainly used to prepare formaldehyde, dimethyl terephthalate, halomethane, explosives, medicine, dyes, pesticides and other organic chemical products. With the increasing consumption of energy in the world, natural gas and oil resources are becoming increasingly tight. Many new fields have been developed in the application of methanol. For example, methanol has rapidly entered the fuel market as a non-petroleum based fuel, and has become a substitute fuel for gasoline, gasoline seep-burning has been rapidly developed. Methanol can be directly synthesized into gasoline, and methyl tert-butyl ether (MTBE) can also be synthesized from methanol as a high-quality additive for lead-free gasoline, which has important economic and social benefits. In addition, methanol can also be used as a carbon source for protein synthesis, which also has broad prospects.

The raw gas of methanol synthesis is CO and H_2, which can be made from coal, natural gas, light oil, heavy oil, cracking gas and coke oven gas. In recent years, it can be seen from the development that the raw material route of methanol production from natural gas has been attached great importance, and its investment cost and consumption index are lower than coal and oil, while natural gas reserves are more abundant than oil. At present, the capacity of methanol synthesis from natural gas accounts for more than 80% of the total capacity.

6.4.1　Basic principle of methanol synthesis

1. Reaction thermodynamics of methanol synthesis

The reaction formula for the hydrogenation of carbon monoxide to methanol is

$$CO + 2H_2 \Longrightarrow CH_3OH(g) \tag{6-27}$$

Methanol can also be synthesized when CO_2 is present in the syngas.

$$CO_2 + 3H_2 \Longrightarrow CH_3OH(g) + H_2O \tag{6-28}$$

In addition to methanol formation, many side reactions occur in CO hydrogenation, such as

$$2CO + 4H_2 \Longrightarrow (CH_3)_2O + H_2O \tag{6-29}$$

$$CO + 3H_2 \Longrightarrow CH_4 + H_2O \tag{6-30}$$

$$4CO + 8H_2 \Longrightarrow C_4H_9OH + 3H_2O \tag{6-31}$$

$$CO_2 + H_2 \Longrightarrow CO + H_2O \tag{6-32}$$

The by-products are mainly dimethyl ether, isobutanol and methane gas, in addition to a small amount of ethanol and trace aldehydes, ketones, ethers and esters. Therefore, the condensation product is crude methanol containing impurities, which requires a refining process.

2. Reaction kinetics of methanol synthesis

The reaction mechanism of methanol synthesis has been studied and reported by many scholars, which can be summarized as three hypotheses.

The first assumption is that methanol is generated by direct hydrogenation of CO, and CO_2 need to be inverted into CO first.

The second hypothesis holds that methanol is directly synthesized from CO_2, while CO should first through transformation reaction to synthetic methanol.

The third hypothesis holds that methanol is formed directly from both CO and CO_2.

The first and second hypotheses assume that methanol synthesis is a series of reactions, while the third assumes that methanol synthesis is a parallel reaction. All the hypotheses are based on certain experimental data and need to be further studied and explored. As for the active center and adsorption type, there is still no consensus. For Cu-based catalysts, there are several views: the surface zero-valent copper Cu is the active center; Cu^+ dissolved in ZnO is the active center; $Cu - Cu^+$ is the active center. The adsorption centers of CO and CO_2 are related to Cu, while the adsorption centers of H_2 and H_2O are independent of ZnO.

6.4.2 Methanol synthesis catalyst

At present, the catalysts used in industrial production can be roughly divided into zinc-chromium series and copper-zinc (or aluminum) series. Different types of catalysts have different properties and require different reaction conditions.

1. Zinc-Chromium catalyst

Zinc-Chromium catalyst is an early methanol synthesis catalyst, which has low activity and requires a higher reaction temperature (380℃ ~ 400℃). Due to the limitation of equilibrium conversion at high temperature, the pressure (30 MPa) must be increased to meet the requirement. Therefore, the catalyst is characterized by the requirement of high temperature and high pressure. Besides, the catalyst has good mechanical strength and heat resistance, with long service life, generally 2~3 years.

2. Copper based catalyst

The active components are Cu and ZnO, and some cocatalysts need to be added to promote the activity of the catalyst. The activity is higher when aluminum and chromium are added. The addition of Cr_2O_3 can improve the dispersion of copper in the catalyst, and prevent the dispersed copper grains from being sintered and grown up when heated, which can prolong the life of the catalyst. Adding Al_2O_3 cocatalyst makes the catalyst activity higher, and Al_2O_3 is cheap and non-toxic, so it is better to use Al_2O_3 instead of Cr_2O_3 copper-based catalyst.

6.4.3 Process conditions of methanol synthesis

Synthesis of methanol reaction is a plurality of reactions at the same time. In addition to the main reaction, there are other side reactions generating dimethyl ether, isobutanol, methane and so on. Therefore, how to improve the selectivity of methanol synthesis and methanol yield is the core problem, which involves the selection of catalysts and the control of operating conditions, such as reaction temperature, pressure, space velocity and the composition of raw gas.

(1) Reaction temperature and pressure

Methanol synthesis is a reversible exothermic reaction, and the equilibrium yield is related to temperature and pressure. As the temperature increases, the reaction rate increases while the equilibrium constant decreases. As with ammonia synthesis, there is an optimum temperature. The temperature distribution of the catalyst bed should be as close as possible to the optimum temperature curve. For this reason, the internal structure of the reactor is complicated to remove the reaction heat in time.

Cold excitation type and indirect type are generally used. From the thermodynamic anlysis, methanol synthesis is a reaction of volume reduction, and increasing pressure is conducive to the improvement of methanol equilibrium yield.

(2) Space velocity

Theoretically speaking, high space velocity leads to a short contact time between reaction gas and catalyst, and the conversion rate decreases, while low space velocity leads to an increase in conversion rate. For the synthesis of methanol, due to the number of side reactions, if the space velocity is too low, it will promote the increase of side reactions, reduce the selectivity and production capacity of methanol synthesis. Of course, too high space velocity is also unfavorable. Too low methanol content will increase the separation difficulty of the product. Choosing an appropriate space velocity is favorable, which can improve the production capacity, reduce side reactions and improve the purity of methanol products. For Zn – Cr catalyst, the space velocity is $20\ 000 \sim 40\ 000\ h^{-1}$, while for Cu – Zn – Al catalyst, the space velocity is $10\ 000\ h^{-1}$.

(3) Raw gas ratio of methanol synthesis

The stoichiometric ratio (mole ratio) of H_2 to CO is 2 : 1. Besides H_2 and CO, there is also a certain amount of CO_2 in the raw gas of industrial production. $H_2 - CO_2 / (CO + CO_2) = 2.1 \pm 0.1$ is usually used as the composition of fresh raw gas for methanol synthesis. In fact, the value of H_2 to CO in the mixture entering the tower is always greater than 2. The reason is that the high hydrogen content can improve the reaction rate, reduce the generation of side reactions, and the thermal conductivity of hydrogen (thermal conductivity) is large, which is conducive to the derivation of reaction heat, and easy to control the reaction temperature.

6.4.4　Methanol synthesis process and reactor

The process flow of methanol synthesis is similar to that of ammonia synthesis. Due to the low synthesis rate, the unreacted syngas must be recycled, and the circulating compressor is used to boost the pressure, the product methanol must be separated from the reaction tail gas. Due to the high boiling point of methanol, methanol can be condensed by cooling with water under the pressure of synthesis. In order to keep the inert gas content within a certain range, the circulating gas needs to be partially emptied. As there are many side reactions in methanol synthesis, crude methanol must be refined to meet the requirements of the product.

The early high pressure methanol synthesis process has many defects. Due to the advanced technical and economic indicators of the low pressure method, the power consumption of the low pressure method is only about 60% of that of the high pressure method, so the high pressure method is gradually replaced by the low pressure method.

A schematic diagram of the low pressure methanol separation process is shown in Figure 6 – 2.

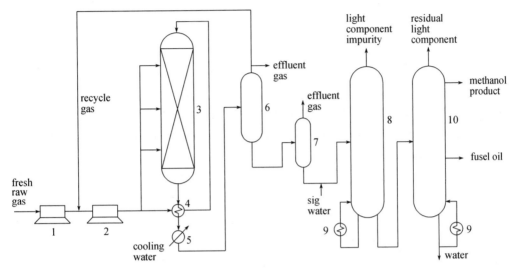

1 – Synthetic compressor; 2 – Cycle compressor; 3 – Synthesis tower;
4 – Heat exchanger; 5 – Condenser; 6 – Separator; 7 – Flash tank;
8 – Light component removal tower; 9 – Reboiler; 10 – Rectification tower.

Figure 6 – 2   Low pressure methanol separation process diagram

6.4.5   Progress of methanol synthesis technology

Although the copper-based catalyst with high activity has been developed and methanol synthesis technology has changed from high pressure to low pressure, there are still many problems: reactor structure is complex; the one-way conversion rate is low, the energy consumption of gas compression and circulation is large, the reaction temperature is not easy to control and the thermal stability of the reactor is poor. All these problems reveal to people that there is still great potential in methanol synthesis technology, newer and higher technologies are waiting for us to develop. The new achievements made since 1980s are introduced below.

(1) Gas-liquid-solid three-phase methanol synthesis process

This process is first proposed by American Chemical Systems Company, using a three-phase fluidized bed. The liquid phase is inert medium, the catalyst is ICI Cu-Zn improved catalyst. For liquid medium, it should have good thermal and chemical stability under methanol synthesis conditions. Methanol is not only the fluidization medium of catalyst, but also the reaction heat absorption medium. The smaller the solubility of methanol in the liquid medium the better, and the product methanol leaves the reactor in the form of gas phase.

（2）Methanol synthesis process by liquid phase method

The process of methanol synthesis in liquid phase is characterized by the use of transition metal coordination catalysts with higher activity. The catalyst is evenly distributed in the liquid medium, so there is no influence of catalyst surface heterogeneity and internal diffusion. The reaction temperature is low, generally not more than 200℃.

（3）Novel GSSTFR and RSIPR reaction system

The system adopts a new reaction device of reaction, adsorption and product exchange alternately. GSSTFR refers to a gas-solid-solid dripflow reaction system in which CO and H_2 react to synthesize methanol under the action of catalyst. The methanol is immediately adsorbed by a solid powder adsorbent and dripped out of the reaction system. RSIPR is interstage product removal reaction system, when the powdery adsorbent adsorption gas methanol flows into the system, it is exchanged with the liquid phase of tetraethylene glycol dimethyl ether in the system, and the gaseous methanol is absorbed by the liquid phase （gaseous methanol dissolved in liquid tetraethylene glycol dimethyl ether）, then separate tetraethylene glycol dimethyl ether in methanol. In this way, methanol synthesis continues to the right, the one-way conversion of CO can reach 100%, and the gas phase reactants do not cycle. The new process is still under research and has not yet been put into industrial production. There are still many technical problems to be solved and perfected.

6.5　Dehydrogenation of ethylbenzene to styrene

Styrene is an unsaturated aromatic hydrocarbon, colorless liquid, insoluble in water, soluble in methanol, ethanol, carbon tetrachloride, ether and other solvents. The boiling point is 145℃

Styrene is an important monomer of polymer composite material, its homopolymerization can obtain polystyrene resin, its usage is very extensive, its copolymerization with other monomers can make a variety of valuable copolymer, such as with acrylonitrile can make SAN resin, with acrylonitrile and butadiene can make ABS resin, and with butadiene can make styrene butadiene rubber and SBS plastic rubber, etc. In addition, styrene is also widely used in pharmaceutical, coating, textile and other industries.

Styrene polymers were discovered in 1827, and Berthelot discovered in 1867 that ethylbenzene produced styrene when passed through a red-heated porcelain tube. In 1930, Dow Chemical Company in the United States pioneered the styrene production process by thermal dehydrogenation of ethylbenzene and realized the industrial production of styrene in 1945. Over the past years, the production technology of styrene has been improved and perfected. In 2000, the world styrene production capacity exceeded 22. 3 million tons. Due to the strong market demand, styrene production will develop rapidly.

6.5.1　Brief introduction to the method of preparing styrene

1. Ethylbenzene dehydrogenation

Ethylbenzene dehydrogenation is the main method of producing styrene at present. The production of ethylbenzene in industry mainly adopts alkylation method, using benzene and ethylene to generate under the action of catalyst. Ethylbenzene can also be separated by distillation from reforming oil in refinery, cracking gasoline in the cracking process of alkane and coal tar in coking plant.

$$\text{(6-33)}$$

$$\text{(6-34)}$$

2. Ethylbenzene cooxidation

The method is divided into three steps, the first is the generation of ethylbenzene hydrogen peroxide, and then the reaction of ethylbenzene hydrogen peroxide with propylene to generate α-methylbenzyl alcohol and propylene oxide, and then the dehydration of α-methylbenzyl alcohol to styrene.

$$\text{(6-35)}$$

$$\text{(6-36)}$$

$$\text{(6-37)}$$

3. Synthesis of styrene with toluene as raw material

In the first step, PbO · MgO/Al$_2$O$_3$ catalyst is used to dehydrogenate toluene into styrene benzene in the presence of water vapor. In the second step, styrene and ethylene are synthesized under the action of WO · K$_2$O/SiO$_2$ catalyst.

$$\text{(6-38)}$$

$$\text{(6-39)}$$

Another method is to synthesize styrene directly from toluene and methanol.

$$CH_3OH \longrightarrow HCHO + H_2 \qquad (6-40)$$

$$HCHO + \underset{}{\bigcirc}-CH_3 \longrightarrow \underset{}{\bigcirc}-CH_2CH_2OH \qquad (6-41)$$

$$\underset{}{\bigcirc}-CH_2CH_2OH \longrightarrow \underset{}{\bigcirc}-CH=CH_2 + H_2O \qquad (6-42)$$

$$\underset{}{\bigcirc}-CH_2CH_2OH + H_2 \longrightarrow \underset{}{\bigcirc}-C_2H_5 + H_2O \qquad (6-43)$$

4. Direct synthesis of styrene from ethylene and benzene

$$\underset{}{\bigcirc} + C_2H_4 + \frac{1}{2}O_2 \longrightarrow \underset{}{\bigcirc}-CH=CH_2 + H_2O \qquad (6-44)$$

5. Oxydehydrogenation of ethylbenzene

$$\underset{}{\bigcirc}-C_2H_5 + \frac{1}{2}O_2 \longrightarrow \underset{}{\bigcirc}-CH=CH_2 + H_2O \qquad (6-45)$$

6.5.2 Basic principle of ethylbenzene catalytic dehydrogenation

1. Main and side reactions of ethylbenzene catalytic dehydrogenation

Main reaction

$$\underset{}{\bigcirc}-C_2H_5 \rightleftharpoons \underset{}{\bigcirc}-CH=CH_2 + H_2 \qquad (6-46)$$

Side reactions mainly include cracking of ethylbenzene, hydrocracking, steam cracking, polymerization and condensation to form tar, etc.

$$\underset{}{\bigcirc}-C_2H_5 \rightleftharpoons \underset{}{\bigcirc} + C_2H_4 \qquad \Delta H(873\ K) = -102\ kJ/mol \quad (6-47)$$

$$\underset{}{\bigcirc}-C_2H_5 + H_2 \rightleftharpoons \underset{}{\bigcirc}-CH_3 + CH_4 \qquad \Delta H(873\ K) = 64.4\ kJ/mol \qquad (6-48)$$

$$\underset{}{\bigcirc}-C_2H_5 + H_2 \rightleftharpoons \underset{}{\bigcirc} + C_2H_6 \qquad \Delta H(298\ K) = 41.8\ kJ/mol \qquad (6-49)$$

$$\underset{}{\bigcirc}-C_2H_5 \rightleftharpoons 8C + 5H_2 \qquad \Delta H(873\ K) = 1.72\ kJ/mol \qquad (6-50)$$

$$\underset{}{\bigcirc}-C_2H_5 + 16H_2O \rightleftharpoons 8CO_2 + 21H_2 \qquad \Delta H(873\ K) = 793\ kJ/mol \qquad (6-51)$$

2. Ethylbenzene dehydrogenation catalyst

Thermodynamically, cracking is more favorable than dehydrogenation, and hydrocracking is also more favorable than dehydrogenation, with a large equilibrium constant even at 700℃. Therefore, when ethylbenzene is dehydrogenated at high temperature, the main product is benzene. To make the main reaction smoothly, it is necessary to use a catalyst with high activity and high selectivity.

In addition to activity and selectivity, dehydrogenation is carried out at high temperature in the presence of hydrogen and a large amount of water vapor, which also requires good thermal and chemical stability of the catalyst, as well as coking resistance and easy regeneration.

The active component of dehydrogenation catalyst is iron oxide, and the cocatalyst is oxides of potassium, vanadium, molybdenum, tungsten, cerium and so on. Such as the composition of catalyst is $Fe_2O_3 : K_2O : Cr_2O_3 = 87 : 10 : 3$, it can make ethylbenzene conversion reach 60%, the selectivity is 87%.

Cocatalyst K_2O can change the surface acidity of catalyst, reduce the occurrence of pyrolysis reaction, and can improve the anti-coking performance of catalyst and carbon elimination, promote the regeneration of catalyst, prolong the regeneration cycle.

Cocatalyst Cr_2O_3 is a high melting point metal oxide, which can improve the heat resistance of catalyst, stable iron valence state. However, chromium oxide is toxic to human body and the environment, so we should use chrome free catalyst, such as $Fe_2O_3 - Mo_2O_3 - CeO - K_2O$ catalyst, as well as domestic XH－02 and 335 chrome free catalyst.

6.5.3　Selection of dehydrogenation conditions for ethylbenzene

1. Temperature

Ethylbenzene dehydrogenation is a reversible endothermic reaction. The increase of temperature is beneficial to the improvement of equilibrium conversion rate and reaction rate. The pyrolysis and hydrocracking of ethylbenzene are both favorable with the increase of temperature. As a result, the conversion of ethylbenzene increases and the selectivity of styrene decreases with the increase of temperature. When the temperature decreases, the side reaction decreases, which is beneficial to the improvement of styrene selectivity, but the yield is not high because of the decrease of reaction rate.

2. Pressure

As for the reaction rate, increasing the pressure will accelerate the reaction rate, but it is not good for the equilibrium of dehydrogenation. In industry, water vapor is used to dilute the raw gas to reduce the partial pressure of ethylbenzene and improve the equilibrium conversion rate of ethylbenzene. If the water vapor has an effect on the performance of the catalyst, only depressurization operation can be adopted.

3. Space speed

Ethylbenzene dehydrogenation is a complex reaction with low space speed, increased contact time, intensified side reactions and decreased selectivity. Therefore, it is necessary to use a higher space speed to improve selectivity. Although the conversion rate is not very high, the unreacted raw gas can be used in a cycle, but it will inevitably cause an increase in energy consumption. Therefore, it is necessary to choose the best space speed comprehensively.

(4) Influence of catalyst particle size

The selectivity of dehydrogenation decreases with the increase of particle size. It can be explained that the main reaction is greatly affected by internal diffusion while the side reaction is not. Therefore, catalysts with smaller particle size are commonly used in industry to reduce the internal diffusion resistance of catalysts. At the same time, the catalyst can be modified by roasting at high temperature to reduce the microporous structure of the catalyst.

(5) Amount of water vapor

Using water vapor as a diluent for dehydrogenation has the following advantages: ① The partial pressure of ethylbenzene is reduced, which is beneficial to improve the equilibrium conversion rate of ethylbenzene dehydrogenation. ② It can inhibit the coking of catalyst surface and eliminate carbon. ③ It can provide the heat needed for the reaction, and easy to separate the product. Increasing the amount of water vapor is beneficial to the above three points, but the more the amount of water vapor is not the better, the improvement of equilibrium conversion rate is not obvious after a certain ratio.

6.6　Oxidative dehydrogenation of n-butene to butadiene

Butadiene is the simplest diolefin with conjugated double bonds. It is easy to oligomerize and polymerize with other unsaturated compounds with double bonds. Therefore, it is the most important polymer monomer, mainly used in the production of synthetic rubber, but also used in the synthesis of plastics and resins.

6.6.1　Production method

1. Butadiene is obtained from C_4 fraction of the byproduct of hydrocarbon pyrolysis to lower olefin

This is the most economical and main method to obtain butadiene at present. The yield of C_4 fraction is about $30\% \sim 50\%$ of the yield of ethylene, and the content of butadiene can be as high as 40%. All the butadiene in Western Europe and Japan, and 80% of the Butadiene in the United States are obtained by this way. Due to the similar boiling points of each component of C_4 fraction (the boiling points of n-butene, isobutene and butadiene are $-6.3°C$, $-6.9°C$ and $-4.4°C$ respectively), the industrial extraction distillation method is usually used to separate them, the extraction agent used are N-methyl pyrrolidone, dimethyl formamide and acetonitrile.

2. Dehydrogenation of n-butane and n-butene to produce butadiene

N-butane dehydrogenation is a series of reversible reactions

$$CH_3CH_2CH_2CH_3 \xrightleftharpoons[+H_2]{-H_2} \begin{cases} H_2C\!\!=\!\!CHCH_2CH_3 \\ (\text{anti-form})CH_3HC\!\!=\!\!CHCH_3 \xrightleftharpoons[+H_2]{-H_2} HCH_2C\!\!=\!\!CHCH_2 \\ (\text{cis-form})CH_3HC\!\!=\!\!CHCH_3 \end{cases}$$

$$(6-52)$$

In the first stage of dehydrogenation, three n-butene isomers were obtained, and in the second stage, the three butene isomers were further dehydrogenated to 1,3 - butadiene. The thermal efficiency of the two stages is -126 kJ/mol and -113.7 kJ/mol, respectively. Dehydrogenation reaction is endothermic, it is the amount of substance (mol) increased reaction, so the use of high temperature and low pressure (even negative pressure) dehydrogenation reaction is favorable. Because of the intense side reaction at high temperature, the increase of by-products, we should use catalysts with high catalytic activity and good selectivity.

3. Butadiene by oxidative dehydrogenation of n-butene

Adding an appropriate amount of oxygen in the dehydrogenation reaction gas to quickly remove the hydrogen produced in the dehydrogenation reaction is called oxidative dehydrogenation. It breaks the chemical equilibrium limit and increases the reaction rate. Oxidative dehydrogenation has the following advantages.

(1) The reaction temperature is low, n-butene to butadiene is only $400°C \sim 500°C$, $100°C \sim 200°C$ lower than the usual dehydrogenation reaction.

(2) Usually, dehydrogenation is endothermic reaction (about 126 kJ/mol), which needs to supply heat, and oxidative dehydrogenation is exothermic reaction, which can save the original heating equipment.

(3) Because the catalyst works in a lower temperature and oxygen atmosphere, coking is very little, so the catalyst can run for a long time.

(4) In general dehydrogenation reaction, pressure has a great effect on the equilibrium conversion, but in oxidative dehydrogenation, pressure has little effect, so decompression or water vapor dilution is not necessary.

(5) As the reaction temperature is low, it is possible to improve the conversion rate and selectivity. In the usual dehydrogenation process, the conversion rate can be increased by increasing the temperature, but the selectivity is often decreased due to the violent side reactions.

Therefore, the oxidative dehydrogenation of hydrocarbons has a bright prospect. For example, oxidative dehydrogenation of n-butene has gradually replaced dehydrogenation and become an important method for the manufacture of butadiene.

6.6.2 Process principle

1. Chemical reaction

The main reaction of oxidative dehydrogenation of n-butene to butadiene is an exothermic reaction.

$$C_4H_8 + \frac{1}{2}O_2 \longrightarrow H_2C =CHCH =CH_2 + H_2O(g) \quad \Delta H(773 \text{ K}) = 134.31 \text{ kJ/mol}$$

$$(6-53)$$

The main side effects are:

(1) The oxidation degradation of n-butene into saturated and unsaturated small molecule aldehydes, ketones, acids and other oxygen-containing compounds, such as formaldehyde, acetaldehyde, acrolein, acetone, saturated and unsaturated low organic acids.

(2) The oxidation of n-butene to furan, butenal and butanone.

(3) Complete oxidation to carbon monoxide, carbon dioxide and water.

(4) Oxidation dehydrogenation of n-butene to aromatic hydrocarbons.

(5) Deep oxidative dehydrogenation to vinyl acetylene, methyl acetylene, etc.

(6) The polymerization of product and by-product coking.

2. Catalyst and catalytic mechanism

There are two series of catalysts for oxidative dehydrogenation of n-butene in industrial application.

(1) Bismuth molybdate series catalyst

It is a two-component or multi-component catalyst based on Mo-Bi oxide. Mo-Bi-O two-component and Mo − Bi − P − O three-component catalyst was used in the early stage, but the activity and selectivity are low. After improvement, it is developed into a six-component, seven-component or more component mixed oxide catalyst, such as Mo − Bi − P − Fe − Ni − K − O. The catalytic activity and selectivity of Mo − Bi − P − Fe − Co − Ni − Ti − O were improved obviously.

(2) Ferrite spinel series catalyst

$ZnFe_2O_4$, $MnFe_2O_4$, $MgFe_2O_4$, $ZnCrFeO_4$ and $Mg_{0.1}Zn_{0.9}Fe_2O_4$ (atomic ratio) ferrates are oxides with spinel type ($A^{2+}B_2^{3+}O_4$) structure. They are a kind of oxidation dehydrogenation catalysts of n-butene developed in the late 1960s.

(3) Other types

There are mainly mixed oxide catalysts based on Sb or Sn oxides.

6.6.3   Selection of process conditions

The technological conditions are related to the catalyst and reactor used. Taking ferrite spinel catalyst and adiabatic reactor as an example, the selection of technological conditions for the oxidation dehydrogenation of n-butene to butadiene is discussed.

(1) Requirements of raw material purity

The dehydrogenation rates and selectivity of the three isomers of n-butene over ferrite spinel catalysts are not very different. Therefore, the composition distribution of the three isomers in the raw material has little influence on the selection of technological conditions.

The amount of isobutene in the raw material should be strictly controlled, because isobutene is easy to oxidize, it will increase the consumption of oxygen, and affect the control of reaction temperature.

Alkane C_3 or below C_3 is stable and will not be oxidized, but its high content will affect the production capacity of the reactor, and a small amount of it may be oxidized into CO_2 and water under operating conditions.

(2) Dosage ratio of oxygen to n-butene

Generally, air is used as an oxidant. Because the yield of butadiene is directly related to the oxygen used, the ratio of oxygen and n-butene dosage should be strictly controlled.

(3) Dosage ratio of water vapor to n-butene

The selectivity of butadiene increases with the increase of $H_2O/n - C_2H_8$ until it reaches a maximum.

(4) Reaction temperature

Because oxidative dehydrogenation is exothermic, the outlet temperature is significantly higher than the inlet temperature, and the temperature difference between the two can reach 220℃ or greater. The suitable reaction temperature range is 327℃～547℃.

(5) Space velocity of n-butene

The space velocity of n-butene varies within a certain range and has little effect on selectivity. Generally, when the space velocity increases, the inlet temperature should be increased accordingly to maintain a certain conversion rate. Industrial mass space velocity (GHSV) of n-butene is 600 h^{-1} or higher.

(6) Pressure

Although the inlet pressure of the reactor has little effect on the conversion rate, it has an effect on the selectivity. As the inlet pressure increases, the selectivity decreases and so does the yield. Therefore, it is desirable to operate at low pressure and the resistance drop of catalyst bed should be as small as possible, so the radial adiabatic bed reactor is more suitable. The decrease of selectivity may be due to the degradation or complete oxidation of raw materials, intermediates and butadiene after long time residence on the catalyst surface.

6.6.4 Technological process

The process of oxidative dehydrogenation of n-butene to produce butadiene is divided into three parts: reaction, separation and refining of butadiene, and recovery of unconverted n-butene. The partial process of oxidative dehydrogenation of n-butene to butadiene is shown in Figure 6 – 3.

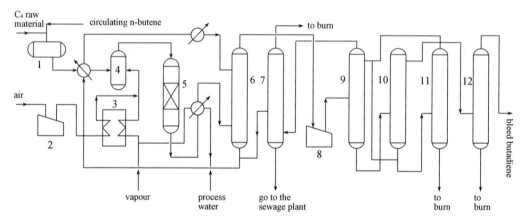

1 – C$_4$ raw material tank; 2 – Air compressor; 3 – Heating furnace; 4 – Mixer; 5 – Reactor;

6 – Quencher; 7 – Stripping tower; 8 – Compressor; 9 – Absorption tower; 10 – Desorption tower;

11 – Oil regeneration tower; 12 – Dehydration and reconstitution tower.

Figure 6 – 3　Partial process of oxidative dehydrogenation of n-butene to butadiene

Questions

6 – 1　What are the requirements of catalysts for hydrogenation and dehydrogenation?

6 – 2　What are the differences between the production process of ammonia synthesis using coal as raw material and natural gas as raw material?

6 – 3　Analyze and compare three refining methods of syngas.

6 – 4　What are the factors that affect the equilibrium concentration of ammonia?

6 – 5　Explain the influence of temperature and pressure on equilibrium ammonia concentration and reaction rate of ammonia synthesis reaction.

6 – 6　Explain the influence of inert gases on equilibrium concentration and reaction rate of ammonia synthesis.

6 – 7　Explain the role of active components and auxiliaries in ammonia synthesis

catalysts.

6 - 8   In the ammonia synthesis process, why is the exhaust gas in front of the circulating compressor, while the ammonia cooling is behind the circulating compressor?

6 - 9   What are the similarities between ammonia synthesis and methanol synthesis?

6 - 10   Compare the high and low pressure methanol synthesis.

6 - 11   Explain the effect of temperature and space velocity on selectivity in the production of ethylbenzene dehydrogenation to styrene.

第7章 烃类选择性氧化

7.1 概 述

化学工业中氧化反应是一大类重要化学反应,它是生产大宗化工原料和中间体的重要反应过程。有机物氧化反应中烃类的氧化最有代表性。烃类氧化反应可分为完全氧化和部分氧化两大类型。完全氧化是指反应物中的碳原子与氧化合生成 CO_2,氢原子与氧结合生成水的反应过程;部分氧化,又称选择性氧化,是指烃类及其衍生物中少量氢原子(有时还有少量碳原子)与氧化剂(通常是氧)发生作用,而其他氢和碳原子不与氧化剂反应的过程。据统计,全球生产的主要化学品中 50% 以上和选择性氧化过程有关。烃类选择性氧化可生成比原料价值更高的化学品,在化工生产中有广泛的应用。选择性氧化不仅能生产含氧化合物,如醇、醛、酮、酸、酸酐、环氧化物、过氧化物等,还可生产不含氧化合物,如丁烯氧化脱氢制丁二烯、丙烷(丙烯)氨氧化制丙烯腈、乙烯氧氯化制二氯乙烷等,这些产品有些是有机化工的重要原料和中间体,有些是三大合成材料的单体,还有些是用途广泛的溶剂,在化学工业中占有重要地位。

7.1.1 氧化过程特点和氧化剂选择

1. 氧化反应的特征

(1) 反应放热量大

氧化反应是强放热反应,氧化深度越大,放出的反应热越多,完全氧化时的热效应约为部分氧化时的 8～10 倍。因此,在氧化反应过程中,反应热的及时转移非常重要,否则会造成反应温度迅速上升,促使副反应增加,反应选择性显著下降,严重时可能导致反应温度无法控制,甚至发生爆炸。利用氧化反应的反应热可副产蒸汽,一般说来,气-固相催化氧化反应温度较高,可回收得到高、中压蒸汽;气-液相氧化反应温度较低,只能回收低品位的能量,如低压蒸汽和热水。

(2) 反应不可逆

对于烃类和其他有机化合物而言,氧化反应的 $\Delta G^* \ll 0$,因此为热力学不可逆反应,不受化学平衡限制,理论上可达 100% 的单程转化率。但对许多反应,为了保证较高的选择性,转化率须控制在一定范围内,否则会造成深度氧化而降低目的产物的产率,如丁烷氧化制顺酐,一般控制丁烷的转化率在 85%～90%,以保证生成的顺酐不继续深度氧化。

(3) 氧化途径复杂多样

烃类及其绝大多数衍生物均可发生氧化反应,且氧化反应多为串联、并联或两者组合而形成的复杂网络。由于催化剂和反应条件的不同,氧化反应可经过不同的反应路径,转化为不同的反应产物,而且这些产物往往比原料的反应性更强,更不稳定,易于发生深度氧化,最终生成二氧化碳和水。因此反应条件和催化剂的选择非常重要,其中催化剂的选用是决定氧化路径的关键。

（4）过程易燃易爆

烃类与氧或空气容易形成爆炸混合物,因此氧化过程在设计和操作时应特别注意其安全性。

2. 氧化剂的选择

要在烃类或其他化合物分子中引入氧,需采用氧化剂,比较常见的有空气和纯氧、过氧化氢和其他过氧化物等,空气和纯氧使用最为普遍。空气比纯氧便宜,但氧分压小,含大量的惰性气体,因此生产过程中动力消耗大,废气排放量大。用纯氧作氧化剂则可降低废气排放量,减小反应器体积。究竟是使用空气还是纯氧,要视技术经济分析而定。

用空气或纯氧对某些烃类及其衍生物进行氧化,生成的烃类过氧化物或过氧酸也可用作氧化剂进行氧化反应,如乙苯经空气氧化生成过氧化氢乙苯,将其与丙烯反应,可制得环氧丙烷。

近年来,过氧化氢作为氧化剂发展迅速,使用过氧化氢氧化条件温和,操作简单,反应选择性高,不易发生深度氧化反应,对环境友好,可实现清洁生产。

7.1.2　烃类选择性氧化分类

就反应类型而言,选择性氧化可分为以下三种。

（1）碳链不发生断裂的氧化反应。如烷烃、烯烃、环烷烃和烷基芳烃的饱和碳原子上的氢原子与氧进行氧化反应,生成新的官能团,烯烃氧化生成二烯烃、环氧化物等。

（2）碳链发生断裂的氧化反应。包括产物碳原子数比原料少的反应,如异丁烷氧化生成乙醇的反应,以及产物碳原子数与原料相同的开环反应,如环己烷氧化生成己二醇等。

（3）氧化缩合反应。在反应过程中,这类反应发生分子之间的缩合,如丙烯氨氧化生成丙烯腈、苯和乙烯氧化缩合生成苯乙烯等。

就反应相态而言,可分为均相催化氧化和非均相催化氧化。均相催化氧化体系中反应组分与催化剂的相态相同,而非均相催化氧化体系中反应组分与催化剂以不同相态存在。目前,化学工业中采用的主要是非均相催化氧化过程,均相催化氧化过程的应用还是少数。

7.2　均相催化氧化

近 40 年,在金属有机化学发展的推动下,均相催化氧化过程以其高活性和高选择性引起人们的关注。均相催化氧化通常指气-液相氧化反应,习惯上称为液相氧化反应。液相催化氧化一般具有以下特点。

（1）反应物与催化剂同相，不存在固体表面活性中心性质及分布不均匀的问题，作为活性中心的过渡金属活性高、选择性好。

（2）反应条件不太苛刻，反应比较平稳，易于控制。

（3）反应设备简单，容积较小，生产能力较高。

（4）反应温度通常不太高，反应热利用率较低。

（5）在腐蚀性较强的体系中要采用特殊材质。

（6）催化剂多为贵金属，必须分离回收。

如今，均相催化氧化技术在高级烷烃氧化制仲醇，环烷烃氧化制醇、酮混合物，Wacker 法制醛或酮，烃类过氧化氢的制备，烃类过氧化氢对烯烃进行的环氧化反应，芳烃氧化制芳香酸等过程中已成功地应用。随着化工产品向精细化方向发展，它在精细化学品合成领域中显示越来越重要的作用。用较廉价的过渡金属代替贵金属作催化剂、催化剂的回收和固载化研究也不断取得进展。

均相催化氧化反应有多种类型，工业上常用催化自氧化和络合催化氧化两类反应。此外，还有烯烃的液相环氧化反应。

7.2.1 催化自氧化

1. 催化自氧化反应

自氧化反应是指具有自由基链式反应特征，能自动加速的氧化反应。非催化自氧化反应的开始阶段，由于没有足够浓度的自由基诱发链反应，因此具有较长的诱导期。催化剂能加速链的引发，促进反应物引发、生成自由基，缩短或消除反应诱导期，因此可大大加速氧化反应，称为催化自氧化。工业上常用此类反应生产有机酸和过氧化物，在适宜的条件下，也可获得醇、酮、醛等中间产物。反应所用催化剂多为 Co、Mn 等过渡金属离子的盐类，如醋酸盐和环烷酸盐等，钴盐的催化效果一般较好，通常溶解在液态介质中形成均相。

2. 自氧化反应机理

经过大量实验，已确定烃类及其他有机化合物的自氧化反应是按自由基链式反应机理进行，但有些过程（如链的引发）的机理尚未完全弄清楚，下面以烃类的液相自氧化为例，对其自氧化的基本步骤进行简单介绍。

$$RH + O_2 \longrightarrow \dot{R} + HO_2 \tag{7-1}$$

$$\dot{R} + O_2 \longrightarrow RO\dot{O} \tag{7-2}$$

$$RO\dot{O} + RH \longrightarrow ROOH + \dot{R} \tag{7-3}$$

$$\dot{R} + \dot{R} \longrightarrow R—R \tag{7-4}$$

上述步骤中，决定性步骤是链的引发过程，也就是烃分子发生均裂反应转化为自由基的过程，需要很大的活化能。所需能量与碳原子的结构有关。已知 C—H 键能大小为

$$\text{叔 C—H} < \text{仲 C—H} < \text{伯 C—H}$$

故叔 C—H 键均裂的活化能最小,其次是仲 C—H。

要使键反应开始,还必须有足够的自由基浓度,因此从链引发到链反应开始,必然有自由基浓度的积累阶段。在此阶段,观察不到氧的吸收,一般称为诱导期,需数小时或更长的时间,诱导期后,反应很快加速而达到最大值。可以采用催化剂和引发剂加速自由基的生成,缩短反应诱导期,如 Co、Mn 等过渡金属离子盐类及易分解生成自由基的过氧化氢异丁烷、偶氮二异丁腈等化合物,通常可在链引发阶段发挥作用。而在链传递阶段,作为载体的是由作用物生成的自由基。

链的传递反应是自由基-分子反应,所需活化能较小。这一过程包括氧从气相到反应区域的传质过程和化学反应过程。各参数的影响甚为复杂,在氧的分压足够高时,式(7-2)反应速率很快,链传递反应速率是由式(7-3)控制。

式(7-3)生成的产物 ROOH,性能不稳定,在温度较高或有催化剂存在下,会进一步分解而生成新的自由基,发生分支反应,生成不同氧化物。

$$ROOH \longrightarrow R\dot{O} + \dot{O}H \qquad (7-5)$$

$$R\dot{O} + RH \longrightarrow ROH + \dot{R} \qquad (7-6)$$

$$\dot{O}H + RH \longrightarrow H_2O + \dot{R} \qquad (7-7)$$

$$2ROOH \longrightarrow RO\dot{O} + R\dot{O} + H_2O \qquad (7-8)$$

$$RO\dot{O} \longrightarrow R'\dot{O} + R''CHO(或酮) \qquad (7-9)$$

$$R\dot{O}(或 R'\dot{O}) + RH \longrightarrow ROH(或 R'OH) + \dot{R} \qquad (7-10)$$

分支反应的结果是生成不同碳原子数的醇和醛,醇和醛又可进一步氧化生成酮和酸,使产物组成甚为复杂。

3. 自氧化反应过程的影响因素

(1) 溶剂的影响

在均相催化氧化体系中,经常要使用溶剂。溶剂的选择非常重要,它不仅能改变反应条件,还会对反应历程产生一定的影响。

(2) 杂质的影响

自氧化反应是自由基链式反应,体系中引发的自由基的数量和链的传递过程,对反应的影响至关重要。杂质的存在有可能使体系中的自由基失活,从而破坏正常的链引发和传递,导致反应速率显著下降甚至终止反应。由于自氧化反应体系的自由基浓度一般较小,因此,对杂质一般非常敏感,有时少量杂质就会产生相当大的影响。杂质对自由基连锁反应的影响称为阻化作用,杂质则为阻化剂。不同的反应体系阻化剂不尽相同,常见的有水、硫化物、酚类等。

(3) 温度和氧气分压的影响

氧化反应伴随大量的反应热,在自氧化反应体系中,由于自由基链式反应特点,保持体系的放热和移出热量平衡非常重要。氧化反应需要氧源,在体系供氧能力足够时,反应由动力学控制,保持较高的反应温度有利于反应的进行,但温度也不宜过高,以免副产物增多,选择性降低,甚至反应失去控制。当氧浓度较低,系统供氧能力不足时,反应由传质控制,此时增大氧分压,可促进氧传递,提高反应速率,但也需要根据设备耐压能力和经济核算而定。若供氧速度在两者之间,传质和动力学因素均有影响,应综合考虑。

另外,由于氧化反应的目的产物为氧化过程的中间产物,因此,氧分压的改变会影响反应的选择性,从而对产物的构成产生影响。

（4）氧化剂用量和空速的影响

氧化剂（空气或氧气）用量的上限由反应排出的尾气中氧的爆炸极限确定,应避开爆炸范围。氧化剂用量的下限为反应所需的理论耗氧量,此时尾气中氧含量为零。在工业实践中,一般尾气中氧含量控制在 $2\%\sim6\%$,以 $3\%\sim5\%$ 为佳。氧化剂的空速定义为空气或氧气的流量和反应器中液体体积之比。空速提高,有利于气液相接触,加速氧的吸收,促进反应进行;但过高的空速会使气体在反应器中停留时间缩短,氧的吸收不完全,利用率降低,导致尾气中氧含量过高,对安全和经济性都有影响。空速的大小受尾气中氧含量要求约束。

4. 对二甲苯氧化制备对苯二甲酸

对苯二甲酸主要用于生产聚对苯二甲酸二乙酯（PET）,也可生产聚对苯二甲酸二丁酯（PBT）和聚对苯二甲酸二丙酯（PPT）,进一步生产聚酯纤维、薄膜和工程塑料。其中聚酯纤维占世界合成纤维产量的 50% 以上。除广泛用于生活中衣物外,聚酯纤维还用于轮胎帘子布、运输带、灭火水管等。

目前,对苯二甲酸的主要生产方法是对二甲苯氧化法,这是一个典型的均相催化自氧化反应,主要包括氧化和加氢精制两部分,采用工艺多数为高温氧化法,其中,以美国 Amoco、英国 ICI 和日本三井油化三家公司技术有代表性,应用广泛。

（1）氧化过程

以对二甲苯（PX）为原料,醋酸钴、醋酸锰为催化剂,四溴乙烷为促进剂,在一定的压力和温度下,用空气于醋酸溶剂中把对二甲苯连续地氧化成粗对苯二甲酸,反应方程式如下。

主反应

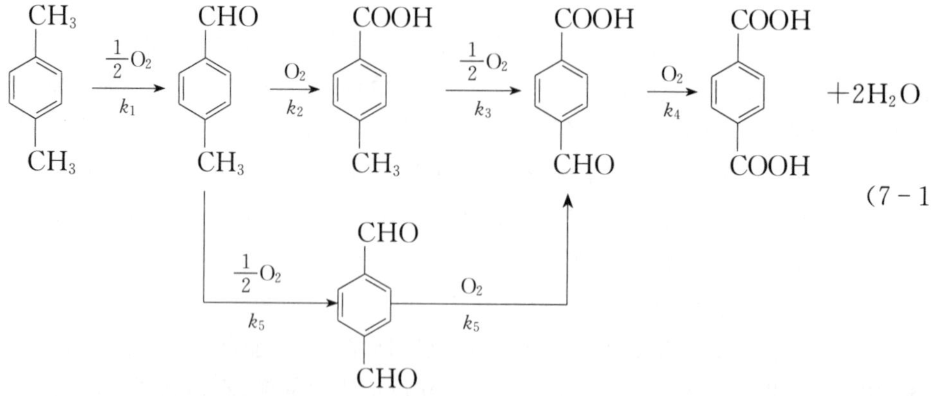

$$(7-11)$$

除以上主反应外,还伴随着一些副反应。例如,溶剂醋酸和对二甲苯会产生部分深度氧化,生成 CO 和 CO_2,氧化反应的配比不当,或原料不纯,带入某些杂质时,也会发生一些副反应。

副反应

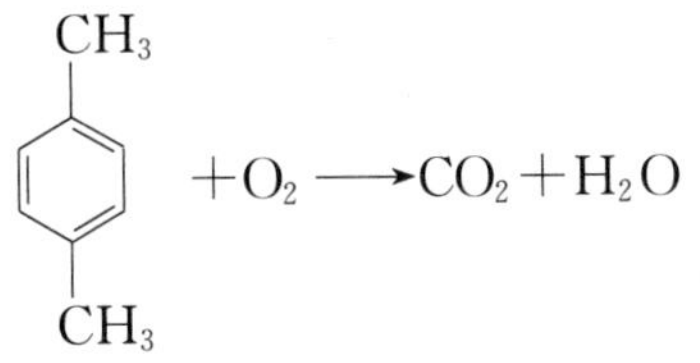

$$\text{（对二甲苯结构，上下两个 CH}_3\text{）} + O_2 \longrightarrow CO_2 + H_2O \tag{7-12}$$

$$CH_3COOH + O_2 \longrightarrow CO_2 + H_2O \tag{7-13}$$

（2）氧化机理

对二甲苯高温氧化采用 Co‑Mn‑Br 三元混合催化剂，其催化剂主体是 Co 和 Mn，但仅用 Co、Mn 并不能完成其反应，因为对二甲苯的第二个甲基很难氧化。加入溴化物，利用溴离子基的强烈吸氢作用，使得对二甲苯的另一个甲基分子中的氢很容易被取代，而使分子活化。

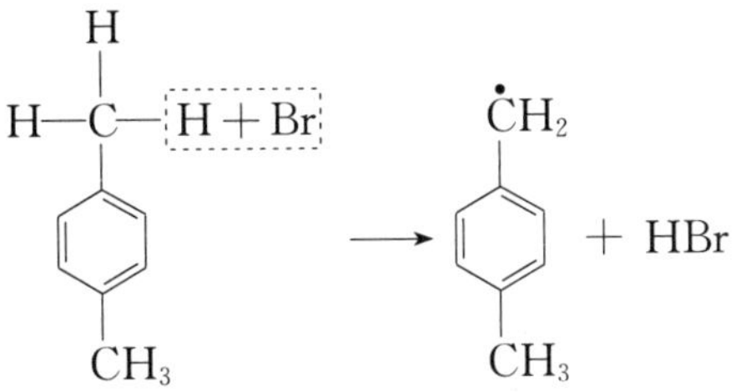

$$\text{（甲苯基上的 H—C—H + Br）} \longrightarrow \dot{C}H_2\text{（对甲苯基）} + HBr \tag{7-14}$$

因此，对二甲苯的反应历程可用下列反应式表示。

$$R-CH_3 + \dot{B}r \longrightarrow R-\dot{C}H_2 + HBr \tag{7-15}$$

$$R-\dot{C}H_2 + O_2 \longrightarrow R-CH_2OO\cdot \tag{7-16}$$

$$R-CH_2OO\cdot + HBr \longrightarrow R-CH_2OOH + Br\cdot \tag{7-17}$$

$$R-CH_2OOH + Me^{2+} \longrightarrow R-CH_2O\cdot + Me^{3+} + OH^- \tag{7-18}$$

$$R-CH_2O\cdot + Me^{3+} \longrightarrow R-CHO + Me^{2+} + H^+ \tag{7-19}$$

$$R-CHO + Br\cdot \longrightarrow R-CO\cdot + HBr \tag{7-20}$$

$$R-CO\cdot + O_2 \longrightarrow R-COOO\cdot \tag{7-21}$$

$$R-COOO\cdot + HBr \longrightarrow R-COOOH + Br\cdot \tag{7-22}$$

$$R-COOOH + Me^{2+} \longrightarrow R-COO\cdot + Me^{3+} + OH^- \tag{7-23}$$

$$R-COO\cdot + HBr \longrightarrow R-COOH + Br\cdot \tag{7-24}$$

式中，R—代表 （苯环，CH₃）、（苯环，CHO）或（苯环，COOH）；Me^{2+} 代表 Co^{2+}、Mn^{2+}；Me^{3+} 代表 Co^{3+}、Mn^{3+}。

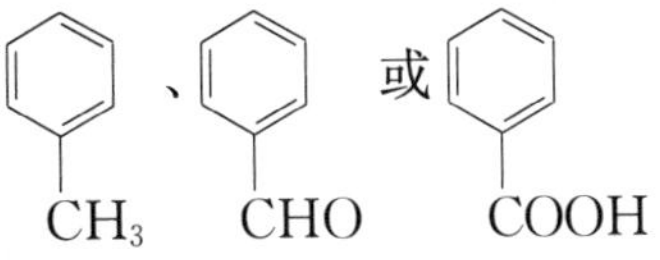

（3）加氢精制过程

加氢精制工艺是利用氧化反应的逆反应原理，在 6.9 MPa 压力和 281℃高温条件下，将粗对苯二甲酸充分溶解于脱盐水中，然后通过钯-碳催化剂床层，进行加氢反应，使粗对苯二甲酸产品中的杂质——对羧基苯甲醛（4 - CBA）还原为易溶于水的对甲基苯甲酸（即 PT 酸），其他有色杂质也同时被分解。反应方程式如下。

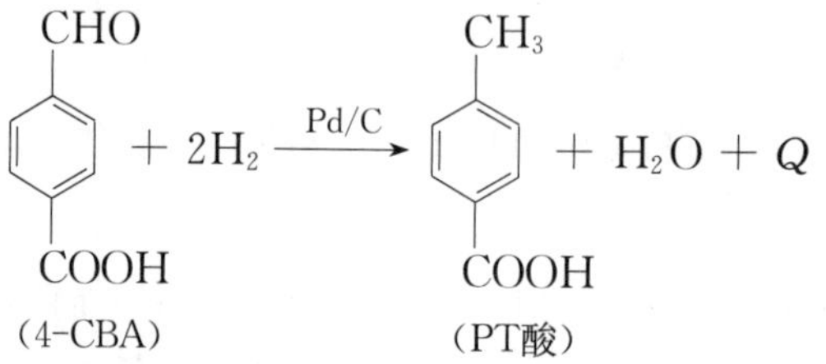

$$\text{(4-CHO,COOH)} + 2H_2 \xrightarrow{Pd/C} \text{(CH_3,COOH)} + H_2O + Q \qquad (7-25)$$

（4-CBA）　　　　　　　（PT酸）

生成的水溶性对甲基苯甲酸经过多级结晶从母液中分出，多次结晶的产物即为高纯度对苯二甲酸。

7.2.2　配位催化氧化

1. 配位催化氧化反应

均相配位催化氧化与催化自氧化反应的机理不同。在配位催化氧化反应中，催化剂由中心金属离子与配位体构成。过渡金属离子与反应物形成配位键并使反应物氧化，而金属离子或配位体被还原，然后，还原态的催化剂再被分子氧氧化成初始状态，完成催化循环过程。而催化自氧化是通过金属离子的单电子转移引起链引发和氢化过氧化物的分解来实现氧化的过程。

2. 乙烯配位催化氧化制乙醛

（1）反应原理

Wacker 法乙烯氧化制乙醛是一个典型的配位催化氧化反应，其过程包括以下三个基本化学反应。

① 烯烃的羰化反应。烯烃在氯化钯水溶液中氧化成醛，并析出金属钯。

$$H_2C{=\!=}CH_2 + PdCl_2 + H_2O \longrightarrow CH_3CHO + Pd + 2HCl \qquad (7-26)$$

② 金属钯氧化反应。式（7-26）析出的金属钯由系统内的氯化铜氧化，转变成二价钯。

$$Pd + 2CuCl_2 \rightleftharpoons PdCl_2 + 2CuCl \qquad (7-27)$$

③ 氯化亚铜的氧化。被还原的氯化亚铜，在盐酸溶液中通入氧气就可迅速氧化转变成氯化铜。

$$2CuCl + \frac{1}{2}O_2 + 2HCl \longrightarrow 2CuCl_2 + H_2O \qquad (7-28)$$

这样，第一个反应中被还原的钯，通过第二个反应转变成二价钯，而后被还原的一价铜，在第三个反应中被氧气氧化成二价铜，由此构成了系统内的催化剂循环。在此，氯化钯和氯化铜称为共催化剂。虽然前两个反应不需要氧气，但系统中氧气的存在是必要的,其作用是将低价的铜氧化,重新转变成高价的铜,这样就实现了乙烯氧化生产乙醛的完整过程。

$$H_2C{=}CH_2 + \frac{1}{2}O_2 \xrightarrow[\text{水溶液}]{PdCl_2 - CuCl_2 - HCl} CH_3CHO \qquad \Delta H_{298}^{\ominus} \qquad (7-29)$$

三步反应中羰化反应速率最慢,是控制步骤。

(2) 工艺流程

乙烯均相氧化制乙醛的过程包括三个基本反应。三个反应在同一反应器进行的是 Hoechst 公司开发的一段法;乙烯羰化与钯的氧化在一台反应器中,Cu^+ 的氧化在另一反应器中的工艺是 Wacker-Chemie 公司开发的二段法。

一段法生产乙醛的工艺流程如图 7-1 所示。

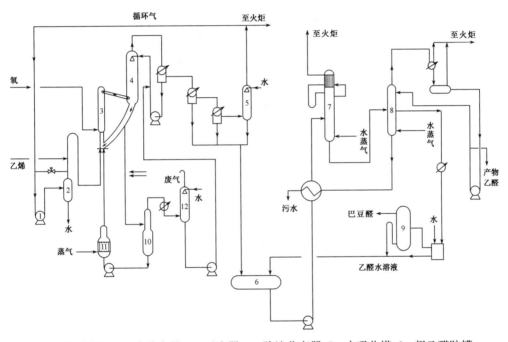

1—水环压缩机;2—水分离器;3—反应器;4—除沫分离器;5—水吸收塔;6—粗乙醛贮糟;
7—脱轻组分塔;8—精馏塔;9—乙醛水溶液分离器;10—分离器;11—分解器;12—水洗涤塔。

图 7-1　一段法乙烯直接氧化生产乙醛的工艺流程

7.2.3　烯烃液相环氧化

除乙烯外,丙烯和其他高级烯烃的气相环氧化法转化率不高,选择性很低,因此,常采用液相环氧化法生产,其中环氧丙烷的生产具有代表性。环氧丙烷是重要的有机化工中间体,主要用于生产聚氨酯泡沫塑料、非离子表面活性剂、乳化剂、破乳剂等,在丙烯衍生物中,仅次于聚丙烯和丙烯腈而居第三位。目前工业上采用的环氧丙烷的生产方法有氯醇法和有机过氧化物法。

氯醇法是生产环氧丙烷的最古老的方法,以丙烯和氯气为原料,基本原理是丙烯先经氯醇化反应生成氯丙醇,然后氯丙醇经皂化反应生成环氧丙烷。

$$CH_3CH=\!\!=CH_2+Cl_2+H_2O \xrightarrow{100℃左右} CH_3CH(OH)CH_2Cl+HCl \quad (7-30)$$

$$2CH_3CH(OH)CH_2Cl+Ca(OH)_2 \longrightarrow 2CH_3CH\!\!-\!\!CH_2+CaCl_2+2H_2O$$

(7-31)

氯醇法生产环氧丙烷的特点是流程短、操作负荷大、选择性好、收率高、生产比较安全、对丙烯纯度要求不高、建厂投资较少,但设备腐蚀性大,生产过程中每生产1 t环氧丙烷产生的含氯化钙的废水量多达 $40\sim60$ t,环境污染严重,并需要有充足的氯源。因此,该方法现已被有机过氧化物法逐渐取代,但对于环氧丁烷等的生产仍采用氯醇法。

有机过氧化物制环氧化烯烃的方法又称共氧化法,目前工业上仅采用异丁烷和乙苯的两种有机氢过氧化物进行共氧化法生产环氧丙烷。

环氧化反应催化剂常选用能溶于反应介质的过渡金属,如钼、钒、钨、钛等有机酸盐类或配位化合物。反应转化率和选择性与所用金属的氧化还原电位及 L 酸酸度有密切关系,以具有低的氧化还原电位和高的 L 酸的钼配位化合物效果最佳,如环烷酸钼、乙酰丙酮钼和六羰基钼等,每摩尔过氧化氢有机物对应的催化剂通常用量为 $0.001\sim0.03$ mol。

烃类过氧化物 ROOH 基团的空间位阻和电子效应是影响环氧化反应的重要因素。一般地,以过氧化氢乙苯为原料的反应速率较过氧化氢异丁烷快,活化能也较低,但前者的稳定性比后者低,因此,有时环氧化收率要低一些。

烯烃与 ROOH 的配比也对反应的选择性有一定的影响,工业上丙烯与 ROOH 的配比一般在 $2:1$ 至 $10:1$(摩尔比)之间。

7.2.4 均相催化氧化过程反应器类型

均相催化氧化反应如果使用空气或氧气作氧源,则属于气-液两相反应体系,氧气通过气液相界面进行传质,进入液相进行氧化反应。通常液相一侧的传质阻力较大,为减少该部分阻力,常用的方法是让液相在反应器内呈连续相,同时反应器必须能提供充分的氧接触表面,并具有较大的持液量。因此,多采用搅拌鼓泡釜式反应器和各种形式的鼓泡反应器(如连续鼓泡床塔式反应器等)。搅拌鼓泡釜式反应器使用范围较广,在搅拌桨的作用下,气泡被破碎和分散,液体高度湍动,缺点是机械搅拌的耗能和动密封问题。连续鼓泡床塔式反应器不采用机械搅拌,气体由分布器以鼓泡的方式通过液层,使液体处于湍动状态,从而达到强化相间传质和传热的目的,结构比较简单。根据反应热的大小,可设置内冷却管或外循环冷却器等来除去反应热,对于反应速率较快的体系,为避免在入口附近发生飞温,还可加入循环导流筒等来快速移走反应热。

7.3 非均相催化氧化

通常涉及的非均相催化氧化是气-固相催化氧化,即原料和氧或空气均以气态形式通过固体催化剂床层,在固体表面发生氧化反应。近年来,液-固相催化氧化反应也有所发展。与均相催化氧化相比,非均相催化氧化过程具有以下特点。

(1) 固体催化剂的活性温度较高,因此,气-固相催化氧化反应通常在较高的反应温度下进行,一般高于150℃,这有利于能量的回收和节能。

(2) 反应物料在反应器中流速快,停留时间短,单位体积反应器的生产能力高,适于大规模连续生产。

(3) 由于反应过程要经历扩散、吸附、表面反应、脱附和扩散等多个步骤,因此,反应过程的影响因素较多,反应不仅与催化剂的组成有关,还与催化剂的结构(如比表面、孔结构等)有关;同时,催化剂床层间传热、传质过程复杂,对目标产物的选择性和设备的正常运作有着不可忽略的影响。

(4) 反应物料与空气或氧的混合物存在爆炸极限问题,因此,在工艺条件的选择和控制方面,以及在生产操作上必须特别关注生产安全。实践中已有许多措施来保证氧化过程安全地进行。

由于固体催化剂的特点,特别是近几十年来高效催化剂(高选择性、高转化率、高生产能力)的相继研制成功,非均相催化氧化在烃类选择性氧化过程中得以广泛应用。目前工业上非均相催化氧化使用的有机原料主要有两类:一类是具有π电子的化合物,如烯烃和芳烃,其氧化产品占总氧化产品的80%以上;另一类是不具有π电子的化合物,如醇类和烷烃等。以前对低碳烷烃的利用较少,是因其氧化的选择性不高,但近年来,随着高选择性催化剂的开发成功、环保意识的提高,以及烷烃价格低廉的优势,低碳烷烃的选择性氧化已逐渐受到重视,有的已工业化,比较典型的有以丁烷代替价高且污染大的苯氧化制顺酐,以丙烷代替价格较高的丙烯为原料的氨氧化制丙烯腈。另外,一些特殊的氧化反应如氨氧化、乙酰基氧化、氧氯化、氧化脱氢等也是常见的非均相催化氧化过程。

7.3.1 重要的非均相催化氧化反应

1. 烷烃的催化氧化反应

工业上成功利用的烷烃催化氧化的典型是正丁烷气相催化氧化制顺丁烯二酸酐(简称顺酐),可用来代替苯法制顺酐,以减少环境污染,目前该法已在顺酐生产中占主导地位。顺酐主要用于制备不饱和聚酯,还可用来生产增塑剂、杀虫剂、涂料和1,4-丁二醇及其下游产品。

$$C_4H_{10} + \frac{7}{2}O_2 \xrightarrow[400℃\sim500℃]{V-P-O} \begin{matrix} CHCO \\ \parallel \quad \rangle O \\ CHCO \end{matrix} + 4H_2O \quad \Delta H = -1\ 265\ kJ/mol \quad (7-32)$$

2. 烯烃的直接环氧化

烯烃的直接环氧化的工业化范例是乙烯环氧化制环氧乙烷。

$$H_2C=CH_2 + \frac{1}{2}O_2 \xrightarrow[220℃\sim260℃]{Ag/\alpha-Al_2O_3} C_2H_4O \quad (7-33)$$

3. 烯丙基催化氧化反应

三个碳原子以上的烯烃,如丙烯、正丁烯、异丁烯等,其α碳原子的C—H键解离能比普通的 C—H 键小,易于断裂,在催化剂的作用下,可在碳原子上发生选择性氧化反应。这些氧化反应都经历烯丙基 H_2C ＝ CH — CH_2 — 反应历程,因此统称为烯丙基氧化反应。使用不同的原料和反应条件,利用烯丙基氧化反应,可生成 α,β-不饱和醛或酮、α,β-不饱和酸和酸酐、α,β-不饱和腈和二烯烃等诸多重要的氧化产物。这些氧化产物中仍保留有双键,具有共轭体系特性,因此易于聚合,是高分子材料的重要单体。丙烯的烯丙基催化氧化反应可简单表示如下。

$$\text{CH}_3\text{CH}=\text{CH}_2 \longrightarrow \begin{array}{l} \xrightarrow[+O_2]{\text{Mo}-\text{Bi}-\text{Co}-\text{O}/\text{SiO}_2} H_2C=\text{CHCHO} \\ \xrightarrow[+O_2]{\text{Co}-\text{Mo}-\text{O}/\text{SiO}_2} H_2C=\text{CHCOOH} \xrightarrow[H_2O]{\xrightarrow{\text{Mo}-\text{V}-\text{Cu}-\text{O}/\text{SiO}_2}{+O_2} \quad \text{ROH}} H_2C=\text{CHCOOR} \\ \xrightarrow[+NH_3+O_2]{\text{P}-\text{Mo}-\text{Bi}-\text{O}/\text{SiO}_2} H_2C=\text{CHCN} \end{array}$$

$$(7-34)$$

4. 芳烃催化氧化反应

芳烃气-固相催化氧化主要用来生产酸酐,比较典型的应用有苯氧化生产顺酐、萘和邻二甲苯氧化生产邻苯二甲酸酐(简称苯酐)、均四甲苯氧化生产均苯四酸酐等。尽管这些酸酐产物为固体结晶,但挥发性大,能升华,因此可采用气-固相催化氧化来生产。

$$\bigcirc + \frac{9}{2}O_2 \xrightarrow[400℃]{\text{V}-\text{M}-\text{O}/\text{SiO}_2} \begin{array}{c}\text{CHCO}\\ \| \\ \text{CHCO}\end{array}\!\!O + 2CO_2 + 2H_2O \quad \Delta H = -1\,850 \text{ kJ/mol}$$

$$(7-35)$$

$$\bigcirc\!\!\bigcirc + \frac{9}{2}O_2 \xrightarrow{\text{V}_2\text{O}_5-\text{K}_2\text{SO}_4/\text{SiO}_2} \begin{array}{c}\text{CO}\\ \text{CO}\end{array}\!\!O + 2H_2O + 2CO_2$$

$$\Delta H = -1\,792 \text{ kJ/mol} \qquad (7-36)$$

$$\bigcirc\!\!\begin{array}{c}\text{CH}_3\\\text{CH}_3\end{array} + 3O_2 \xrightarrow[400℃]{\text{V}_2\text{O}_5-\text{TiO}_2/载体} \begin{array}{c}\text{CO}\\\text{CO}\end{array}\!\!O + 3H_2O \quad \Delta H = -1\,109 \text{ kJ/mol}$$

$$(7-37)$$

$$\begin{array}{c}\text{H}_3\text{C}\\\text{H}_3\text{C}\end{array}\!\!\bigcirc\!\!\begin{array}{c}\text{CH}_3\\\text{CH}_3\end{array} + 6O_2 \xrightarrow[440℃]{\text{V}-\text{Ti}-\text{O}/载体} O\!\!\begin{array}{c}\text{OC}\\\text{OC}\end{array}\!\!\bigcirc\!\!\begin{array}{c}\text{CO}\\\text{CO}\end{array}\!\!O + 6H_2O$$

$$\Delta H = -2\,700 \text{ kJ/mol} \qquad (7-38)$$

5. 醇的催化氧化反应

醇类氧化经过不稳定的过氧化物中间体可生产醛或酮,比较重要的是甲醇氧化制甲醛,还有乙醇氧化制乙醛、异丙醇氧化制丙酮等。甲醇氧化制甲醛可使用电解银作催化剂,于 620℃ 左右进行反应,或采用 Mo-Fe-O、Mo-Bi-O 催化剂,在 200℃～300℃ 进行反应。甲醛主要用来生产脲醛树脂、酚醛树脂、聚甲醛、季戊四醇等。

6. 烯烃乙酰基氧化反应

在催化剂的作用下,氧与烯烃或芳烃和有机酸反应生成酯类的过程,称为乙酰基氧化反应。这类反应中,以乙烯和醋酸进行乙酰基氧化反应生产醋酸乙烯最为重要,目前乙烯法已基本取代乙炔法生产醋酸乙烯。醋酸乙烯可用来生产维尼龙纤维,聚醋酸乙烯广泛用于生产聚乙烯醇、水溶性涂料和黏结剂,醋酸乙烯还可与氯乙烯、乙烯等共聚,形成共聚物。丙烯和醋酸乙酰基氧化生成醋酸丙烯,丁二烯的乙酰基氧化产物主要用来生产 1,4 -丁二醇。

$$H_2C{=}CH_2 + CH_3COOH + \frac{1}{2}O_2 \xrightarrow[160℃\sim180℃,0.8\ MPa\sim1.2\ MPa]{Pd-Au-CH_3COOK/SiO_2} CH_3COOCH{=}CH_2 + H_2O$$

$$\Delta H = -147\ kJ/mol \tag{7-39}$$

$$CH_3CH{=}CH_2 + CH_3COOH + \frac{1}{2}O_2 \xrightarrow{Pd/Al_2O_3} CH_3COOC_3H_5 + H_2O$$

$$\Delta H = -167\ kJ/mol \tag{7-40}$$

$$H_2C{=}CH{-}CH{=}CH_2 + 2CH_3COOH + \frac{1}{2}O_2 \xrightarrow{Pd/C}$$

$$CH_3COO{-}CH_2CH{=}CHCH_2{-}OOCCH_3 + H_2O \tag{7-41}$$

7. 氧氯化反应

典型的氧氯化反应是以金属氯化物为催化剂,如乙烯氧氯化制二氯乙烷,二氯乙烷高温裂解可生产重要的有机单体氯乙烯,并副产 HCl。

$$C_2H_4 + 2HCl + \frac{1}{2}O_2 \xrightarrow[240℃]{CuCl_2/载体} ClH_2C{-}CH_2Cl + H_2O \tag{7-42}$$

$$ClH_2C{-}CH_2Cl \xrightarrow{裂解} H_2C{=}CHCl + HCl \tag{7-43}$$

其他氧氯化技术,如甲烷氧氯化制氯甲烷、二氯乙烷氧氯化制三氯乙烯和四氯乙烯都已工业化。

$$8C_2H_4Cl_2 + 6Cl_2 + 7O_2 \xrightarrow[420℃]{CuCl_2-KCl/载体} 4C_2HCl_3 + 4C_2Cl_4 + 14H_2O \tag{7-44}$$

7.3.2 非均相催化氧化反应机理

尽管烃类的气-固相催化氧化反应过程复杂,系统内可以存在多个相互独立的反应,并以串联或并联的形式相互关联,但还是可以对过程进行简化,建立比较符合实际的反应网络,研究其反应机理。常见的烃类气-固相催化氧化的反应机理有三种。

1. 氧化还原机理

氧化还原机理又称晶格氧作用机理。该机理认为晶格氧参与了反应,其模型描述是:反应物首先和催化剂的晶格氧结合,生成氧化产物,催化剂变成还原态,接着还原态的活性组分再与气相中的氧气反应,重新成为氧化态催化剂,由此,氧化还原循环构成了有机物在催化剂上的氧化过程。

2. 化学吸附氧化机理

化学吸附氧化机理以 Langmuir 化学吸附模型为基础,假定氧是以吸附态形式化学吸附在催化剂表面的活性中心上,再与烃分子反应。该模型简明并便于数学处理,因此,在气-固相催化反应中广为应用,对于具有复杂反应网络的体系也可较方便地推导出反应速率方程。

3. 混合反应机理

混合反应机理是化学吸附和氧化还原机理的综合,假定反应物首先化学吸附在催化剂表面含晶格氧的氧化态活性中心上,然后与氧化态活性中心在表面反应生成产物,同时氧化态的活性中心变为还原态,它们再与气相中的氧发生表面氧化反应,重新转化为氧化态活性中心。

7.3.3 非均相氧化催化剂和反应器

非均相氧化催化剂的活性组分主要是具有可变价的过渡金属钼、铋、钒、钛、钴、锑等的氧化物,如 $MoO_3 \cdot Bi_2O_3$、$Co_2O_3 \cdot MoO_3$、$V_2O_5 \cdot TiO_2$、$V_2O_5 \cdot P_2O_5$、$CoO \cdot WO_3$等;一些能化学吸附氧的金属(如银等)在环氧化反应、醇的氧化中也成功地得以应用。近年来,杂多酸和新型分子筛催化剂的开发应用也十分活跃。

变价过渡金属氧化物作催化剂时,单一氧化物对特定的氧化反应而言,常表现为活性很高时选择性较差。为了使活性和选择性恰当而获得较高收率,工业催化剂常采用两种或两种以上的金属氧化物,以产生协同效应,这些氧化物可以形成复合氧化物、固溶体或以混合物的形式存在。同时,催化剂中变价金属离子处于氧化态和还原态的比例应保持在一合适的范围内,以保持催化剂的氧化还原能力适当,如丁烷氧化制顺酐所用的 V-P-O 催化剂,其中既有 V^{3+},又有 V^{5+},合适的催化剂应保持平均钒价态为 $4.0\sim4.1$。

有些氧化催化剂是负载型的,常用的载体有氧化铝、硅胶、活性炭等。载体的品种和性能对催化剂的催化作用常有相当大的影响。

常用的烃类气-固相催化氧化反应器有固定床反应器、流化床反应器。由于氧化反应放热量很大,需要及时移出,故一般采用换热式反应器。

7.4 乙烯环氧化制环氧乙烷

7.4.1 环氧乙烷的性质与用途

环氧乙烷(简称 EO)是最简单、最重要的环氧化物,在常温下为气体,沸点为 $10.4℃$,可与水、醇、醚及大多数有机溶剂以任意比例混合,在空气中的爆炸极限(体积分数)为 $2.6\%\sim100\%$,有毒。环氧乙烷易自聚,当有铁、酸、碱、醛等杂质或在高温下时更是如此,自聚时会放出大量热,甚至发生爆炸,因此存放环氧乙烷的贮槽必须清洁,并保持在 $0℃$ 以下。

由于环氧乙烷具有含氧三元环结构,性质非常活泼,极易发生开环反应,在一定条件下,可与水、醇、氢卤酸、氨及氨的化合物等发生加成反应,其通式为

$$\text{H}_2\text{C}\underset{\text{O}}{-}\text{CH}_2 \ +XY \longrightarrow \ \text{H}_2\text{C}\underset{\text{OX}}{-}\underset{\text{Y}}{\text{CH}_2} \qquad\qquad (7-45)$$

环氧乙烷与水发生水合反应生成乙二醇,是制备乙二醇的主要方法;与氨反应可生成乙醇胺、二乙醇胺和三乙醇胺。环氧乙烷本身还可开环聚合生成聚乙二醇。

环氧乙烷是以乙烯为原料的产品中的第三大品种,仅次于聚乙烯和苯乙烯。环氧乙烷的主要用途是生产乙二醇,其消费量约占全球环氧乙烷总消费量的 60%,乙二醇是生产聚酯纤维的主要原料之一。环氧乙烷还可用于生产非离子表面活性剂以及乙醇胺类、乙二醇醚类、二甘醇、三甘醇等。

7.4.2　环氧乙烷生产方法

环氧乙烷的工业生产采用乙烯直接氧化法。直接氧化法又分为空气氧化法和氧气氧化法。1931 年,法国催化剂公司的 Lefort 发现乙烯在银催化剂作用下可以直接氧化成环氧乙烷,经过进一步的研究与开发形成了乙烯-空气直接氧化法制环氧乙烷技术。1937 年,美国 UCC 公司首次采用此法建厂生产。1958 年,美国 Shell 公司首次建成了氧气直接氧化法工业装置,氧气直接氧化法技术先进,适宜大规模生产,生产成本低,产品纯度可达 99.99%,设备体积小,放空量少,氧气氧化法排出的废气量只相当于空气氧化法的 2%,相应的乙烯损失也少;另外,氧气氧化法流程比空气氧化法短,设备少,建厂投资可减少 15%~30%,考虑空分装置的投入,总投资会比空气氧化法高一些,但用纯氧作氧化剂可提高进料浓度和选择性,生产成本大约为空气氧化法的 90%;同时,氧气氧化法比空气氧化法反应温度低,有利于延长催化剂的使用寿命。因此,近年来新建的大型装置均采用纯氧作氧化剂,氧气氧化法逐渐成为占绝对优势的工业生产方法。

7.4.3　乙烯直接氧化法制环氧乙烷的反应

乙烯在银催化剂上的氧化反应包括选择氧化和深度氧化,除生成目的产物环氧乙烷外,还生成副产物二氧化碳、水及少量甲醛和乙醛。

$$\text{C}_2\text{H}_4+\frac{1}{2}\text{O}_2 \longrightarrow \text{C}_2\text{H}_4\text{O} \quad \Delta H=-103.4 \text{ kJ/mol} \qquad (7-46)$$

$$\text{C}_2\text{H}_4+3\text{O}_2 \longrightarrow 2\text{CO}_2+2\text{H}_2\text{O(g)} \quad \Delta H=-1\ 324.6 \text{ kJ/mol} \qquad (7-47)$$

$$\text{C}_2\text{H}_4\text{O}+\frac{5}{2}\text{O}_2 \longrightarrow 2\text{CO}_2+2\text{H}_2\text{O(g)} \quad \Delta H=-1\ 221.2 \text{ kJ/mol} \qquad (7-48)$$

研究表明,二氧化碳和水主要由乙烯直接氧化生成,反应的选择性主要取决于平行副反应的竞争,环氧乙烷串联副反应是次要的。环氧乙烷的氧化可能是先异构化为乙醛,再氧化为二氧化碳和水,而乙醛在反应条件下易氧化,所以产物中只有少量乙醛存在。由于这些氧化反应都是强放热反应,具有较大的平衡常数,尤其是深度氧化,为选择性氧化反应放热的十余倍,因此为减少副反应的发生,提高选择性,催化剂的选择非常重要。选择不当会因副反应进行而引起操作条件的恶化,甚至变得无法控制,造成反应器内发生"飞温"事故。

7.4.4　乙烯直接环氧化机理

1. 催化剂

乙烯直接氧化法生产环氧乙烷的工业催化剂为银催化剂。在乙烯直接氧化制环氧乙烷的生产过程中，原料乙烯消耗的费用占 EO 生产成本的 70% 左右，因此，降低乙烯单耗是提高经济效益的关键，最佳措施是开发高性能催化剂。工业上使用的银催化剂由活性组分银、载体和助催化剂组成。

（1）载体

载体的主要功能是提高活性组分银的分散度，防止银的微小晶粒在高温下烧结。载体的表面结构、孔结构及导热性能，对催化剂颗粒内部的温度分布、催化剂上银晶粒的大小及分布、反应原料气体及生成气体的扩散速率等有非常大的影响，从而显著影响其活性和选择性。载体比表面积大，有利于银晶粒的分散，催化剂初始活性高，但比表面积大的催化剂孔径较小，反应产物环氧乙烷难以从小孔中扩散出来，脱离表面的速度慢，从而造成环氧乙烷深度氧化，选择性下降。

载体的形状对催化剂的催化性能也有影响。为了提高载体性能，应尽量把载体制成传质传热性能良好的形状，如环形、马鞍形、阶梯形等。同时，载体形状选择应保证反应过程中气流在催化剂颗粒间有强烈搅动，不发生短路，床层阻力小。

（2）助催化剂

只含活性组分银的催化剂并不是最好的，必须添加助催化剂。研究表明，碱金属、碱土金属和稀土元素等具有助催化作用，两种或两种以上的助催化剂有协同作用，效果优于单一组分。碱金属助催化剂的主要作用是使载体表面酸性中心中毒，以减少副反应的发生。

此外，还可添加活性抑制剂。抑制剂的作用是使催化剂表面部分可逆中毒，从而使活性适当降低，减少深度氧化，提高选择性，见诸报道的有二氯乙烷、氯乙烯、氮氧化物、硝基烷烃等。工业生产中常添加微量二氯乙烷，二氯乙烷热分解生成乙烯和氯，氯被吸附在银表面，影响氧在催化剂表面的化学吸附，减少乙烯的深度氧化。

（3）银含量

增加催化剂的银含量，可提高催化剂的活性，但会使选择性降低，因此，目前工业催化剂的银的质量含量基本在 20% 以下。但最近的研究结果表明，只要选择合适的载体和助催化剂，高银含量的催化剂也能保证选择性基本不变，而活性明显提高。

（4）催化剂制备

银催化剂的制备有两种方法，早期采用粘接法或称涂覆法，现在采用浸渍法。粘接法使用黏结剂将活性组分、助催化剂和载体粘接在一起，制得的催化剂银的分布不均匀，易剥落，催化性能差，寿命短。浸渍法一般采用水或有机溶剂溶解有机银，如羧酸银及有机胺构成的银铵配位化合物作银浸渍液，该浸渍液中也可溶有助催化剂组分，将载体浸渍其中，经后处理制得催化剂。银盐的选择、银盐和助催化剂浸渍次序和方法、还原剂的选择和制备过程工艺条件等，都对银粒在载体表面上的大小和分布有影响，从而影响催化剂的催化性能。

2. 反应机理

乙烯在银催化剂上直接氧化制环氧乙烷的反应机理至今尚无定论。P. A. Kilty 等根据氧在银催化剂表面的吸附、乙烯和吸附氧的作用以及选择性氧化反应,提出了氧在银催化剂表面上存在两种化学吸附态,即原子吸附态和分子吸附态。当由四个相邻的银原子簇组成吸附位时,氧便解离形成原子吸附态 O^{2-},这种吸附的活化能低,在任何温度下都有较高的吸附速度,且易与乙烯发生深度氧化。

活性抑制剂的存在,可使催化剂的银表面部分被覆盖,如添加二氧乙烷时,若银表面的 1/4 被氯覆盖,则无法形成四个相邻银原子簇组成的吸附位,从而抑制氧的原子态吸附和乙烯的深度氧化。

$$O_2 + 4Ag(相邻) \longrightarrow 2O^{2-}(吸附态) + 4Ag^+ \tag{7-49}$$

$$12Ag^+ + 6O^{2-}(吸附态) + C_2H_4 \longrightarrow 2CO_2 + 12Ag + 2H_2O \tag{7-50}$$

在较高温度下,在不相邻的银原子上也可产生氧解离形成的原子态吸附,但这种吸附需较高的活化能,因此不易形成。

$$O_2 + 4Ag(不相邻) \longrightarrow 2O^{2-}(吸附态) + 4Ag^+ \tag{7-51}$$

当没有由四个相邻银原子簇构成的吸附位时,可发生氧的分子态吸附,即氧的非解离吸附,形成活化了的离子化氧分子,乙烯与此种分子氧反应生成环氧乙烷,同时产生一个吸附的原子态氧。此原子态的氧与乙烯反应,则生成二氧化碳和水。

$$O_2 + Ag \longrightarrow Ag - O_2^-(吸附态) \tag{7-52}$$

$$C_2H_4 + Ag - O_2^-(吸附态) \longrightarrow C_2H_4O + Ag - O^-(吸附态) \tag{7-53}$$

$$C_2H_4 + 6Ag - O^-(吸附态) \longrightarrow 2CO_2 + 6Ag + 2H_2O \tag{7-54}$$

$$7C_2H_4 + 6Ag - O_2^-(吸附态) \longrightarrow 6C_2H_4O + 2CO_2 + 6Ag + 2H_2O \tag{7-55}$$

7.4.5　反应条件对乙烯环氧化的影响

1. 反应温度

乙烯环氧化过程中存在着平行的完全氧化副反应,反应温度是影响选择性的主要因素。尽管催化反应机理和动力学还未取得一致的认识,但研究表明环氧化反应的活化能小于完全氧化反应的活化能。反应温度升高,两个反应的速率都加快,但完全氧化反应的速率增加更快。

2. 空速

空速是影响乙烯转化率和环氧乙烷选择性的另一因素,但与反应温度相比,该因素是次要的。空速减小,会导致转化率提高,选择性下降,但影响不如温度显著。空速不仅影响转化率和选择性,还影响催化剂的空时收率和单位时间的放热量,应全面考虑。空速提高,可增大反应器中气体流动的线速度,减小气膜厚度,有利于传热。工业上采用的空速与选用的催化剂有关,还与反应器和传热速率有关,一般在 $4\,000 \sim 8\,000\ \mathrm{h}^{-1}$ 左右。催化剂活性高、反应热可及时移出时,可选择高空速,反之选择低空速。

3. 反应压力

乙烯直接氧化的主、副反应在热力学上都不可逆,因此压力对主、副反应的平衡和选择性影响不大。但加压可提高乙烯和氧的分压,加快反应速率,提高反应器的生产能力,也有利于采用加压吸收法回收环氧乙烷,故工业上大都采用加压氧化法。但压力也不能太高,否则设备耐压要求提高,费用增大,环氧乙烷也会在催化剂表面产生聚合和积碳,影响催化剂寿命。一般工业上采用的压力在 2.0 MPa 左右。

4. 原料配比及致稳气

对于具有循环的乙烯环氧化过程,进入反应器的混合气由循环气和新鲜原料气混合形成,它的组成不仅影响过程的经济性,也与安全生产息息相关。实际生产过程中乙烯与氧的配比一定要在爆炸限以外,同时必须控制乙烯和氧的浓度在合适的范围内,过低时催化剂的生产能力小,过高时反应放出的热量大,易造成反应器的热负荷过大,产生飞温。乙烯与空气混合物的爆炸极限(体积分数)为 $2.7\%\sim36\%$,与氧的爆炸极限(体积分数)为 $2.7\%\sim80\%$,实际生产中因循环气带入二氧化碳等,爆炸限也有所改变。为了提高乙烯和氧的浓度,可以用加入第三种气体来改变乙烯的爆炸极限,这种气体通常称为致稳气。致稳气是惰性的,能减小混合气的爆炸极限,增加体系安全性;具有较高的比热容,能有效地移出部分反应热,增加体系稳定性。工业上曾广泛采用的致稳气是氮气,近年来常采用甲烷作致稳气。

5. 原料气纯度

许多杂质对乙烯环氧化过程都有影响,必须严格控制。主要有害物质及危害如下:① 催化剂中毒。如硫化物、砷化物、卤化物等能使催化剂永久中毒,乙炔会使催化剂中毒并能与银反应生成有爆炸危险的乙炔银。② 增大反应热效应。氢气、乙炔、C_3 以上的烷烃和烯烃可发生燃烧反应放出大量热,使过程难以控制,乙炔、高碳烯烃的存在还会加快催化剂表面的积碳失活。③ 影响爆炸限。氩气和氢气是空气和氧气中带来的主要杂质,其含量过高会改变混合气体的爆炸限,降低氧的最大容许浓度。④ 选择性下降。原料气及反应器管道中带入的铁离子会使环氧乙烷重排为乙醛,导致生成二氧化碳和水,使选择性下降。

6. 乙烯转化率

单程转化率的控制与氧化剂的种类有关,用纯氧作氧化剂时,单程转化率一般应控制在 $12\%\sim15\%$,选择性可达 $83\%\sim84\%$;用空气作氧化剂时,单程转化率一般应控制在 $30\%\sim35\%$,选择性达 70% 左右。单程转化率过高时,由于放热量大,温度升高快,会加快深度氧化,使环氧乙烷的选择性明显降低。为了提高乙烯的利用率,工业上采用循环流程,即将环氧乙烷分离后未反应的乙烯再送回反应器,所以单程转化率不能过低,否则会因循环气量过大而导致能耗增加。同时,生产中要引出 $10\%\sim15\%$ 的循环气以除去有害气体(如二氧化碳、氩气等),单程转化率过低也会造成乙烯的损失增加。

乙烯直接氧化法生产环氧乙烷的工艺流程包括反应部分和环氧乙烷回收、精制两大部分。氧气法生产环氧乙烷的工艺流程示意图见图 7-2。

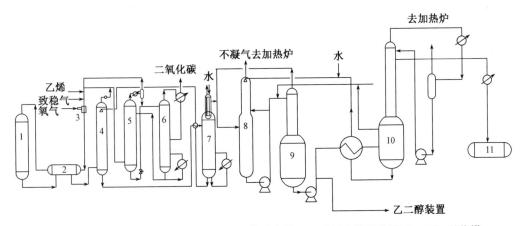

1—环氧乙烷反应器；2—热交换器；3—气体混合器；4—环氧乙烷吸收塔；5—CO$_2$吸收塔；
6—CO$_2$吸收液再生塔；7—解析塔；8—再吸收塔；9—脱气塔；10—精馏塔；11—环氧乙烷贮槽。

图7-2 氧气法生产环氧乙烷工艺流程示意图

7.5 丙烯氨氧化制丙烯腈

7.5.1 丙烯腈概况

烃类的氨氧化是指用空气或氧气对烃类及氨进行共氧化生成腈或有机氮化物的过程。烃类可以是烷烃、环烷烃、烯烃、芳烃等，其中最有工业价值的是丙烯氨氧化。在烯丙基氧化过程中，丙烯氨氧化制丙烯腈（AN）可以作为此类过程的典型实例。

丙烯腈是重要的有机化工产品，在丙烯系列产品中居第二位，仅次于聚丙烯。在常温常压下，丙烯腈是无色液体，味甜，微臭，沸点为 77.3℃。丙烯腈有毒，室内允许浓度为 0.002 mg/L，在空气中爆炸极限（体积分数）为 3.05%～17.5%，与水、苯、四氯化碳、甲醇、异丙醇等可形成二元共沸物。丙烯腈分子中含有碳碳双键和氰基，化学性质活泼，能发生聚合、加成、氰基和氰乙基等反应，可制备出各种合成纤维、合成橡胶、塑料、涂料等。

7.5.2 丙烯氨氧化制丙烯腈的化学反应

丙烯氨氧化过程中，除生成主产物丙烯腈外，还有多种副产物。
主反应

$$C_3H_6 + NH_3 + \frac{3}{2}O_2 \longrightarrow H_2C{=\!=}CH{-\!-}CN(g) + 3H_2O(g) \tag{7-56}$$

副反应

$$C_3H_6 + \frac{3}{2}NH_3 + \frac{3}{2}O_2 \longrightarrow \frac{3}{2}CH_3CN(g) + 3H_2O(g) \tag{7-57}$$

$$C_3H_6 + 3NH_3 + 3O_2 \longrightarrow 3HCN + 6H_2O(g) \tag{7-58}$$

$$C_3H_6+O_2 \longrightarrow H_2C=CHCHO(g)+H_2O(g) \qquad (7-59)$$

$$C_3H_6+\frac{3}{2}O_2 \longrightarrow H_2C=CHCOOH(g)+H_2O(g) \qquad (7-60)$$

$$C_3H_6+O_2 \longrightarrow CH_3CHO(g)+HCHO(g) \qquad (7-61)$$

$$C_3H_6+\frac{1}{2}O_2 \longrightarrow CH_3COCH_3(g) \qquad (7-62)$$

$$C_3H_6+3O_2 \longrightarrow 3CO+3H_2O(g) \qquad (7-63)$$

$$C_3H_6+\frac{9}{2}O_2 \longrightarrow 3CO_2+3H_2O(g) \qquad (7-64)$$

7.6 芳烃氧化制邻苯二甲酸酐

7.6.1 邻苯二甲酸酐的性质、用途及工艺概况

邻苯二甲酸酐简称苯酐,沸点为 284.5℃,凝固点(干燥空气中)为 131.11℃,有刺激性。苯酐主要用来生产增塑剂邻苯二甲酸二辛酯、邻苯二甲酸二丁酯及其他酯类,还可用于制造不饱和聚酯树脂和染料、医药、农药等。

7.6.2 邻二甲苯制苯酐的反应机理

邻二甲苯气相催化氧化制苯酐的反应历程复杂,包括一系列平行和串联反应,反应均为不可逆放热过程。

主反应

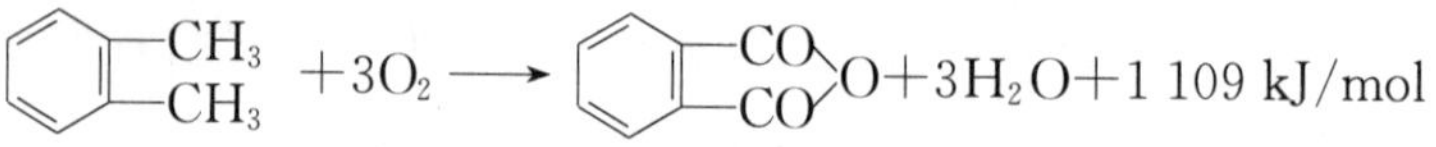

$$\text{（结构式）} + 3O_2 \longrightarrow \text{（结构式）} + 3H_2O + 1\ 109\ \text{kJ/mol} \qquad (7-65)$$

副反应

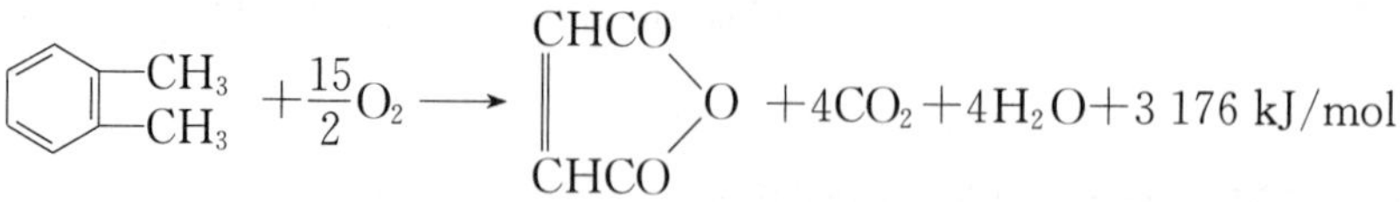

$$\text{（结构式）} + \frac{15}{2}O_2 \longrightarrow \text{（结构式）} + 4CO_2 + 4H_2O + 3\ 176\ \text{kJ/mol} \qquad (7-66)$$

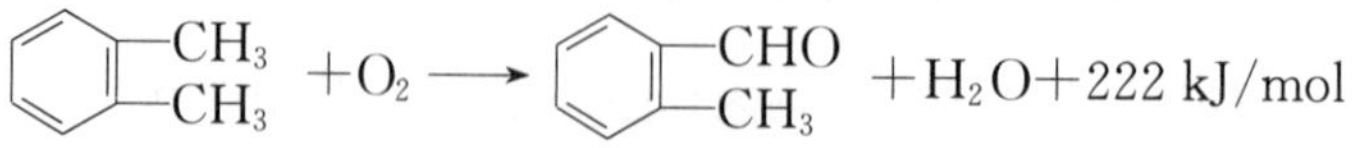

$$\text{（结构式）} + O_2 \longrightarrow \text{（结构式）} + H_2O + 222\ \text{kJ/mol} \qquad (7-67)$$

$$\text{（结构式）} + 2O_2 \longrightarrow \text{（结构式）} + 2H_2O + 874\ \text{kJ/mol} \qquad (7-68)$$

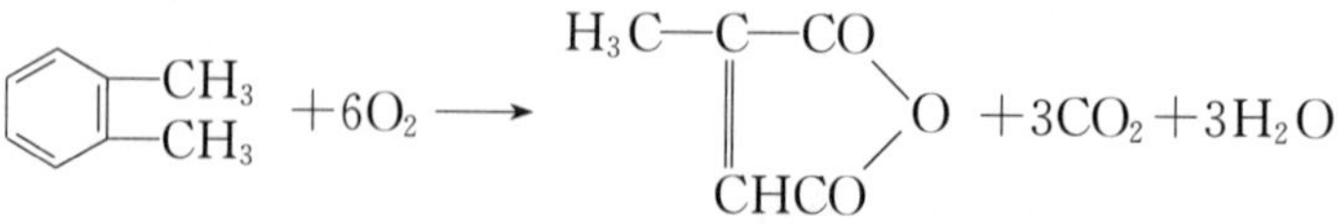

$$\text{（结构式）} + 6O_2 \longrightarrow \text{（结构式）} + 3CO_2 + 3H_2O \qquad (7-69)$$

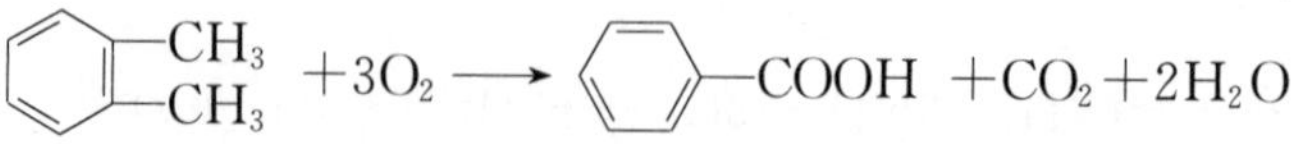

$$\text{（结构式）} + 3O_2 \longrightarrow \text{（结构式）}-COOH + CO_2 + 2H_2O \qquad (7-70)$$

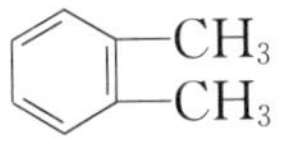

$$\begin{matrix} & CH_3 \\ & CH_3 \end{matrix} + \frac{21}{2}O_2 \longrightarrow 8CO_2 + 5H_2O + 4\,380\text{ kJ/mol} \qquad (7-71)$$

7.7 氧化操作安全技术

7.7.1 爆炸极限

选择性氧化过程中,烃类及其衍生物的气体或蒸气与空气或氧气形成混合物,在一定浓度范围内,由于引火源如明火、高温或静电火花等因素的作用,该混合物会自动迅速发生支链型连锁反应,导致极短时间内体系温度和压力急剧上升,火焰迅速传播,最终发生爆炸。该浓度范围称为爆炸极限,简称爆炸限,一般以体积分数表示,其最低浓度为爆炸下限,最高浓度为爆炸上限。不同的体系有不同的爆炸限。常温常压下,空气中,邻二甲苯的爆炸限为 $1.0\% \sim 6.0\%$,萘的爆炸限为 $0.9\% \sim 7.8\%$,丙烯的爆炸限为 $2.4\% \sim 11.0\%$,丙烯腈的爆炸限为 $3.05\% \sim 17.5\%$,乙烯的爆炸限为 $2.7\% \sim 36\%$,环氧乙烷的爆炸限为 $2.6\% \sim 100\%$,环氧丙烷的爆炸限为 $1.9\% \sim 24.0\%$。

7.7.2 防止爆炸的工艺措施

爆炸极限的存在限制了反应原料浓度的提高,对反应速率、选择性、能量利用和设备投资等都不利,但为了安全起见,在确定进料浓度和配比时,大部分工业氧化反应还是在爆炸极限以外(通常在爆炸下限以下)操作。由于惰性气体的存在可改变体系的爆炸限,对于一些爆炸威力较大的物系,不仅要在爆炸极限以外操作,工业生产中还加入惰性气体作致稳气,如乙烯为原料制环氧乙烷时,N_2、CO_2、甲烷等都有致稳作用。

反应原料和空气(尤其是氧气)在混合时最容易发生事故,因此,混合器的设计和混合顺序的选择也非常重要。混合应尽量在接近反应器入口处,氧气或空气在喷嘴出口处的速度要大大高于原料的火焰传播速度,以利于快速混合。

7.8 催化氧化技术进展

近年来,均相催化氧化技术的进展主要在以下三个方面。

(1) 新反应的开发

如使用 $PdCl_2 - CuCl - LiCl - CH_3COOLi$ 为催化剂,可使乙烯、CO 和氧直接进行羰基氧化,一步合成丙烯酸;以 $PdCl_2 - CuCl_2$ 或 $PdCl_2 - FeCl_3$ 为催化剂,可进行一系列羰基氧化反应,如以 CO、O_2 和醇为原料生产草酸酯等;环己烷两步氧化生成尼龙-66 的主要单体己二酸,首先用环烷酸钴作催化剂,空气氧化环己烷生成环己酮和环己醇,然后再以 $Cu(II)/V$(体积比)盐为催化剂,硝酸为氧化剂,可得己二酸,收率大于 90%,副产的氮化物可在系统内再氧化成硝酸。

(2) 催化剂的改进

如用较廉价的过渡金属代替贵金属作催化剂;贵金属催化剂的回收和固定化等。

（3）不对称催化氧化

此反应可用来生产光学活性物质，其中研究比较多的是不对称环氧化反应、双羟基化反应等。

目前，非均相催化氧化技术的进展除了低碳烷烃的选择性氧化和氨氧化以外，最引人注目的是钛硅沸石的发现及其在合成领域的应用。

思考题

7-1　分析氧化过程的作用及其特点。

7-2　分析催化自氧化反应的特点并给出在化工应用中的实例。

7-3　分析比较非均相催化氧化和均相催化氧化的特点。

7-4　非均相催化氧化反应机理有几种？各自描述的特点是什么？

7-5　掌握乙烯环氧化制环氧乙烷的原理、催化体系和反应主流程。

7-6　乙烯环氧化反应工艺条件选择的依据是什么？

7-7　致稳气的作用是什么？

7-8　分析共沸精馏的原理及其在丙烯腈精制过程中的应用。

7-9　从原料的来源和价格，谈丙烯氨氧化制丙烯腈的前景。

7-10　分析苯酐生产的基本原理、催化体系的构成及工艺技术走向。

7-11　何谓爆炸极限？其主要影响因素有哪些？

7-12　请列举一些化工生产中常用的防爆措施。

Chapter 7   Selective oxidation of hydrocarbons

7.1   Overview

Oxidation reaction in the chemical industry is a large class of important chemical reactions, and it is an important reaction process for the production of bulk chemical raw materials and intermediates. The oxidation of hydrocarbons is the most representative in the oxidation of organic matter. Hydrocarbon oxidation reactions can be divided into two types: complete oxidation and partial oxidation. Complete oxidation refers to the reaction process in which carbon atoms in the reactants combine with oxygen to form CO_2, and hydrogen atoms combine with oxygen to form water. Partial oxidation, also known as selective oxidation, refers to a process in which a small amount of hydrogen atoms (sometimes with a small amount of carbon atoms) in hydrocarbons and their derivatives react with an oxidizing agent (usually oxygen), while other hydrogen and carbon atoms do not. According to statistics, more than 50% of the major chemicals produced in the world are related to the selective oxidation process. The selective oxidation of hydrocarbons can generate chemicals of higher value than the raw materials and has a wide range of applications in chemical production. Selective oxidation can not only produce oxygen-containing compounds, such as alcohols, aldehydes, ketones, acids, acid anhydrides, epoxides, peroxides, but also produce oxygen-free compounds, such as oxidative dehydrogenation of butene to butadiene, propane (propylene) ammoxidation to acrylonitrile, ethylene oxychlorination to dichloroethane. Some of these products are important raw materials and intermediates of organic chemical industry, some are monomers of three major synthetic materials, and some are widely used solvents. They occupy an important position in the chemical industry.

7.1.1   Characteristics of the oxidation process and oxidizers selection

1. Characteristics of oxidation reactions

(1) Large exothermic reaction heat

The oxidation reaction is a strong exothermic reaction. The deeper the oxidation, the more reaction heat is released. The thermal effect of complete oxidation is about 8 to 10 times than that of partial oxidation. Therefore, in the process of oxidation reaction, the timely transfer of reaction heat is very important. Otherwise, the reaction temperature will rise rapidly, which will promote the increase of side reactions, and the reaction selectivity will drop significantly. In severe cases, the reaction temperature may be uncontrollable, and even explosion may occur. Utilizing the reaction heat of the oxidation reaction, steam can be by-produced. Generally speaking, the gas-solid phase catalytic oxidation reaction temperature is high, high and medium pressure steam can be recovered. The gas-liquid phase oxidation reaction temperature is low, and only low-grade energy can be recovered, such as low-pressure steam and hot water.

(2) Irreversible reaction

For hydrocarbons and other organic compounds, the oxidation reaction has $\Delta G^* \ll 0$. Therefore, it is a thermodynamic irreversible reaction, which is not limited by chemical equilibrium, and can theoretically achieve a single-pass conversion rate of 100%. But for many reactions, in order to ensure high selectivity, the conversion rate must be controlled within a certain range, otherwise it will cause deep oxidation and reduce the yield of the target product. For example, when butane is oxidized to maleic anhydride, the conversion rate of butane is generally controlled at about 85% to 90% to ensure that the generated maleic anhydride does not continue to be deeply oxidized.

(3) Complex and diverse oxidation pathways

Hydrocarbons and most derivatives can undergo oxidation reactions, and oxidation reactions are mostly complex networks formed by series reactions, parallel reactions or a combination of the two. Due to different catalysts and reaction conditions, oxidation reactions can go through different reaction paths, leading to different reaction products. These products tend to be more reactive and less stable than the starting materials, and are prone to deeper oxidation, ultimately producing carbon dioxide and water. Therefore, the selection of reaction conditions and catalysts is very important, and the selection of catalysts is the key to determining the oxidation path.

(4) Flammable and explosive process

Hydrocarbons can easily form explosive mixtures with oxygen or air, so the oxidation process should be designed and operated with special attention to ensure its safety.

2. Selection of oxidant

To introduce oxygen into the molecules of hydrocarbons or other compounds, an oxidant is required. The common ones are air and pure oxygen, hydrogen peroxide and other peroxides, etc. Air and pure oxygen are the most commonly used. Air is cheaper than pure oxygen, but the oxygen partial pressure is small and it contains a lot of inert gases. So, the power consumption and the exhaust gas emission in the production process is large. Using pure oxygen as an oxidant can reduce exhaust emissions and reactor volume. Whether to use air or pure oxygen depends on the technical and economic analysis.

Hydrocarbon peroxides or peroxy acids generated by oxidation of some hydrocarbons and their derivatives with air or pure oxygen can also be used as oxidants for oxidation reactions. For example, hydrogen peroxide ethylbenzene from ethylbenzene by air oxidation can be reacted with propylene to produce propylene oxide.

In recent years, hydrogen peroxide has developed rapidly as an oxidant. The use of hydrogen peroxide has mild conditions, simple operation, high reaction selectivity, and is not prone to deeper oxidation reactions. It is environmentally friendly, and clean production can be achieved.

7.1.2 Classification of hydrocarbons selective oxidation

In terms of reaction types, selective oxidation can be classified into the following three genres.

(1) Oxidation reaction in which the carbon chain is not broken. For example, the hydrogen atoms on the saturated carbon atoms of alkanes, alkenes, cycloalkanes and alkyl aromatics undergo oxidation reaction with oxygen to generate new functional groups, and alkenes are oxidized to form diolefin and epoxides, etc.

(2) Oxidation reaction in which the carbon chain is broken. It includes the reactions in which the carbon number of the product is less than that of the raw material, such as the oxidation of isobutane to produce ethanol. It also includes the ring-opening reaction in which the carbon number of the product is the same as that of the raw material, such as the oxidation of cyclohexane to hexanediol, etc.

(3) Oxidative condensation reaction. In the course of the reaction, condensation between molecules occurs in this type of reaction, such as the ammoxidation of propylene to produce acrylonitrile, the oxidative condensation of benzene and ethylene to produce styrene, and the like.

As far as the reaction phase is concerned, it can be divided into homogeneous catalytic oxidation and heterogeneous catalytic oxidation. In the homogeneous catalytic oxidation system, the reaction components and the catalyst are in the same phase, while in the heterogeneous catalytic oxidation system, the reaction components and the catalyst exist in different phases. At present, heterogeneous catalytic oxidation process is mainly used in chemical industry, and the application of homogeneous catalytic oxidation process is still few.

7.2 Homogeneous catalytic oxidation

In the past 40 years, driven by the development of metal-organic chemistry, the homogeneous catalytic oxidation process has attracted attention due to its high activity and high selectivity. Homogeneous catalytic oxidation generally refers to a gas-liquid phase oxidation reaction. It is customarily called the liquid phase oxidation reaction. Liquid phase catalytic oxidation generally has the following characteristics.

(1) The reactant is in the same phase as the catalyst. There is no issue of uneven distribution of the active center on the solid surface, and the transition metal as the active center has high activity and good selectivity.

(2) The reaction conditions are not too harsh, and the reaction is relatively stable and easy to control.

(3) The reaction equipment is simple, the volume is small, and the production capacity is high.

(4) Reaction temperature is usually not too high, so reaction heat utilization rate is relatively low.

(5) Special materials should be used in highly corrosive systems.

(6) Catalysts are mostly precious metals, they must be separated and recovered.

Nowadays, homogeneous catalytic oxidation technology has been successfully applied, such as oxidation of higher alkanes to secondary alcohols, the oxidation of naphthene to alcohols and ketone mixtures, the production of aldehydes or ketones by the Wacker process, the preparation of hydrocarbon hydrogen peroxide, epoxidation of olefins by hydrocarbon hydrogen peroxide, oxidation of aromatic hydrocarbons to aromatic acids and other processes. With the development of chemical products towards refinement, it plays an increasingly important role in the field of fine chemical synthesis. Research on the use of cheaper transition metal catalyst instead of precious metals, catalyst recovery, and catalyst immobilization have also made progress.

There are many types of homogeneous catalytic oxidation reactions, and catalytic auto-oxidation and complex catalytic oxidation are commonly used in industry. In addition, there is the liquid phase epoxidation of olefins.

7.2.1 Catalytic auto-oxidation

1. Catalytic auto-oxidation reaction

Auto-oxidation reaction refers to an oxidation reaction that has the characteristics of a free radical chain reaction and can be automatically accelerated. The initial stage of the non-catalytic auto-oxidation reaction has a longer induction period, because there is no sufficient concentration of free radicals to induce the chain reaction.

Catalysts can accelerate the initiation of the chain, promote the initiation of the reactant to generate free radicals, shorten or eliminate the reaction induction period, so it can greatly accelerate the oxidation reaction, which is called catalytic auto-oxidation. Such reactions are commonly used in industry to produce organic acids and peroxides. Under suitable conditions, intermediate products such as alcohols, ketones, and aldehydes can also be obtained. The catalysts used in the reaction are mostly salts of transition metal ions such as Co and Mn, like acetate and naphthenate. The catalytic effect of cobalt salts is generally good, and it is usually dissolved in a liquid medium to form a homogeneous phase.

2. Mechanism of auto-oxidation reaction

After a lot of experiments, it has been determined that the auto-oxidation of hydrocarbons and other organic compounds is carried out according to the mechanism of free radical chain reaction, but some processes (such as chain initiation) have not been fully clarified. Next, the basic steps of auto-oxidation will be briefly introduced with the example of liquid-phase auto-oxidation of hydrocarbons.

$$RH + O_2 \longrightarrow \dot{R} + HO_2 \qquad (7-1)$$

$$\dot{R} + O_2 \longrightarrow RO\dot{O} \qquad (7-2)$$

$$RO\dot{O} + RH \longrightarrow ROOH + \dot{R} \qquad (7-3)$$

$$\dot{R} + \dot{R} \longrightarrow R\!-\!R \qquad (7-4)$$

Among the above three steps, the decisive step is the initiation process of the chain, that is, the process in which the hydrocarbon molecules undergo a homolytic reaction and are converted into free radicals, which requires a large activation energy. The energy required is related to the structure of the carbon atom. The known C—H bond energy is

tertiary C—H $<$ secondary C—H $<$ primary C—H

Therefore, the activation energy of homolytic cleavage of tertiary C—H bond is the smallest, followed by secondary C—H.

For the bond reaction to start, there must also be a sufficient free radical concentration. So, from the initiation of the chain to the start of the chain reaction, there must be an accumulation stage of the free radical concentration. At this stage, no oxygen absorption is observed. It is generally called the induction period, which takes several hours or longer. After the induction period, the reaction accelerates quickly and reaches the maximum value. Catalysts and initiators can be used to accelerate the generation of free radicals and shorten the induction period of the reaction, such as Co and Mn transition metal ion salts, and compounds such as hydrogen peroxide, isobutane and 2,2′- azobis(2 - methylpropionitrile) are easily decomposed to generate free radicals. These substances usually act during the chain initiation stage. In the chain transfer stage, the free radicals generated by the substrate act as carriers.

The chain transfer reaction is a free radical-molecular reaction and requires a small activation energy. This process includes the mass transfer process and the chemical reaction process of oxygen from the gas phase to the reaction zone. The influence of each parameter is very complicated. When the partial pressure of oxygen is high enough, the reaction rate of formula (7-2) is very fast, and the reaction rate of chain transfer is controlled by the reaction formula (7-3).

The product ROOH generated by the reaction formula (7-3) has unstable performance. In the presence of a higher temperature or a catalyst, it will be further decomposed to generate new free radicals, and branching reactions will occur to generate different oxides.

$$ROOH \longrightarrow R\dot{O} + \dot{O}H \tag{7-5}$$

$$R\dot{O} + RH \longrightarrow ROH + \dot{R} \tag{7-6}$$

$$\dot{O}H + RH \longrightarrow H_2O + \dot{R} \tag{7-7}$$

$$2ROOH \longrightarrow RO\dot{O} + R\dot{O} + H_2O \tag{7-8}$$

$$RO\dot{O} \longrightarrow R'\dot{O} + R''CHO(\text{or ketone}) \tag{7-9}$$

$$R\dot{O}(\text{or } R'\dot{O}) + RH \longrightarrow ROH(\text{or } R'OH) + \dot{R} \tag{7-10}$$

As a result of the branching reaction, alcohols and aldehydes with different carbon numbers are generated, and they can be further oxidized to generate ketones and acids, making the product composition very complicated.

3. Influencing factors of auto-oxidation reaction process

(1) Influence of solvent

In homogeneous catalytic oxidation systems, solvents are often used. The choice of solvent is very important. It can not only change the reaction conditions, but also have a certain influence on the reaction process.

(2) Influence of impurities

The auto-oxidation reaction is a free radical chain reaction, the number of free radicals initiated in the system and the chain transfer process are very important to the reaction. The presence of impurities may deactivate the free radicals in the system, thereby disrupting the normal chain initiation and transfer, resulting in a significant decrease in the reaction rate or even termination of the reaction. Since the free radical concentration of the auto-oxidation reaction system is generally small, it is very sensitive to the influence of impurities, and sometimes even a small number of impurities will have a considerable impact. The influence of impurities on the free radical chain reaction is called inhibition, and impurities are inhibitor. Different reaction systems have different inhibitors, the common ones are water, sulfide, phenol, etc.

(3) Influence of temperature and oxygen partial pressure

The oxidation reaction is accompanied by large reaction heat. In the auto-oxidative reaction system, due to the characteristics of the free radical chain reaction, it is significant to maintain the balance of the exothermic heat and the heat removal of the system. The oxidation reaction requires an oxygen source. When the oxygen supply capacity of the system is sufficient, the reaction is controlled by kinetics. Keeping a high reaction temperature is conducive to the progress of the reaction, but it should not be too high, so as to avoid the increase of by-products, the decrease of selectivity, and even the loss of the reaction control. When the oxygen concentration is low and the oxygen supply capacity of the system is insufficient, the reaction is controlled by mass transfer. At this time, increasing the oxygen partial pressure can promote the oxygen transfer and improve the reaction rate, but it also needs to be determined according to the pressure resistance capacity of the equipment and economic accounting. If the oxygen supply rate is between the two, both mass transfer and kinetic factors have an impact, which should be considered comprehensively.

In addition, since the target product of the oxidation reaction is an intermediate product of the oxidation process, the change of the oxygen partial pressure will affect the selectivity of the reaction, thereby affecting the composition of the product.

(4) Influence of oxidant dosage and space velocity

The upper limit of the amount of oxidant (air or oxygen) is determined by the explosion limit of oxygen in the exhaust gas discharged from the reaction, and the explosion range should be avoided. The lower limit of the amount of oxidant is the theoretical oxygen consumption required for the reaction, and the oxygen content in the exhaust gas is zero at this time. In industrial practice, the oxygen content in the exhaust gas is generally controlled at 2% to 6%, preferably 3% to 5%. The space velocity of the oxidant is defined as the ratio of the flow rate of air or oxygen to the liquid volume in the reactor. The increase of the space velocity is beneficial to the contact between the gas and the liquid phase. The absorption of oxygen is accelerated, and the reaction speed is promoted. However, an excessively high space velocity will shorten the residence time of the gas in the reactor, make oxygen absorption incomplete, and the utilization rate will be reduced, resulting in an excessively high oxygen content in the tail gas, which will affect both safety and economy. The space velocity is constrained by the oxygen content in the exhaust.

4. Terephthalic acid preparation by paraxylene oxidation

Terephthalic acid is mainly used in the production of polyethylene terephthalate (PET), and also in the production of polybutylece terephthalate (PBT) and polypropylene terephthalate (PPT) for further preparation of polyester fibers, films and engineering plastics. Among them, polyester fiber accounts for more than 50% of the world's synthetic fiber output. In addition to being widely used in daily life clothes, it is also used in tire cord fabrics, conveyor belts, fire extinguishing water pipes, etc.

At present, the main production method of terephthalic acid is the p-xylene oxidation method, which is a typical homogeneous catalytic auto-oxidation reaction. It mainly includes two parts: oxidation and hydrofining. The three technologies of Amoco in the United States, ICI in the United Kingdom and Mitsui Petrochemical in Japan are representative and widely used.

(1) Oxidation process

Taking p-xylene (PX) as raw material, using cobalt acetate and manganese acetate as catalyst, and tetrabromoethane as accelerator, under certain pressure and temperature, p-xylene is continuously oxidized into crude terephthalic acid in acetic acid solvent with air. The reaction equation is as follows.

Main reactions

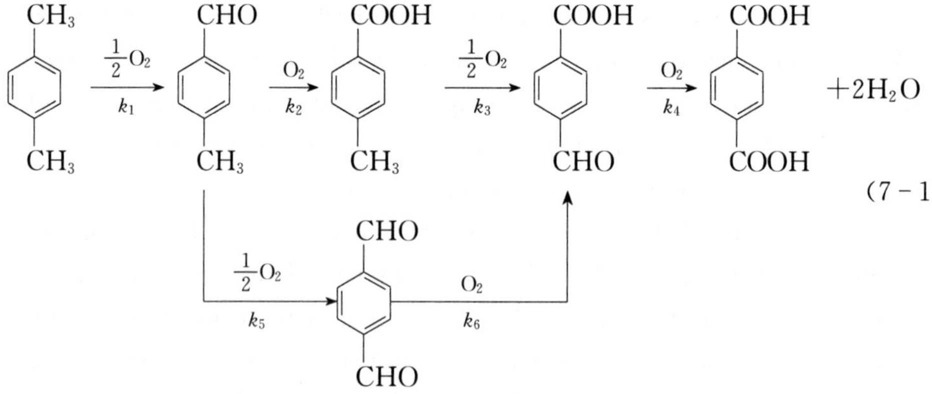

$$(7-11)$$

In addition to the above main reactions, there are also some side reactions. For example, the solvent acetic acid and p-xylene will undergo partial deeper oxidation to generate CO and CO_2. When the ratio of the oxidation reaction is improper, or the raw materials are not pure, side reactions will also occur.

Side reactions

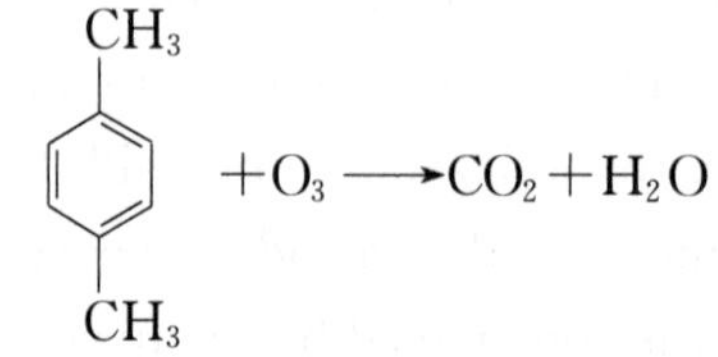

$$(7-12)$$

$$CH_3COOH + O_2 \longrightarrow CO_2 + H_2O \qquad (7-13)$$

(2) Oxidation mechanism

The high-temperature oxidation of p-xylene uses a Co - Mn - Br ternary mixed catalyst. The main catalysts are Co and Mn, but only Co and Mn cannot complete the reaction. This is because the second methyl group of p-xylene is difficult to get oxidized. Therefore, adding bromide and using the strong hydrogen absorption effect of bromide ion group makes the hydrogen in another methyl molecule of p-xylene easily replaced, and the molecule is activated.

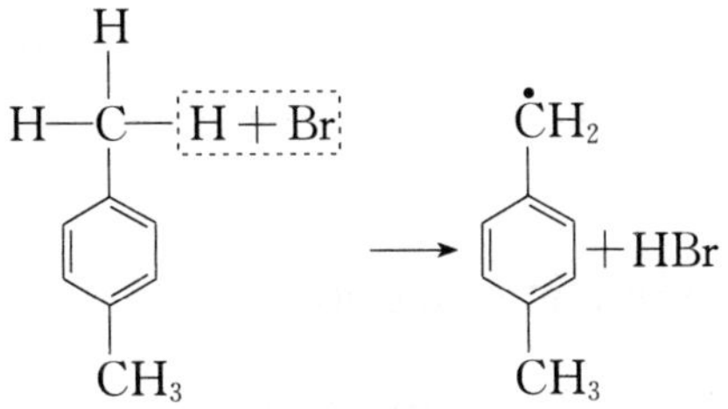

$$\underset{\underset{CH_3}{|}}{\overset{\overset{H}{|}}{H-C-[H+Br]}} \quad \longrightarrow \quad \underset{\underset{CH_3}{|}}{\overset{\dot{C}H_2}{}} +HBr \qquad (7-14)$$

Therefore, the reaction scheme of p-xylene can be represented by the following reaction formula.

$$R-CH_3+\dot{B}r \longrightarrow R-\dot{C}H_2+HBr \qquad (7-15)$$

$$R-\dot{C}H_2+O_2 \longrightarrow R-CH_2OO\cdot \qquad (7-16)$$

$$R-CH_2OO\cdot+HBr \longrightarrow R-CH_2OOH+Br\cdot \qquad (7-17)$$

$$R-CH_2OOH+Me^{2+} \longrightarrow R-CH_2O\cdot+Me^{3+}+OH^- \qquad (7-18)$$

$$R-CH_2O\cdot+Me^{3+} \longrightarrow R-CHO+Me^{2+}+H^+ \qquad (7-19)$$

$$R-CHO+Br\cdot \longrightarrow R-CO\cdot+HBr \qquad (7-20)$$

$$R-CO\cdot+O_2 \longrightarrow R-COOO\cdot \qquad (7-21)$$

$$R-COOO\cdot+HBr \longrightarrow R-COOOH+Br\cdot \qquad (7-22)$$

$$R-COOOH+Me^{2+} \longrightarrow R-COO\cdot+Me^{3+}+OH^- \qquad (7-23)$$

$$R-COO\cdot+HBr \longrightarrow R-COOH+Br\cdot \qquad (7-24)$$

In the formula, R— represents $\underset{CH_3}{\bigcirc}$, $\underset{CHO}{\bigcirc}$ or $\underset{COOH}{\bigcirc}$; Me^{2+} stands for Co^{2+}, Mn^{2+} ; Me^{3+} stands for Co^{3+}, Mn^{3+}.

(3) Hydrorefining process

The hydrorefining process utilizes the reverse reaction mechanism of the oxidation reaction. At a pressure of 6.9 MPa and under temperature of 281℃, the crude terephthalic acid is fully dissolved in the desalinated water, and then through the palladium carbon catalyst bed, hydrogenation reaction is carried out. The impurity p-carboxy benzaldehyde (4 – CBA) in the crude terephthalic acid product is reduced to water-soluble p-toluic acid (PT acid), other colored impurities are also decomposed at the same time. The reaction equation is as follows.

$$\underset{\underset{(4\text{-CBA})}{\underset{\text{COOH}}{\bigcirc}}}{\overset{\text{CHO}}{\bigcirc}} + 2H_2 \xrightarrow{\text{Pd/C}} \underset{\underset{(\text{PT acid})}{\underset{\text{COOH}}{\bigcirc}}}{\overset{\text{CH}_3}{\bigcirc}} + H_2O + Q \qquad (7-25)$$

The generated water-soluble p-toluic acid is separated from the mother liquor through multi-stage crystallization, and the product of repeated crystallization is high-purity terephthalic acid.

7.2.2 Coordination catalytic oxidation

1. Coordination catalytic oxidation reactions

The mechanism of homogeneous coordination catalytic oxidation is different from that of catalytic auto-oxidation. In coordination catalytic oxidation, the catalyst is composed of central metal ions and ligands. The transition metal ion forms a coordination bond with the reactant and activates it, so that the reactant is oxidized, and the metal ion or ligand is reduced. Then, the catalyst in the reduced state is oxidized to the initial state by oxygen to complete the catalytic cycle process. Catalytic auto-oxidation is a process in which chain initiation and decomposition of hydroperoxides are achieved by single electron transfer of metal ions.

2. Coordination catalytic oxidation of ethylene to acetaldehyde

(1) Reaction mechanism

The Wacker method of ethylene oxidation to acetaldehyde is a typical coordination catalytic oxidation reaction, and the process includes the following three basic chemical reactions.

① Carbonylation of olefins. Olefins are oxidized to aldehydes in aqueous palladium chloride solution, and palladium is precipitated.

$$H_2C{=}CH_2 + PdCl_2 + H_2O \longrightarrow CH_3CHO + Pd + 2HCl \qquad (7-26)$$

② Palladium oxidation reaction. The palladium precipitated by the formula (7-26) is oxidized by cupric chloride in the system and converted into divalent palladium.

$$Pd + 2CuCl_2 \Longleftrightarrow PdCl_2 + 2CuCl \qquad (7-27)$$

③ Oxidation of cuprous chloride. The reduced cuprous chloride can be rapidly oxidized into cupric chloride when oxygen is introduced into the hydrochloric acid solution.

$$2CuCl + \frac{1}{2}O_2 + 2HCl \longrightarrow 2CuCl_2 + H_2O \qquad (7-28)$$

In this way, the reduced palladium in the first reaction is converted into divalent palladium through the second reaction, and then the reduced monovalent copper is oxidized to divalent copper by oxygen in the third reaction, thus constituting the catalyst cycle inside the system. Here, palladium chloride and copper chloride are referred to as co-catalysts. Although the first two reactions do not require oxygen, the presence of oxygen in the system is necessary to re-oxidize low-valent copper into high-valent copper, thus realizing the complete process of ethylene oxidation to acetaldehyde.

$$H_2C\!\!=\!\!CH_2+\frac{1}{2}O_2 \xrightarrow[\text{aqueous solution}]{\text{PdCl}_2-\text{CuCl}_2-\text{HCl}} CH_3CHO \qquad \Delta H^{\ominus}_{298} \qquad (7-29)$$

In the three-step reaction, carbonylation reaction rate is the slowest, which is the rate-controlling step.

(2) Process flow

The process of homogeneous oxidation of ethylene to acetaldehyde includes three basic reactions. The three reactions are carried out in the same reactor by the one-stage method developed by Hoechst Company. The process of ethylene carbonylation and palladium oxidation in one reactor and Cu^+ oxidation in another reactor is a two-stage process developed by Wacker-Chemie Company.

The technological process of producing acetaldehyde by one-stage method is shown in Figure 7 − 1.

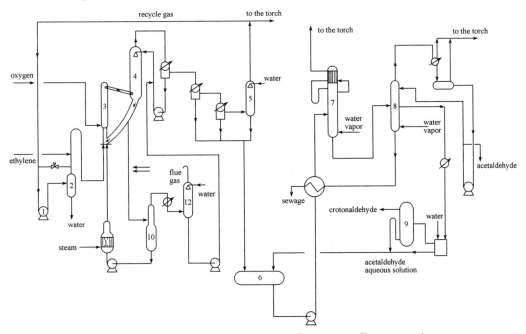

1 − Water ring compressor; 2 − Water separator; 3 − Reactor; 4 − Foam removing separator; 5 − Water absorption tower; 6 − Crude acetaldehyde storage tank; 7 − Removing light component tower; 8 − Rectifying column; 9 − Acetaldehyde aqueous solution separator; 10 − Separator; 11 − Decomposer; 12 − Water washing tower.

Figure 7 − 1　Production process of acetaldehyde by one-stage process direct oxidation of ethylene

7.2.3 Liquid-phase epoxidation of olefins

In addition to ethylene, the gas-phase epoxidation of propylene and other higher olefins has low conversion rate and low selectivity. Therefore, liquid-phase epoxidation is often used for production, and the production of propylene oxide is representative. Propylene oxide is an important organic chemical intermediate. It is mainly used in the production of polyurethane foam, nonionic surfactants, emulsifiers, demulsifiers, etc. Among propylene derivatives, it ranks third right after polypropylene and acrylonitrile. Production methods currently used in industry include chlorohydrin method and organic peroxide method.

The chlorohydrin method is the oldest method for producing propylene oxide. The basic principle is to use propylene and chlorine as raw materials. First, propylene undergoes chlorohydrin reaction to generate chloropropanol, and then chlorohydrin is saponified to generate propylene oxide.

$$CH_3CH\!=\!CH_2 + Cl_2 + H_2O \xrightarrow{\sim 100℃} CH_3CH(OH)CH_2Cl + HCl \quad (7-30)$$

$$2CH_3CH(OH)CH_2Cl + Ca(OH)_2 \longrightarrow 2CH_3\underset{\diagdown \;\; \diagup}{\underset{O}{CH\!-\!CH_2}} + CaCl_2 + 2H_2O$$

$$(7-31)$$

Features of chlorohydrin method for producing propylene oxide include short process flow, large operating load, good selectivity, high yield, relatively safe production, low requirements for propylene purity, and less investment in plant construction. However, the equipment suffers from highly corrosive materials. In the production process, every ton of propylene oxide produced produces $40\sim60$ tons of waste water containing calcium chloride, which causes serious environmental pollution and requires sufficient chlorine sources. Therefore, this method has been gradually replaced by the organic peroxide method, but the chlorohydrin method is still used for the production of butylene oxide and the like.

The method of epoxidizing olefin by organic peroxide is also called co-oxidation method. At present, the co-oxidation method to produce propylene oxide adopts organic hydroperoxide, and only two kinds of organic hydroperoxide (isobutane and ethylbenzene) are used in industry.

Epoxidation catalysts are often selected from transition metals that are soluble in the reaction medium, such as organic acid salts or coordination compounds of molybdenum, vanadium, tungsten, titanium and other. The reaction conversion rate and selectivity are related to the redox potential of the used metal and the L acid acidity. Molybdenum complexes with low redox potential and high L acid have the best effect, such as molybdenum naphthenate, molybdenum acetylacetonate and molybdenum hexacarbonyl, etc. The usual dosage of catalyst is 0. 001~0. 03 mol/ mol (hydrogen peroxide organic compound).

The steric hindrance and electronic effect of the ROOH group in hydrocarbon peroxides are important factors affecting the epoxidation reaction. Generally, the reaction rate of ethylbenzene hydrogen peroxide is faster than that of isobutane hydrogen peroxide. The activation energy is also low, but the stability of the former is lower. Therefore, sometimes the epoxidation yield is low.

The ratio of olefin to ROOH also has a certain influence on the selectivity of the reaction. In industry, the ratio of propylene to ROOH is generally between 2:1 and 10:1 (by moles).

7.2.4   Reactor type of homogeneous catalytic oxidation process

If the homogeneous catalytic oxidation reaction uses air or oxygen as the oxygen source, it belongs to a gas-liquid two-phase reaction system. The mass transfer of oxygen occurs through the gas-liquid phase interface and oxygen enters the liquid phase for the oxidation reaction. Usually, the mass transfer resistance on the liquid phase side is relatively large. In order to reduce this part of the resistance, the common method is to make the liquid phase a continuous phase in the reactor. At the same time, the reactor must be able to provide sufficient oxygen contact surface and have a large liquid holdup. Therefore, agitated bubbling tank reactors and various forms of bubbling reactors such as continuous bubbling bed tower reactors are mostly used. The stirred bubbling tank reactor has a wide range of applications. Under the action of the stirring paddle, the bubbles are broken and dispersed, and the liquid is highly turbulent. The disadvantage is the energy consumption of mechanical stirring and dynamic sealing issues. The continuous bubbling bed tower reactor does not use mechanical stirring, and the gas is bubbled through the liquid layer by the distributor to keep the liquid in a turbulent state, so as to achieve the purpose of strengthening mass transfer and heat transfer between the phases, and the structure is relatively simple. According to the magnitude of the reaction heat, an inner cooling tube or an outer circulation cooler can be set up to remove the reaction heat. For a system with a fast reaction rate, in order to avoid the occurrence of overheating near the inlet, measures such as adding a circulating guide tube can also be used to quickly remove reaction heat.

7.3   Heterogeneous catalytic oxidation

The heterogeneous catalytic oxidation usually involved is gas-solid catalytic oxidation, that is, the raw material and oxygen or air pass through the solid catalyst bed in gaseous form, and the oxidation reaction occurs on the solid surface. In recent years, the liquid-solid catalytic oxidation reaction has also been developed. Compared with homogeneous catalytic oxidation, the heterogeneous catalytic oxidation process has the following characteristics.

(1) The solid catalyst is usually high temperature catalyst. Therefore, the gas-solid phase catalytic oxidation reaction is often carried out at a relatively high reaction temperature, usually higher than 150℃, which facilitates energy recovery and energy saving.

(2) The flow rate of the reaction material in the reactor is fast, and the residence time is short. The production capacity per unit volume of the reactor is high, which is suitable for large-scale continuous production.

(3) Since the reaction process has to go through multiple steps such as diffusion, adsorption, surface reaction, desorption and diffusion, there are many factors affecting the reaction process. The reaction is not only related to the composition of the catalyst, but also to the structure of the catalyst such as specific surface area and pore structure. Also, the process of heat transfer and mass transfer between catalyst beds is quite complex, which has a non-negligible impact on the selectivity of the target product and the normal operation of the equipment.

(4) The mixture of reaction material and air or oxygen has the problem of explosion limit. Therefore, special attention must be paid to production safety in the selection and control of process conditions, as well as in production operation. There are many measures in practice to ensure that the oxidation process can be carried out safely.

Due to the characteristics of solid catalysts, especially the successful development of high-efficiency catalysts (high selectivity, high conversion, high productivity) in recent decades, heterogeneous catalytic oxidation has been widely used in the selective oxidation of hydrocarbons. At present, there are two main types of organic raw materials used in industrial heterogeneous catalytic oxidation: one is compounds with π electrons, such as olefins and aromatic hydrocarbons, whose oxidation products account for more than 80% of the total oxidation products; the other is compounds without π electrons, such as alcohols and alkanes. In the past, there was little use of low-carbon alkanes because of its low oxidation selectivity. However, in recent years, due to the successful development of high-selectivity catalysts, the advantages of low prices and the improvement of environmental awareness, low-carbon alkanes' selective oxidation has gradually drawn attention, and some have been industrialized. The typical ones are the production of maleic anhydride by butane instead of expensive and polluting benzene, and the ammoxidation of acrylonitrile by propane instead of high-priced propylene as raw material. In addition, some special oxidation reactions such as ammonia oxidation, acetyl oxidation, oxychlorination, oxidative dehydrogenation are also common heterogeneous catalytic oxidation processes.

7.3.1   Important heterogeneous catalytic oxidation reactions

1. Catalytic oxidation of alkanes

A typical example of successful industrial use is the gas-phase catalytic oxidation of n-butane to produce maleic anhydride, which can be used to replace the benzene method and reduce environmental pollution. At present, this method has dominated the production of maleic anhydride. Maleic anhydride is mainly used in the preparation of unsaturated polyesters, and can also be used in the production of plasticizers, pesticides, coatings, 1,4 - butanediol and its downstream products.

$$C_4H_{10}+\frac{7}{2}O_2 \xrightarrow[\text{400℃}\sim\text{500℃}]{\text{V-P-O}} \begin{matrix} CHCO \\ \\ CHCO \end{matrix}\!\!\!\!>\!\!O \;+4H_2O \quad \Delta H=-1\,265\text{ kJ/mol} \qquad (7-32)$$

2. Direct epoxidation of olefins

An industrialization example is the epoxidation of ethylene to ethylene oxide.

$$H_2C\!\!=\!\!CH_2+\frac{1}{2}O_2 \xrightarrow[\text{220℃}\sim\text{260℃}]{\text{Ag}/\alpha-\text{Al}_2\text{O}_3} C_2H_4O \qquad\qquad (7-33)$$

3. Allyl-catalyzed oxidation reaction

Olefins with more than three carbon atoms, such as propylene, n-butene, isobutene, have a C—H bond of α-carbon atom with dissociation energy smaller than that of ordinary C—H bonds, and the bond is easy to break. With catalysts, selective oxidation occurs on these carbon atoms. These oxidation reactions all go through the allyl ($H_2C\!=\!CH\!-\!CH_2\!-$) reaction process, so they are collectively referred to as allyl oxidation reactions. By allyl oxidation reaction with different raw materials and reaction conditions, many important oxidation products such as α, β-unsaturated aldehydes or ketones, α, β-unsaturated acids and acid anhydrides, α, β-unsaturated nitriles and dienes can be generated. These oxidation products still retain double bonds and have the characteristics of conjugated systems, so they are easy to polymerize and are important monomers for polymer materials. The allyl-catalyzed oxidation of propylene can be simply represented as follows.

$$CH_3CH\!=\!CH_2 \longrightarrow \begin{cases} \xrightarrow[+O_2]{\text{Mo}-\text{Bi}-\text{Co}-\text{O}/\text{SiO}_2} H_2C\!=\!CHCHO \\[2mm] \xrightarrow[+O_2]{\text{Co}-\text{Mo}-\text{O}/\text{SiO}_2} H_2C\!=\!CHCOOH \xrightarrow{ROH} H_2C\!=\!CHCOOR \\[2mm] \xrightarrow[+NH_3+O_2]{\text{P}-\text{Mo}-\text{Bi}-\text{O}/\text{SiO}_2} H_2C\!=\!CHCN \end{cases}$$

$$\;+O_2\!\!\downarrow\text{Mo}-\text{V}-\text{Cu}-\text{O}/\text{SiO}_2 \qquad H_2O\!\!\uparrow$$

$$(7-34)$$

4. Aromatic catalytic oxidation reaction

Aromatic gas-solid phase catalytic oxidation is mainly used to produce acid anhydrides. The typical ones are: benzene oxidation to produce maleic anhydride, naphthalene and o-xylene oxidation to produce phthalic anhydride, and 1, 2, 4, 5 – tetramethylbenzene oxidation to produce pyromellitic dianhydride. Although these acid anhydride products are solid crystals, they are volatile and can be sublimated. Therefore, they can be produced by gas-solid phase catalytic oxidation.

$$\text{C}_6\text{H}_6 + \frac{9}{2}\text{O}_2 \xrightarrow[400℃]{\text{V－M－O/SiO}_2} \begin{array}{c}\text{CHCO}\\ \|\quad\quad \text{O}\\ \text{CHCO}\end{array} +2\text{CO}_2+2\text{H}_2\text{O} \quad \Delta H=-1\ 850\ \text{kJ/mol}$$

(7 – 35)

$$\text{C}_{10}\text{H}_8 + \frac{9}{2}\text{O}_2 \xrightarrow{\text{V}_2\text{O}_5－\text{K}_2\text{SO}_4/\text{SiO}_2} \begin{array}{c}\text{CO}\\ \quad\quad \text{O}\\ \text{CO}\end{array}+2\text{H}_2\text{O}+2\text{CO}_2$$

$$\Delta H=-1\ 792\ \text{kJ/mol} \quad\quad\quad (7 – 36)$$

$$\begin{array}{c}\text{CH}_3\\ \text{CH}_3\end{array} +3\text{O}_2 \xrightarrow[400℃]{\text{V}_2\text{O}_5－\text{TiO}_2/\text{support}} \begin{array}{c}\text{CO}\\ \quad\quad \text{O}\\ \text{CO}\end{array}+3\text{H}_2\text{O} \quad \Delta H=-1\ 109\ \text{kJ/mol}$$

(7 – 37)

$$\begin{array}{c}\text{H}_3\text{C}\quad\text{CH}_3\\ \text{H}_3\text{C}\quad\text{CH}_3\end{array} +6\text{O}_2 \xrightarrow[440℃]{\text{V－Ti－O/support}} \text{O}\begin{array}{c}\text{OC}\quad\text{CO}\\ \quad\quad\quad\quad \text{O}\\ \text{OC}\quad\text{CO}\end{array}+6\text{H}_2\text{O}$$

$$\Delta H=-2\ 700\ \text{kJ/mol} \quad\quad\quad (7 – 38)$$

5. Catalytic oxidation of alcohols

The oxidation of alcohols can produce aldehydes or ketones through unstable peroxide intermediates. The more important ones are the oxidation of methanol to formaldehyde, the oxidation of ethanol to acetaldehyde, and the oxidation of isopropanol to acetone. Electrolytic silver can be used as a catalyst for methanol oxidation to formaldehyde at about 620℃, or Mo-Fe-O and Mo-Bi-O can be used as catalysts at 200℃ ～ 300℃. Formaldehyde is mainly used to produce urea-formaldehyde resin, phenolic resin, polyoxymethylene, pentaerythritol, etc.

6. Acetyloxidation of olefin

With a catalyst, the ester formation process in which oxygen reacts with olefins or aromatic hydrocarbons react with organic acids is called acetyl oxidation. In this kind of reaction, the acetyl oxidation of ethylene and acetic acid is the most important to produce vinyl acetate. At present, the ethylene method has basically replaced the acetylene method to produce vinyl acetate. Vinyl acetate can be used to produce vinylon fibers, and polyvinyl acetate is widely used in the production of polyvinyl alcohol, water-soluble coatings and adhesives. Vinyl acetate can also be copolymerized with vinyl chloride and ethylene. The acetyl oxidation reaction of propylene and acetic acid produces propylene acetate, and the acetyl oxidation product of butadiene is mainly used to produce 1,4 - butanediol.

$$H_2C\!=\!CH_2+CH_3COOH+\frac{1}{2}O_2 \xrightarrow[160℃\sim180℃,0.8\,MPa\sim1.2\,MPa]{Pd-Au-CH_3COOK/SiO_2} CH_3COOCH\!=\!CH_2+H_2O$$

$$\Delta H\!=\!-147\ kJ/mol \qquad\qquad (7-39)$$

$$CH_3CH\!=\!CH_2+CH_3COOH+\frac{1}{2}O_2 \xrightarrow{Pd/Al_2O_3} CH_3COOC_3H_5+H_2O$$

$$\Delta H\!=\!-167\ kJ/mol \qquad\qquad (7-40)$$

$$H_2C\!=\!CH\!-\!CH\!=\!CH_2+2CH_3COOH+\frac{1}{2}O_2 \xrightarrow{Pd/C}$$

$$CH_3COO\!-\!CH_2CH\!=\!CHCH_2\!-\!OOCCH_3+H_2O \qquad (7-41)$$

7. Oxychlorination reaction

Typical oxychlorination reactions use metal chloride as catalysts. Oxychlorination of ethylene produces dichloroethane, and high temperature cracking of dichloroethane can produce important organic monomer vinyl chloride, and by-product HCl.

$$C_2H_4+2HCl+\frac{1}{2}O_2 \xrightarrow[240℃]{CuCl_2/support} ClH_2C\!-\!CH_2Cl+H_2O \qquad (7-42)$$

$$ClH_2C\!-\!CH_2Cl \xrightarrow{cracking} H_2C\!=\!CHCl+HCl \qquad (7-43)$$

Other oxychlorination technologies such as methane oxychlorination to produce methyl chloride, dichloroethane oxychlorination to produce trichloroethylene and tetrachloroethylene have been industrialized.

$$8C_2H_4Cl_2+6Cl_2+7O_2 \xrightarrow[420℃]{CuCl_2-KCl/support} 4C_2HCl_3+4C_2Cl_4+14H_2O$$

$$(7-44)$$

7.3.2　Mechanism of heterogeneous catalytic oxidation reaction

Although the process of gas-solid phase catalytic oxidation of hydrocarbons is complex, and there can be multiple independent reactions in the system, which are connected in series or in parallel, the process can still be simplified and a more realistic reaction network can be established so that its reaction mechanism can be studied. There are three common reaction mechanisms for the gas-solid phase catalytic oxidation of hydrocarbons.

1. Redox mechanism

Redox mechanism is also known as lattice oxygen mechanism. This mechanism believes that lattice oxygen participates in the reaction, and its model description is that the reactant first combines with the lattice oxygen of the catalyst to form an oxidation product, the catalyst becomes a reduced state, and then the active component in the reduced state reacts with oxygen in the gas phase, which becomes the oxidized catalyst again. This is how the redox cycle constitutes the oxidation process of organics on the catalyst.

2. Chemical adsorption oxidation mechanism

This mechanism is based on the Langmuir chemisorption model, which assumes that oxygen is chemisorbed on the active center of the catalyst surface in an adsorbed state, and then reacts with hydrocarbon molecules. This model is concise and easy to handle mathematically. Therefore, it is widely used in gas-solid phase catalytic reactions, and the reaction rate equation can be easily derived for systems with complex reaction networks.

3. Mixed reaction mechanism

This mechanism is a combination of chemisorption mechanism and redox mechanism. It is assumed that the reactants are first chemisorbed on the oxidized active centers containing lattice oxygen on the surface of the catalyst, and then react with the oxidized active centers on the surface to form products. At the same time, the oxidized active centers change to reduced state. They undergo surface oxidation reaction with oxygen in the gas phase and re-convert to the oxidized active center.

7.3.3　Heterogeneous oxidation catalysts and reactors

The active components of heterogeneous oxidation catalysts are mainly oxides of transition metals (molybdenum, bismuth, vanadium, titanium, cobalt, antimony, etc.) with variable valence, such as $MoO_3 \cdot Bi_2O_3$, $Co_2O_3 \cdot MoO_3$, $V_2O_5 \cdot TiO_2$, $V_2O_5 \cdot P_2O_5$ and $CoO \cdot WO_3$; some metals that can chemically adsorb oxygen, such as silver, have also been successfully applied in epoxidation and alcohol oxidation. In recent years, the development and application of heteropolyacids and new molecular sieve catalysts are also very active.

When a valence transition metal oxide is used as a catalyst, a single oxide for a specific oxidation reaction often exhibits poor selectivity when the activity is high, and low activity when the selectivity is guaranteed. In order to obtain high yield with proper activity and selectivity, industrial catalysts are often composed of two or more metal oxides to produce a synergistic effect. These oxides can form complex oxides, solid solutions or in the form of mixtures. At the same time, the ratio of the oxidized and reduced metal ions in the catalyst should be kept within a suitable range to keep appropriate redox ability of the catalyst. For example, the V-P-O catalyst used in the oxidation of butane to maleic anhydride has both V^{3+} and V^{5+}. A suitable catalyst should keep the average vanadium valence between 4.0 and 4.1.

Some oxidation catalysts are supported, and the commonly used supports are alumina, silica gel, activated carbon, etc. The type and performance of the carrier often have a considerable influence on the catalytic effect of the catalyst.

Commonly used hydrocarbon gas-solid phase catalytic oxidation reactors are fixed bed reactors and fluidized bed reactors. Since the exothermic heat of the oxidation reaction is large and needs to be removed in time, a heat exchange reactor is usually used.

7.4　Epoxidation of ethylene to ethylene oxide

7.4.1　Properties and uses of ethylene oxide

Ethylene oxide (EO for short) is the simplest and most important epoxide. It is a gas at room temperature and has a boiling point of 10.4℃. It can be mixed with water, alcohol, ether and most organic solvents in any proportion. The explosive limit (volume fraction) is 2.6%~100%. Ethylene oxide is toxic, and is easy to self-polymerize, especially when there are impurities such as iron, acid, alkali, aldehyde or at high temperature. During self-polymerization, a great amount of heat is released, and even explosion occurs. Therefore, the storage tank for storing ethylene oxide must be clean and kept below 0℃.

Since ethylene oxide has an oxygen-containing three-membered ring structure, it is very active and prone to ring-opening reactions. Under certain conditions, it can undergo addition reactions with water, alcohol, hydrohalic acid, ammonia and ammonia compounds. Its general formula is

$$\underset{\diagdown\,\diagup}{\overset{}{H_2C-CH_2}} + XY \longrightarrow \underset{|\quad|}{\overset{}{H_2C-CH_2}} \qquad (7-45)$$
$$\quad O \qquad\qquad\qquad OX\ Y$$

Ethylene oxide can undergo hydration reaction with water to generate ethylene glycol, which is the main method for preparing ethylene glycol. It reacts with ammonia to produce monoethanolamine, diethanolamine and triethanolamine. Ethylene oxide itself can also be polymerized by ring-opening to form polyethylene glycol.

Ethylene oxide is the third largest variety of ethylene-based products after polyethylene and styrene. The main use of ethylene oxide is the production of ethylene glycol, which accounts for about 60% of the total global ethylene oxide consumption. Ethylene glycol is one of the main raw materials for the production of polyester fibers. Ethylene oxide is also used for the production of non-ionic surfactants and ethanolamines, glycol ethers, diethylene glycol, triethylene glycol, etc.

7.4.2　The production method of ethylene oxide

The industrial production of ethylene oxide uses the direct oxidation of ethylene. Direct oxidation can be divided into air oxidation and oxygen oxidation. In 1931, Lefort of the French Catalyst Company found that ethylene could be directly oxidized into ethylene oxide with silver catalyst. After further research and development, the technology of ethylene oxide production by direct air oxidation of ethylene was formed. In 1937, UCC Company in the United States first adopted this technology, built factory and started production. In 1958, Shell Corporation built an industrial plant for the direct oxygen oxidation method for the first time. The direct oxygen oxidation method has advanced technology, and is suitable for large-scale production. This method has low production cost, and the product purity can reach 99.99%. In addition, the equipment is small in size, and the amount of emptying is not much. The amount of exhaust gas discharged by the method is only 2% of that of the air oxidation method, and the corresponding ethylene loss is also less. Besides, the oxygen oxidation method has a shorter process and less equipments than the air oxidation method, so the construction investment can be reduced by 15% to 30%. Taking the cost of air separation unit into consideration, the total investment will be higher than that of the air oxidation method, but the use of pure oxygen as the oxidant can improve the feed concentration and selectivity, so the production cost is about 90% of that of the air oxidation method. The low reaction temperature by oxygen oxidation is beneficial to prolonging the service life of the catalyst. Therefore, pure oxygen is used as the oxidant in newly built large-scale plants in recent years, which gradually replaces air and becomes the dominant method in industrial production.

7.4.3　Reaction of ethylene oxide production by direct oxidation of ethylene

The oxidation reaction of ethylene on silver catalyst includes selective oxidation and deep oxidation. In addition to the target product ethylene oxide, carbon dioxide and water as by-products and a small amount of formaldehyde and acetaldehyde are also generated.

$$C_2H_4 + \frac{1}{2}O_2 \longrightarrow C_2H_4O \quad \Delta H = -103.4 \text{ kJ/mol} \tag{7-46}$$

$$C_2H_4 + 3O_2 \longrightarrow 2CO_2 + 2H_2O(g) \quad \Delta H = -1\,324.6 \text{ kJ/mol} \tag{7-47}$$

$$C_2H_4O + \frac{5}{2}O_2 \longrightarrow 2CO_2 + 2H_2O(g) \quad \Delta H = -1\,221.2 \text{ kJ/mol} \tag{7-48}$$

Studies have shown that carbon dioxide and water are mainly produced by the direct oxidation of ethylene. The selectivity of the reaction mainly depends on the competition of parallel side reactions, and the series side reactions of ethylene oxide are secondary. The oxidation process of ethylene oxide may be that reactant is firstly isomerized to form acetaldehyde, and then oxidized to carbon dioxide and water. Acetaldehyde is easily oxidized under the reaction conditions, so only a small amount of acetaldehyde exists in the product. Since these oxidation reactions are strongly exothermic reactions, and they have large equilibrium constants, especially deep oxidation, which is more than ten times the exothermic heat of selective oxidation reactions, in order to reduce the occurrence of side reactions and improve the selectivity, the selection of catalyst is extremely important. Otherwise, the operation conditions will be deteriorated due to the side reaction, and the reaction could even become uncontrollable, resulting in a "flying temperature" accident in the reactor.

7.4.4   Mechanisms for direct epoxidation of ethylene

1. Catalyst

The industrial catalyst for the direct oxidation of ethylene to produce ethylene oxide is silver catalyst. In the production process of direct oxidation of ethylene to ethylene oxide, the consumption of raw material ethylene accounts for about 70% of the production cost of EO. Therefore, reducing the unit consumption of ethylene is the key to improving economic benefits, and the best measure is to develop high-performance catalysts. Industrially used silver catalysts are composed of active components silver, supports and co-catalysts.

(1) Support

The main function of a support is to improve the dispersion of the active component (silver) and prevent the tiny grains of silver from sintering at high temperature. The surface structure, pore structure and thermal conductivity of support have a great influence on the temperature distribution inside the catalyst particles, the size and distribution of the silver crystallites on the catalyst, the diffusion rate of the reaction raw material gas and the generated gas and so on, thus significantly affecting its activity and selectivity. The large specific surface area of the support is conducive to the dispersion of silver grains, and the initial activity of the catalyst is high. However, the catalyst with a large specific surface area has a small pore size, the reaction product ethylene oxide is difficult to diffuse out of the small pores, and the speed of separation from the surface is slow, resulting in deep oxidation of ethylene oxide and decreased selectivity.

The shape of the support also affects the catalytic performance of the catalyst. In order to improve the performance of the support, it is made into shapes with good mass transfer and heat transfer performance, such as a ring, a saddle, or a ladder. Also, the shape of the support should be selected to ensure that the gas flow is strongly agitated between the catalyst particles during the reaction process, so as to avoid short circuit and keep bed resistance small.

(2) Co-catalyst

Catalysts only containing the active component (silver) are not the best and co-catalysts must be added. Studies have shown that alkali metals, alkaline earth metals and rare earth elements have promoter effects, and two or more co-catalysts have a synergistic effect better than that of a single component. The main function of the alkali metal co-catalyst is to poison the acid center on the support surface to reduce the progress of side reactions.

In addition, activity inhibitors can also be added. The role of the inhibitor is to reversibly poison part of catalyst surface, moderately decrease activity, reduce deep oxidation, and improve selectivity. There are reports of dichloroethane, vinyl chloride, nitrogen oxides, nitroalkanes, etc. In industrial production, a small amount of dichloroethane is often added, which is thermally decomposed to generate ethylene and chlorine, and chlorine is adsorbed on the surface of silver, which affects the chemical adsorption of oxygen on the surface of the catalyst and reduces the deep oxidation of ethylene.

(3) Silver content

Increasing the silver content of the catalyst can improve the activity of the catalyst, but it will reduce the selectivity. Therefore, the mass content of silver in the current industrial catalyst is usually below 20%. However, recent research shows that as long as appropriate supports and co-catalysts are selected, catalysts with high silver content can also ensure that the selectivity is basically unchanged, while the activity is significantly improved.

(4) Catalyst preparation

There are two methods for the preparation of silver catalysts. In the early days, the bonding method or coating method was used, and now the impregnation method was used. The bonding method uses an adhesive to bond the active component, the co-catalyst and the carrier together, and the obtained catalyst silver has issues of uneven distribution, easy peeling, poor catalytic performance and short life. The dipping method uses water or an organic solvent to dissolve organic silver such as silver ammonium complexes composed of silver carboxylates and organic amines as a silver dipping solution. The dipping solution can also dissolve co-catalyst components. The support is dipped in it, the catalyst is obtained by post-treatment. The selection of silver salt, the order and method of impregnation of silver salt and co-catalyst, the selection of reducing agent and the technological conditions of preparation process, all have an impact on the size and distribution of silver particles on the surface of the support, thereby affecting performance of the catalyst.

2. Reaction mechanism

The reaction mechanism of the direct oxidation of ethylene to ethylene oxide over silver catalyst is still inconclusive. According to the adsorption of oxygen on the surface of silver catalyst, the role of ethylene and adsorbed oxygen, and the selective oxidation reaction, P. A. Kilty and others. proposed that oxygen exists in two chemical adsorption states on the surface of silver catalyst, namely atomic adsorption state and molecular adsorption state. When the adsorption site is composed of four adjacent silver atomic clusters, the oxygen will dissociate to form the atomic adsorption state (O^{2-}). The activation energy of this adsorption is low, and the adsorption speed is high at any temperature. The atomic state adsorbed oxygen is prone to deep oxidation with ethylene.

The presence of the activity inhibitor can partially cover the silver surface of the catalyst. For example, when dioxyethane is added, if a quarter of the silver surface is covered by chlorine, the adsorption site consisting of four adjacent silver atom clusters cannot be formed. Thus, the atomic adsorption of oxygen and the deep oxidation of ethylene are suppressed.

$$O_2 + 4Ag(\text{adjacent}) \longrightarrow 2O^{2-}(\text{adsorbed}) + 4Ag^+ \qquad (7-49)$$

$$12Ag^+ + 6O^{2-}(\text{adsorbed}) + C_2H_4 \longrightarrow 2CO_2 + 12Ag + 2H_2O \qquad (7-50)$$

At higher temperatures, atomic adsorption of oxygen by dissociation of oxygen can also occur on non-adjacent silver atoms, but this adsorption requires higher activation energy, so it is not easy to form.

$$O_2 + 4Ag(\text{non-adjacent}) \longrightarrow 2O^{2-}(\text{adsorbed}) + Ag^+ \qquad (7-51)$$

In the absence of adsorption sites composed of four adjacent clusters of silver atoms, molecular adsorption of oxygen, that is, non-dissociative adsorption of oxygen, can occur to form activated ionized oxygen molecules, and ethylene reacts with such molecular oxygen to form ethylene oxide, while producing an adsorbed atomic oxygen. This atomic oxygen reacts with ethylene to form carbon dioxide and water.

$$O_2 + Ag \longrightarrow Ag - O_2^-(\text{adsorbed}) \qquad (7-52)$$

$$C_2H_4 + Ag - O_2^-(\text{adsorbed}) \longrightarrow C_2H_4O + Ag - O^-(\text{adsorbed}) \qquad (7-53)$$

$$C_2H_4 + 6Ag - O^-(\text{adsorbed}) \longrightarrow 2CO_2 + 6Ag + 2H_2O \qquad (7-54)$$

$$7C_2H_4 + 6Ag - O_2^-(\text{adsorbed}) \longrightarrow 6C_2H_4O + 2CO_2 + 6Ag + 2H_2O \qquad (7-55)$$

7.4.5　Influence of reaction conditions on ethylene epoxidation

(1) Reaction temperature

During the epoxidation of ethylene, there are parallel complete oxidation side reactions, and the reaction temperature is the main factor affecting the selectivity. Although a consensus of the catalytic reaction mechanism and kinetics have not yet been reached, studies have shown that the activation energy of the epoxidation reaction is smaller than that of the complete oxidation reaction. As the reaction temperature increases, the rates of both reactions increase, but the rate of complete oxidation increases more rapidly.

(2) Space velocity

Space velocity is another factor that affects ethylene conversion and ethylene oxide selectivity, which is secondary to reaction temperature. A reduction in space velocity leads to an increase in conversion and a decrease in selectivity, but the effect is not as pronounced as temperature. The space velocity not only affects the conversion rate and selectivity, but also affects the space-time yield of the catalyst and the heat release per unit time, which should be considered comprehensively. The increase of the space velocity can increase the linear velocity of the gas flow in the reactor, reduce the thickness of the gas film, and is conducive to heat transfer. The space velocity used in industry is related to the catalyst selected, as well as the reactor and heat transfer rate, generally around $4,000 \sim 8,000 \ h^{-1}$. When the catalyst activity is high and the reaction heat can be removed in time, a high space velocity can be selected, otherwise, a low space velocity should be selected.

(3) Reaction pressure

The main and side reactions of the direct oxidation of ethylene are not thermodynamically reversible, so pressure has little effect on the balance and selectivity of the main and side reactions. However, pressurization can increase the partial pressure of ethylene and oxygen, speed up the reaction rate, improve the production capacity of the reactor, and it is also conducive to the use of pressurized absorption method to recover ethylene oxide. Therefore, pressurized oxidation method is mostly used in industry. The pressure should not be too high, otherwise the pressure resistance requirements and cost of the equipment will increase, and ethylene oxide will also produce polymer and carbon deposition on the surface of the catalyst, which will affect the life of the catalyst. The pressure generally used in industry is about 2.0 MPa.

(4) Raw material ratio and stabilizing gas

For the ethylene epoxidation process with circulation, the mixed gas entering the reactor is formed by mixing the circulating gas and the fresh feed gas, and its composition not only affects the economy of the process, but also is closely related to safe production. In the actual production process, the ratio of ethylene and oxygen must be outside the explosion limit, and at the same time, the concentration of ethylene and oxygen must be controlled within a suitable range. When it is too low, the production capacity of the catalyst is small, and when it is too high, the heat released by the reaction is large, which is easy to cause large thermal load of the reactor, resulting in flying temperature. The explosion limit (by volume) of the mixture of ethylene and air is 2.7%~36%, and the explosion limit (by volume) of ethylene with oxygen is 2.7%~80%. In actual production, the explosion limit is also changed due to the introduction of carbon dioxide from circulating gas. In order to increase the concentration of ethylene and oxygen, the explosion limit of ethylene can be changed by adding a third gas. This gas is usually called stabilizing gas. The stabilizing gas is inert and can reduce the explosion limit of the mixture and increase system safety. With high specific heat capacity, stabilizing gas can effectively remove part of the reaction heat and increase the system stability. The stabilizing gas that was widely used in the industry is nitrogen, and methane has been used as the stabilizing gas in recent years.

(5) Purity of raw material gas

Many impurities have an impact on the ethylene epoxidation process and must be strictly controlled. The main harmful substances and hazards are as follows. ① Catalyst poisoning. Sulfides, arsenides, halides and so on, can permanently poison the catalyst. Acetylene can poison the catalyst and react with silver to generate explosive acetylene silver. ② Increase the reaction heat effect. Hydrogen, acetylene, alkanes and alkenes above C_3 can produce a large amount of heat in the combustion reaction, making the process difficult to control. ③ Affect the explosion limit. Argon and hydrogen are the main impurities brought by air and oxygen. When the value is too high, the explosion limit of the mixed gas will be changed, and the maximum allowable concentration of oxygen will be reduced. ④ Declined selectivity. The brought in iron ions from the feed gas and the reactor pipeline will rearrange ethylene oxide to acetaldehyde, resulting in the production of carbon dioxide and water, which reduces the selectivity.

(6) Ethylene conversion

The control of single-pass conversion is related to the type of oxidant. When pure oxygen is used as the oxidant, the single-pass conversion is generally controlled at 12%~15%, and the selectivity can reach 83%~84%. When air is used as the oxidant, the single-pass conversion is generally controlled at 30%~35%, and the selectivity is about 70%. When the single-pass conversion is too high, due to the large exothermic heat and rapid temperature rise, deep oxidation will be accelerated, and the selectivity of ethylene oxide will be significantly reduced. In order to improve the utilization of ethylene, a recycling process is adopted in industry, that is, the unreacted ethylene after the separation of ethylene oxide is sent back to the reactor. Therefore, the conversion of the single-pass should not be too low, otherwise the energy consumption will increase due to the excessive amount of circulating gas. At the same time, about $10\% \sim 15\%$ of the circulating gas should be extracted in the production to remove harmful gases such as carbon dioxide and argon. If the single-pass conversion is too low, loss of ethylene will be increased.

Ethylene direct oxidation process to produce ethylene oxide includes reaction part and ethylene oxide recovery, refining. The process diagram of ethylene oxide production by oxygen method is shown in Figure 7 - 2.

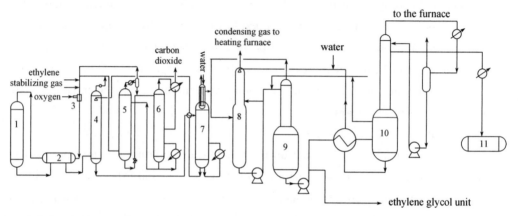

1 - Ethylene oxide reactor; 2 - Heat exchanger; 3 - Gas mixer; 4 - Ethylene oxide absorption tower;
5 - CO₂ absorption tower; 6 - CO₂ absorption liquid regeneration tower; 7 - Stripper tower;
8 - Reabsorption tower; 9 - Degassing tower; 10 - Distillation tower; 11 - Ethylene oxide storage tank.

Figure 7 - 2　Schematic diagram of the production process of ethylene oxide by oxygen method

7.5　Ammoxidation of propylene to acrylonitrile

7.5.1　Characteristics, use and process overview of acrylonitrile

Ammoxidation of hydrocarbons refers to the process of co-oxidizing hydrocarbons and ammonia with air or oxygen to form nitriles or organic nitrogen compounds. Hydrocarbons can be alkanes, naphthenes, alkenes, aromatic hydrocarbons, etc. The most industrially valuable reaction is propylene ammoxidation. In the allyl oxidation process, the ammoxidation of propylene to acrylonitrile (AN) can serve as a typical example of such a process.

Acrylonitrile is an important organic chemical product, ranking second in the propylene series, second only to polypropylene. Acrylonitrile is a colorless and sweet liquid at room temperature and pressure, with a slightly pungent odor. It has a boiling point of 77.3℃. Acrylonitrile is toxic. Its indoor allowable concentration is 0.002 mg/L, and the explosion limit (by volume) in the air is $3.05\% \sim 17.5\%$. It can form binary co-boilings with water, benzene, carbon tetrachloride, methanol, isopropanol, etc. Acrylonitrile molecule contains carbon double bond and cyano group, its chemical properties are active, which can undergo polymerization and addition, and it can react with cyano group and cyanoethyl group to prepare various synthetic fibers, synthetic rubbers, plastics and coatings.

7.5.2 The chemical reaction of propylene ammoxidation to acrylonitrile

In the process of propylene ammoxidation, in addition to the main product (acrylonitrile), there are many by-products.

Main reaction

$$C_3H_6 + NH_3 + \frac{3}{2}O_2 \longrightarrow H_2C =CH—CN(g) + 3H_2O(g) \qquad (7-56)$$

Side reactions

$$C_3H_6 + \frac{3}{2}NH_3 + \frac{3}{2}O_2 \longrightarrow \frac{3}{2}CH_3CN(g) + 3H_2O(g) \qquad (7-57)$$

$$C_3H_6 + 3NH_3 + 3O_2 \longrightarrow 3HCN + 6H_2O(g) \qquad (7-58)$$

$$C_3H_6 + O_2 \longrightarrow H_2C =CHCHO(g) + H_2O(g) \qquad (7-59)$$

$$C_3H_6 + \frac{3}{2}O_2 \longrightarrow H_2C =CHCOOH(g) + H_2O(g) \qquad (7-60)$$

$$C_3H_6 + O_2 \longrightarrow CH_3CHO(g) + HCHO(g) \qquad (7-61)$$

$$C_3H_6 + \frac{1}{2}O_2 \longrightarrow CH_3COCH_3(g) \qquad (7-62)$$

$$C_3H_6 + 3O_2 \longrightarrow 3CO + 3H_2O(g) \qquad (7-63)$$

$$C_3H_6 + \frac{9}{2}O_2 \longrightarrow 3CO_2 + 3H_2O(g) \qquad (7-64)$$

7.6 Oxidation of aromatic hydrocarbons to phthalic anhydride

7.6.1 Characteristics, use and process overview of phthalic anhydride

Phthalic anhydride has a boiling point of 284.5℃ and a freezing point (in dry air) of 131.11℃, and it is irritating. Phthalic anhydride is mainly used to produce dioctyl phthalate, dibutyl phthalate and other esters. It can also be used to produce unsaturated polyester resins, dyes, medicines and pesticides.

7.6.2 Reaction mechanism of o-xylene to phthalic anhydride

The gas-phase catalytic oxidation of o-xylene to phthalic anhydride has a complicated reaction process, including a series of parallel and series reactions, all of which are irreversible exothermic processes.

Main reaction

$$\text{o-xylene(}CH_3, CH_3\text{)} + 3O_2 \longrightarrow \text{(phthalic anhydride)} + 3H_2O + 1\ 109\ \text{kJ/mol} \qquad (7-65)$$

Side reactions

$$\text{(}CH_3, CH_3\text{)} + \frac{15}{2}O_2 \longrightarrow \text{(maleic anhydride, CHCO/CHCO)O} + 4CO_2 + 4H_2O + 3\ 176\ \text{kJ/mol} \qquad (7-66)$$

$$\text{(}CH_3, CH_3\text{)} + O_2 \longrightarrow \text{(}CHO, CH_3\text{)} + H_2O + 222\ \text{kJ/mol} \qquad (7-67)$$

$$\text{(}CH_3, CH_3\text{)} + 2O_2 \longrightarrow \text{(}CH_2/CO\text{)O} + 2H_2O + 874\ \text{kJ/mol} \qquad (7-68)$$

$$\text{(}CH_3, CH_3\text{)} + 6O_2 \longrightarrow \text{(}H_3C-C-CO, CHCO\text{)O} + 3CO_2 + 3H_2O \qquad (7-69)$$

$$\text{(}CH_3, CH_3\text{)} + 3O_2 \longrightarrow \text{(}COOH\text{)} + CO_2 + 2H_2O \qquad (7-70)$$

$$\text{(}CH_3, CH_3\text{)} + \frac{21}{2}O_2 \longrightarrow 8CO_2 + 5H_2O + 4\ 380\ \text{kJ/mol} \qquad (7-71)$$

7.7 Safety techniques for oxidation operations

7.7.1 Explosion limit

During the selective oxidation process, the gas or vapor of hydrocarbons and their derivatives forms a mixture with air or oxygen. Within a certain concentration range, the mixture will automatically and rapidly undergo a branched chain reaction due to the action of ignition sources such as open flame, high temperature or static sparks, resulting in a sharp rise in the temperature and pressure of the system in a very short period of time. The flames spread rapidly and will eventually cause explosion. This concentration range is called the explosion limit, and is generally expressed in volume fraction. The lowest concentration is the lower explosion limit, and the highest concentration is the upper explosion limit. Different systems have different explosion limits. Under normal temperature and pressure and in air, the explosion limit is $1\% \sim 6.0\%$ for o-xylene, $0.9\% \sim 7.8\%$ for naphthalene, $2.4\% \sim 11.0\%$ for propylene, $3.05\% \sim 17.5\%$ for acrylonitrile, $2.7\% \sim 36\%$ for ethylene, $2.6\% \sim 100\%$ for ethylene oxide, and $1.9\% \sim 24.0\%$ for propylene oxide.

7.7.2 Technological measures to prevent explosion

The existence of the explosion limit restricts the increase of the reaction raw material concentration, which is unfavorable to the reaction rate, selectivity, energy utilization and equipment investment. However, for the sake of safety, most industrial oxidation reactions are still operated outside the explosion limit (usually below the lower explosion limit). Since the existence of the inert gas can change the explosion limit of the system, for some systems with high explosive power, not only the operation should be outside the explosion limit, but also the inert gas should be added to stabilize the gas in industrial production. For example, when ethylene is used as raw material to produce ethylene oxide, N_2, CO_2 and methane have a stabilizing effect.

Accidents are most likely to occur when reaction materials and air (especially oxygen) are mixed, so the design of the mixer and the selection of the mixing sequence are also very important. The mixing should be as close as possible to the inlet of the reactor, and the velocity of oxygen or air at the outlet of the nozzle should be much higher than the flame spreading speed of the raw materials to facilitate rapid mixing.

7.8 Advances in catalytic oxidation technology

In recent years, the advances in homogeneous catalytic oxidation technology are mainly in the following three aspects.

(1) Development of new reactions

For example, if $PdCl_2 - CuCl - LiCl - CH_3COOLi$ is used as a catalyst, ethylene, CO and oxygen can directly undergo one-step carbonyl oxidation to synthesize acrylic acid; with $PdCl_2 - CuCl_2$ or $PdCl_2 - FeCl_3$ as a catalyst, a series of carbonyl oxidation reactions can be carried out. For example, CO, O_2 and alcohol are used as raw materials to produce oxalate and related products. Cyclohexane is oxidized in two steps to generate adipic acid, the main monomer of nylon - 66. First, cobalt naphthenate is used as a catalyst, and cyclohexane is oxidized with air to generate cyclohexanone and cyclohexane alcohol. Then using $Cu(II)/V$ (by volume) salt as catalyst and nitric acid as oxidant, adipic acid is obtained with a yield greater than 90%, and nitrides can be re-oxidized into nitric acid in the system.

(2) Improvement of catalysts

For example, inexpensive transition metals can be used to replace precious metals as catalysts; the recovery and immobilization of precious metal catalysts.

(3) Asymmetric catalytic oxidation

This reaction can be used to produce optically active substances, among which asymmetric epoxidation and double hydroxylation are mostly studied.

At present, in addition to the selective oxidation of low-carbon alkanes and ammoxidation, the most remarkable progress in heterogeneous catalytic oxidation technology is the discovery of titanium silicalite and its application in the field of synthesis.

Questions

7 - 1　Analyze the role and characteristics of the oxidation process.

7 - 2　Analyze the characteristics of catalytic auto-oxidation reactions and give examples in chemical applications.

7 - 3　Compare and analyze the characteristics of heterogeneous catalytic oxidation and homogeneous catalytic oxidation.

7 - 4　What are the mechanisms of heterogeneous catalytic oxidation reaction? What are the characteristics of each description?

7 - 5　Master the principle, catalytic system and main flow of the reaction of ethylene epoxidation to ethylene oxide.

7 - 6　What is the basis for the selection of process conditions for ethylene epoxidation?

7 - 7　What is the function of stabilizing gas?

7 - 8　Analyze the principle of azeotropic distillation and its application in the refining process of acrylonitrile.

7 - 9　Talk about the prospect of ammoxidation of propylene to acrylonitrile from the source and price of raw materials.

7 - 10　Analyze the basic principle of phthalic anhydride production, the composition of catalytic system and the trend of process technology.

7 - 11　What is the explosion limit? What are the main influencing factors?

7 - 12　Please list some commonly used explosion-proof measures in chemical production.

第8章 羰基化

8.1 概　述

烯烃与合成气(CO/H₂)或一定配比的一氧化碳及氢气可在过渡金属配位化合物的催化作用下发生加成反应,生成比原料烯烃多一个碳原子的醛。这个反应于 1938 年首先由德国鲁尔化学(Ruhrchime)公司的 O. Röelen 发现,被命名为羰基合成(oxo synthesis),也称 Röelen 反应。

$$2RCH=CH_2+2CO+2H_2 \longrightarrow RCH_2CH_2CHO+RCH(CHO)CH_3 \quad (8-1)$$

由于这一反应的主要工业用途是生产脂肪醇,习惯上又将由烯烃与合成气反应生成醛,然后再加氢(或醛先缩合再加氢)生产醇的过程称作羰基合成(oxo process)。反应式(8-1)可以看作烯烃双键两端的碳原子分别加上一个氢和一个甲酰基(—HCO),因此又称作氢甲酰化(hydroformylation)。

羰基合成的初级产品是醛,在有机化合物中,醛基是最活泼的基团之一,可进行加氢成醇、氧化成酸、氨化成胺以及歧化、缩合、缩醛化等一系列反应;加之原料烯烃的多种多样和醇、酸、胺等产物的后继加工,由此构成以羰基合成为核心,内容十分丰富的产品网络,应用领域涉及化工领域的多个方面。

随着碳一化学的发展,有一氧化碳参与的反应类型逐渐增多,通常将在过渡金属配位化合物(主要是羰基配位化合物)催化剂存在下,有机化合物分子中引入羰基(—CO—)的反应均归入羰化反应的范围,其中主要有两大类,以下分述。

8.1.1 不饱和化合物羰基化反应

1. 烯烃的氢甲酰化

制备多一个碳原子的饱和醛或醇,例如

$$H_2C=CHCH_3+H_2+CO \longrightarrow CH_3CH_2CH_2CHO \xrightarrow{H_2} CH_3CH_2CH_2CH_2OH$$
$$(8-2)$$

2. 烯烃衍生物的氢甲酰化

不饱和的醇、醛、酯、醚、缩醛、卤化物、含氮化合物等中的双键都能进行羰基合成反应,但官能团不参与反应。

烯丙基醇在一定的条件下进行羰基合成反应生成羟基醛,加氢后得到 1,4-丁二醇,后者是一种用途广泛的有机原料。

$$HO-CH_2-CH=CH_2+CO+H_2 \longrightarrow$$

$$HO-CH_2-CH_2-CH_2CHO+HO-CH_2-CH(CHO)-CH_3 \quad (8-3)$$

$$HO-CH_2-CH_2-CH_2CHO+H_2 \longrightarrow HO-CH_2-CH_2-CH_2-CH_2-OH$$

$$(8-4)$$

目前，由日本可乐丽公司开发的以铑膦配位化合物为催化剂，烯丙醇羰基合成法生产 1,4-丁二醇的技术，已被美国 ARCO 公司实现工业化。

另一个有工业意义的反应是丙烯腈的羰基合成。

$$NC-CH=CH_2+CO+H_2 \longrightarrow NC-CH_2-CH_2-CHO \quad (8-5)$$

反应产生的腈基醛经进一步加工可生产 dl-谷氨酸。

羰基合成除可采用上述不饱和化合物为原料外，一些结构特殊的不饱和化合物，甚至某些高分子化合物也能进行羰基合成反应，如萜烯类或甾族化合物的羰基合成产物可用作香料或医药中间体。不饱和树脂的羰基合成是制备特种涂料的一种方法。

3. 不饱和化合物的氢羧基化

不饱和化合物与 CO 和 H_2O 反应。例如

$$H_2C=CH_2+CO+H_2O \longrightarrow CH_3CH_2COOH \quad (8-6)$$

$$HC\equiv CH +CO+H_2O \longrightarrow H_2C=CH-COOH \quad (8-7)$$

由于反应结果是在双键两端或三键两端原子上分别加上一个氢原子和一个羧基，故称氢羧基化反应。利用此反应可制得多一个碳原子的饱和酸或不饱和酸。

以乙炔为原料可制得丙烯酸，由它生产的聚丙烯酸酯被广泛用作涂料。

4. 不饱和化合物的氢酯化反应

不饱和化合物与 CO 和 ROH 反应。例如

$$RCH=CH_2+CO+R'OH \longrightarrow RCH_2CH_2COOR' \quad (8-8)$$

$$HC\equiv CH +CO+ROH \longrightarrow H_2C=CHCOOR \quad (8-9)$$

5. 不对称合成

某些结构的烯烃进行羰基合成反应能生成含有对映异构体的醛。若使用特殊的催化剂，使生成的两种对映体含量不完全相等，理想情况下仅生成某种单一对映体，这样的反应称作不对称催化氢甲酰化反应。

单一对映体在医药、香料、农药、食品添加剂等领域有着广泛的应用前景。

8.1.2 甲醇羰基化反应

1. 甲醇羰基化合成醋酸

采用孟山都法(Monsanto acetic acid process)合成醋酸：

$$CH_3OH+CO \longrightarrow CH_3COOH \quad (8-10)$$

2. 醋酸甲酯羰基化合成醋酐

Tennessce eastman 法合成醋酐：

$$CH_3COOCH_3 + CO \longrightarrow (CH_3CO)_2O \qquad (8-11)$$

醋酸甲酯可由甲醇羰基化再酯化制得：

$$CH_3OH + CO \longrightarrow CH_3COOH \xrightarrow{CH_3OH} CH_3COOCH_3 + H_2O \qquad (8-12)$$

本法实际上是以甲醇为原料。醋酸甲酯对醋酐的选择性为 95%，醋酸甲酯和一氧化碳的转化率为 50%。

醋酸甲酯路线比乙烯酮路线在投资、材质选择上均优越。

3. 甲醇羰基化合成甲酸

$$CH_3OH + CO \longrightarrow HCOOCH_3 \qquad (8-13)$$

$$HCOOCH_3 + H_2O \longrightarrow HCOOH + CH_3OH \qquad (8-14)$$

4. 甲醇羰基化氧化合成碳酸二甲酯、草酸二甲酯或乙二醇

$$2CH_3OH + CO + \frac{1}{2}O_2 \longrightarrow CO(OCH_3)_2 + H_2O \qquad (8-15)$$

$$2CH_3OH + 2CO + \frac{1}{2}O_2 \longrightarrow (COOCH_3)_2 + H_2O \qquad (8-16)$$

$$(COOCH_3)_2 + 2H_2O \longrightarrow (COOH)_2 + 2CH_3OH \qquad (8-17)$$

$$(COOCH_3)_2 + 4H_2 \longrightarrow (CH_2OH)_2 + 2CH_3OH \qquad (8-18)$$

由上述反应可以看出，羰基化的作用物有烯、炔、酸、醇、酯和胺等，羰基化反应已成为获取有机化学品的重要手段。又由于参与羰基化反应的有 CO、H_2 和 CH_3OH 等，它们都是碳一化学工业的主要产品。因而，工业上羰基化往往是碳一化学工业部门开发下游产品的一个重要手段，并已有不少实现工业化，如由甲醇合成醋酸、由二甲胺合成二甲基甲酰胺、由丙烯合成丁醛和丁醇等，其中由甲醇经羰基合成醋酸已成功与乙醛氧化法相竞争，成为生产醋酸的重要方法。煤化工生产的大宗有机化学品能与石油化工产品竞争的不多，到目前为止仅醋酸一个。下一个能与石油化工相竞争的是由甲醇、CO 和氧通过羰基合成反应合成草酸二甲酯，进一步加氢合成的乙二醇。这一原料路线的变更对今后化学工业的发展有重要意义。

8.2 羰基化反应理论基础

在催化反应中，催化剂以配位化合物的形式与反应分子配位使其活化，反应分子在配位化合物体内进行反应形成产物，产物自配位体中解配，最后催化剂还原，这样的催化剂称为配位（络合）催化剂，这样的催化过程称为配位（络合）催化过程。羰基合成反应是典型的配位催化反应。

羰基合成的催化剂往往是"原位"形成的。所谓"原位（in situ）"是指加入反应体系中的化合物或配位化合物在反应条件下就地形成催化剂，同时产生催化作用。这种加

入反应体系的化合物或配位化合物称为催化剂前体或催化剂母体,而真正起作用的则被称为催化剂活性结构。羰基合成催化剂的典型结构是以过渡金属(M)为中心原子的羰基氢化物,它可以被某种配位体(L)所改性,一般形式可表示为 $H_xM_y(CO)_zL_n$。这类催化剂研究的主要对象是中心原子金属(M)、配位体(L),以及它们之间的相互影响和对催化过程的作用。

羰基合成催化剂性能优劣的评价标准包括很多方面,如对反应条件的要求和适用范围、催化剂的稳定性和寿命、耐毒化作用和再生可能性,以及经济方面如催化剂原料的资源和价格、加工和回收方法的难易等,当然最主要的还是催化剂的活性和选择性。羰基合成催化剂的活性,常以单位金属浓度在单位时间内催化产生的目的产物的量来表示;选择性包括化学选择性、区域选择性(醛基的位置)、对映体选择性(不对称合成)。

8.2.1 中心原子

人们最早发现的具有羰基合成催化活性的是金属钴,它可以多种形式加入反应体系,如氧化钴、氢氧化钴、有机酸钴盐或预制成羰基钴$[Co_2(CO)_8]$。大量研究证实,催化剂的活性结构是 $HCo(CO)_4$。而后又发现,铑的羰基合成催化活性是钴的 $10^2\sim10^4$ 倍。基于凡能够形成羰基氢化物的金属都可能具有羰基合成催化活性的认识,对其他金属也进行系统地研究,第Ⅷ族过渡金属的羰基合成催化活性顺序如下:

$$Rh\gg Co>Ir\quad Ru>Os>Pt>Pd>Fe>Ni$$

其他研究过的金属有 Mn、Re、Cr、Cu、Mo,甚至 Na 和 Ca,结果表明它们的羰基合成催化剂活性很低。

8.2.2 配位体

配位化合物中配位体和中心原子之间,以及诸配位体之间是相互影响的。改变配位体必然影响整个配位化合物的电子结构和空间结构,从而影响其催化活性。

经典的羰基合成催化剂是过渡金属的羰基氢化物,其中一个或几个 CO 基团可以被其他配位体所取代。

$$HM(CO)_m+L\longrightarrow HM(CO)_{m-1}L+CO \tag{8-19}$$

$$HM(CO)_{m-1}L+L\longrightarrow HM(CO)_{m-2}L_2+CO \tag{8-20}$$

$$HM(CO)_{m-2}L_2+L\longrightarrow HM(CO)_{m-3}L_3+CO \tag{8-21}$$

这种改变催化剂性能的方法称为催化剂的改性,引入的新配体叫作改性剂。显然,每引入一种新的配体,便产生一种新的催化剂,这种方式为催化剂的不断创新提供了广泛的途径。因此,改变配位体的研究构成了羰基合成催化剂研究的重要方面。

迄今为止,有相当大量的改性被研究过,其中大多是第 V 主族元素的三价化合物。这主要是由于它们可以提供孤对电子与配位化合物的中心原子配位。比较研究显示,三价膦(PR_3)的改性效果最为优越,已被工业采用。

8.2.3 相

配位催化被归于均相催化,因为金属配位化合物常以溶解在溶液中的方式参与反应。这种反应方式的缺点是催化剂与反应产物处于同一相中而难以进行分离。为了克服这一缺点,引出了关于催化剂"应用相"的研究。在羰基合成催化剂研究中,此类关于"相"的研究占有重要地位。

按照"相"的研究的特点,可大体将其分为两种类型。一类是将配位催化剂用各种物理的或化学的方法以固相形式担载在某种固相载体上,使用液体或气体原料进行多相反应,最终实现产物与催化剂分离的目的。在此类方法中,按担载催化剂的固体类型又可分为有机高分子锚定法、无机载体固定法及双重固载方法等。此类方法若在大规模羰基合成工业上使用还要解决诸如催化剂活性组分流失等一系列技术问题。另一类是设法使催化剂和反应产物处于互不相溶的两种液相之中,反应后只需进行简单的相分离,便可达到分离催化剂的目的。这种反应体系又称作两相催化体系。其中最有代表性的是水溶液膦配位体改性的水溶性铑膦催化剂,已经实现了工业化,此类方法中还有非水相的两相催化体系,如 Exxon 公司的氟化物双相体系也在开发中。

8.3 甲醇羰基化合成醋酸

醋酸是重要的有机原料,主要用于生产醋酸乙烯、醋酐、对苯二甲酸、聚乙烯醇、醋酸酯、氯乙酸、醋酸纤维素等。醋酸也用于医药、农药、染料、涂料、合成纤维、塑料和黏合剂等行业。

工业上醋酸的生产方法有乙醛氧化法、丁烷或轻油氧化法以及甲醇羰基化法。以甲醇为原料的羰基合成醋酸工艺,不但原料价廉易得,而且生成醋酸的选择性高达99%以上,基本上无副产物,同时投资省,生产费用低,相对乙醛氧化法有明显的优势。世界醋酸生产正向大型化、规模化方向发展,最大的单套醋酸生产装置已达 1 000 kt/a。

8.3.1 甲醇羰基化合成醋酸的基本原理

BASF 高压法与 Monsanto 低压法通过甲醇羰化反应合成醋酸化学原理基本相同,反应过程大同小异,都有一个催化剂循环和一个助催化剂循环,并且都采用第Ⅷ族元素为催化剂,碘为助催化剂。但因催化剂具体金属元素不同,活性、中间体组成相异,催化效果有差别,反应动力学、反应速率控制步骤也有所不同。

1. 高压法甲醇羰化反应合成醋酸的基本原理

BASF 高压法采用钴碘催化循环,过程如图 8-1 所示。

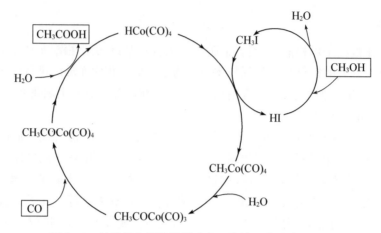

图 8-1 钴碘催化循环的甲醇高压羰基反应过程示意图

整个催化反应方程式如下。

$$Co_2(CO)_8 + H_2O + CO \longrightarrow 2HCo(CO)_4 + CO_2 \qquad (8-22)$$

$$CH_3OH + HI \Longrightarrow CH_3I + H_2O \qquad (8-23)$$

$$HCo(CO)_4 \Longrightarrow H^+ + [Co(CO)_4]^- \qquad (8-24)$$

$$[Co(CO)_4]^- + CH_3I \longrightarrow CH_3Co(CO)_4 + I^- \qquad (8-25)$$

$$CH_3Co(CO)_4 \longrightarrow CH_3CO\!-\!Co(CO)_3 \qquad (8-26)$$

$$CH_3CO\!-\!Co(CO)_3 + CO \Longrightarrow CH_3CO\!-\!Co(CO)_4 \qquad (8-27)$$

$$CH_3CO\!-\!Co(CO)_4 + HI \longrightarrow CH_3COI + H^+ + [Co(CO)_4]^- \qquad (8-28)$$

$$CH_3COI + H_2O \longrightarrow CH_3COOH + HI \qquad (8-29)$$

上述一系列复杂的反应过程要在较高的温度下才能保持合理的反应速率,而为了在较高温度下稳定 $[Co(CO)_4]^-$ 配位化合物,必须提高一氧化碳分压,这决定了高压法生产工艺的苛刻反应条件。该反应副产物有甲烷、二氧化碳、乙醇、乙醛、丙酸、醋酸酯、α-乙基丁醇等,甲醇转化物中甲烷约占 3.55%,液体副产物约 4.5%,废气约 2.0%。一氧化碳约有 10% 通过水蒸气变换反应转化为 CO_2。

2. 低压法甲醇羰基化反应合成醋酸的基本原理

Monsanto 低压法采用铑、碘催化剂体系,主要化学反应如下。

主反应

$$CH_3OH + CO \longrightarrow CH_3COOH \quad \Delta H = -138.6 \text{ kJ/mol} \qquad (8-30)$$

副反应

$$CH_3COOH + CH_3OH \Longrightarrow CH_3COOCH_3 + H_2O \qquad (8-31)$$

$$2CH_3OH \Longrightarrow CH_3OCH_3 + H_2O \qquad (8-32)$$

$$CO + H_2O \Longrightarrow CO_2 + H_2 \qquad (8-33)$$

此外尚有甲烷、丙酸（由原料甲醇中所含乙醇羰化生成）等副产物。由于式(8-31)、式(8-32)是可逆反应,在低压羰化条件下,如将生成的醋酸甲酯和二甲醚循环回反应器,都能羰化生成醋酸,故使用铑催化剂进行低压羰化,副反应很少,以甲醇为基准,生成醋酸的选择性可高达99%。副反应式(8-33)是CO的变换反应,在羰化条件下,此反应也能发生,尤其是在温度高、催化剂浓度高、甲醇浓度下降时。故以一氧化碳为基准,生成醋酸的选择性仅为90%。

8.3.2 甲醇羰基化制醋酸工艺流程

1. BASF 高压法生产工艺流程

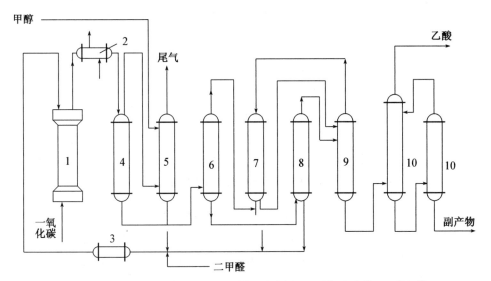

1-反应器;2-冷却器;3-预热器;4-低压分离器;5-尾气洗涤塔;6-脱气塔;

7-分离塔;8-催化剂分离器;9-共沸精馏塔;10-精馏塔。

图 8-2 甲醇高压羰基化合成醋酸工艺流程图(BASF 法)

BASF 高压法生产工艺流程如图 8-2 所示。甲醇经尾气洗涤塔后,与一氧化碳、二甲醚及新鲜补充的催化剂及循环返回的钴催化剂、碘甲烷一起连续加入高压反应器,保持反应温度为250℃,压力为70 MPa。由反应器顶部引出的粗乙酸与未反应的气体经冷却后进入低压分离器,从低压分离器出来的粗酸送至精制工段。在精制工段,粗乙酸经脱气塔脱去低沸点物质,然后在催化剂分离器中脱除碘化钴,碘化钴在乙酸水溶液中作为塔底残余物质除去。脱除催化剂后的粗乙酸在共沸蒸馏塔中脱水并精制,由塔釜得到的不含水与甲酸的乙酸再在两个精馏塔中加工成纯度为99.8%以上的纯乙酸。以甲醇计,乙酸的收率为90%;以一氧化碳计,乙酸的收率为59%。另外,副产物有3.5%的甲烷和4.5%的其他液体副产物。

高压羰基化反应器是 Hastelloy C 合金钢衬里的塔式反应器,反应器内设置循环管,由上升的气体提供能量达到搅拌混合的目的,也借以保持反应器温度的均一。

2. Monsanto 低压法生产工艺流程

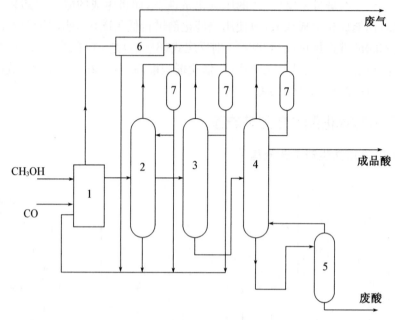

1-反应系统;2-脱轻塔;3-脱水塔;4-脱重塔;5-废酸塔;6-涤气塔;7-蒸馏冷凝液槽。

图 8-3　Monsanto 低压法羰基化合成醋酸工艺流程图

Monsanto 低压法生产流程见图 8-3。甲醇预热后与一氧化碳、返回的含催化剂母液、精制系统返回的轻馏分及含水醋酸一起加入反应器底部,在 175℃～200℃、总压为 3 MPa、一氧化碳分压为 1 MPa～1.5 MPa 下反应,反应后于上部侧线引出反应液,闪蒸压力至 200 kPa 左右,使反应产物与含催化剂的母液分离,后者返回反应器,反应器排出的气体含有一氧化碳、碘甲烷、氢、甲烷,送入涤气塔。精制系统共有四个塔,含粗醋酸、轻馏分的反应混合液以气相送入脱轻塔,在 80℃左右脱出轻馏分,塔顶气含碘甲烷、醋酸甲酯、少量甲醇,送入涤气塔。脱轻塔釜液为含水粗醋酸,送入脱水塔;塔底为无水粗醋酸送入第三个塔,即脱重馏分塔,于塔上部侧线引出成品酸,塔釜液含丙酸约 40% 及其他高级羧酸。第四个塔是废酸蒸馏塔,较小,用于回收脱重塔底部馏分中的醋酸。其塔底排出的重质废酸为产量的 0.2%,可焚烧或回收。四个塔与反应器排出的气体汇总后的组成为 CO 占 40%～80%,其余 20%～60% 为 H_2、CO_2、N_2、O_2 以及微量的醋酸、碘甲烷,它们一起在涤气塔用冷甲醇洗涤回收碘后焚烧放空。

8.3.3　甲醇低压羰基化合成醋酸的优缺点

甲醇低压羰化法制醋酸在技术经济上的优越性很大,其优点如下。

(1) 利用煤、天然气、重质油等为原料,原料路线多样化,不受原油供应和价格波动影响。

(2) 转化率和选择性高,过程能量效率高。

(3) 催化系统稳定,用量少,寿命长。

（4）反应系统和精制系统合为一体，工程和控制都很巧妙，结构紧凑。

（5）虽然醋酸和碘化物对设备腐蚀很严重，但已找到了性能优良的耐腐蚀材料——哈氏合金 C（Hastelloy Alloy C），是一种 Ni‐Mo 合金，解决了设备的材料问题。

（6）用计算机控制反应系统，使操作条件一直保持最佳状态。

（7）副产物很少，三废排放物也少，生产环境清洁。

（8）操作安全可靠。

甲醇低压碳化法的主要缺点是催化剂铑的资源有限，设备用的耐腐蚀材料昂贵。

8.4　丙烯羰基化合成丁醇、辛醇

8.4.1　烯烃氢甲酰化反应的基本原理

1. 反应过程

烯烃氢甲酰化的主反应是生成正构醛，由于原料烯烃和产物醛都具有较高的反应活性，故有连串副反应和平行副反应发生。平行副反应主要是异构醛的生成和原料烯烃的加氢，这两个反应是衡量催化剂选择性的重要指标；主要连串副反应是醛加氢生成醇和缩醛的生成。以丙烯氢甲酰化为例进行说明。

主反应

$$H_2C\!=\!CHCH_3 + CO + H_2 \longrightarrow CH_3CH_2CH_2CHO \qquad (8\text{-}34)$$

副反应

$$H_2C\!=\!CHCH_3 + CO + H_2 \longrightarrow (CH_3)_2CHCHO \qquad (8\text{-}35)$$

$$H_2C\!=\!CHCH_3 + H_2 \longrightarrow CH_3CH_2CH_3 \qquad (8\text{-}36)$$

$$CH_3CH_2CH_2CHO + H_2 \longrightarrow CH_3CH_2CH_2CH_2OH \qquad (8\text{-}37)$$

$$2CH_3CH_2CH_2CHO \longrightarrow CH_3CH_2CH_2CH(OH)CH(CHO)CH_2CH_3 \qquad (8\text{-}38)$$

$$CH_3CH_2CH_2CHO + (CH_3)_2CHCHO \longrightarrow CH_3CH(CH_3)CH(OH)CH(CHO)CH_2CH_3 \qquad (8\text{-}39)$$

当过量丁醛存在时，在反应条件下，缩丁醛又能进一步与丁醛化合，生成环状缩醛、链状三聚物，缩醛很容易脱水生成另一种副产物——烯醛。

$$CH_3CH_2CH_2CH(OH)CH(CHO)CH_2CH_3 \xrightarrow{-H_2O} CH_3CH_2CH_2CH\!=\!C(C_2H_5)CHO \qquad (8\text{-}40)$$

2. 催化剂

各种过渡金属羰基配位化合物催化剂对氢甲酰反应均有催化作用，工业上经常采用的有羰基钴和羰基铑催化剂，现分别讨论如下。

（1）羰基钴和膦羰基钴催化剂

各种形态的钴如粉状金属钴、雷尼钴、氧化钴、氢氧化钴和钴盐均可使用，其中油溶性钴盐和水溶性钴盐用得最多,如环烷酸钴、油酸钴、硬脂酸钴和醋酸钴等,这些钴盐比较容易溶于原料烯烃和溶剂中,使反应在均相系统内进行。

(2) 膦羰基铑催化剂

1952 年,席勒(Schiller)首次报道羰基氢铑 $HRh(CO)_4$ 催化剂可用于氢甲酰化反应。其主要优点是选择性好,产品主要是醛,副反应少,醛-醛缩合和醇-醛缩合等连串副反应很少发生或者根本不发生,活性比羰基氢钴催化剂高 $10^2 \sim 10^4$ 倍,正/异醛比率也高。早期使用的催化剂为 $Rh_4(CO)_{12}$,是由 Rh_2O_3 或 $RhCl_3$ 在合成气存在下于反应系统中形成。羰基铑催化剂的主要缺点是异构化活性很高,正/异醛比率只有 50/50。后来用有机膦配位基取代部分羰基如 $HRh(CO)(PPh_3)_3$(与铑肿羰基配位化合物作用相似),异构化反应可大大被抑制,正/异醛比率达到 15:1,催化剂性能稳定,能在较低 CO 压力下操作,并能耐受 150℃的高温和(1.87×10^3) Pa 的真空蒸馏,并能反复循环使用。

3. 反应热力学、动力学和机理

羰基合成是放热反应,放热量因原料结构的不同而有所不同,反应的平衡常数很大。

影响氢甲酰化反应速率的因素很多,包括反应温度、催化剂浓度、原料烯烃种类和浓度、H_2 和 CO 压力以及配位体浓度、溶剂和所含产物的浓度等。

(1) 双键位置与反应速率密切相关,直链 α-烯烃反应速率最快,当链增长时,反应速率稍有减慢。直链非 α-烯烃反应速率较慢。

(2) 烯烃含支链会降低反应速率,支链愈多,反应速率愈慢,且支链离双键愈近,反应速率减慢愈多。原因可能是支链造成了空间障碍,使烯烃不易与催化剂作用而降低了反应速率。

另外,烯烃结构也影响正/异醛的比例。一般情况下所有烯烃都能进行氢甲酰化反应。除了不因为双键迁移而异构化的烯烃(如环戊烯、环己烯)不产生异构醛外,其他烯烃都得到两个或多个异构体。在通常氢甲酰化条件下,双键位置不同,对正/异醛的比例并无显著影响。由于反应过程中,可能同时有异构化反应发生,所以无论是端烯还是内烯,几乎得到相同的产品组成。带支链的烯烃,双键碳原子因受到支链的空间阻碍,醛基主要加成到 α-原子上,如异丁烯的氢甲酰化,产物中 95％是 3-甲基丁醛。

以羰基铑和三苯基膦改性的羰基铑为催化剂测得的动力学方程,反应速率仍与催化剂中金属浓度和烯烃浓度的一次方成正比,而 $p(H_2)$ 和 $p(CO)$ 的指数则有不同;一些以机理为基础的动力学方程也有不同的方程形式。

4. 影响氢甲酰化反应的因素

(1) 温度的影响

反应温度对反应速率、产物醛的正/异比率和副产物的生成量都有影响。温度升高,反应速率加快,但正/异醛的比率随之降低,重组分和醇的生成量随之增加。

氢甲酰化反应温度不宜过高,使用羰基钴催化剂时,温度一般控制在 140℃ ~ 180℃,使用膦羰基铑催化剂时以 100℃ ~ 110℃ 较宜,并要求反应器有良好的传热条件。

（2）CO、H_2 分压和总压的影响

由烯烃氢甲酰化的动力学方程和反应机理可知，增高一氧化碳分压，会使反应速率减慢，但一氧化碳分压太低，对反应不利，因为金属羰基配位化合物催化剂在一氧化碳分压低于一定值时就会分解，析出金属，而失去催化活性，所需一氧化碳分压与金属羰基配位化合物的稳定性有关，也与反应温度和催化剂的浓度有关。

（3）溶剂影响

氢甲酰化反应常要用溶剂，溶剂的主要作用：① 溶解催化剂；② 当原料是气态烃时，使用溶剂能使反应在液相中进行，对气-液间传质有利；③ 作为稀释剂可以带走反应热。脂肪烃、环烷烃、芳烃、各种醚类、酯、酮和脂肪醇等都可做溶剂。在工业生产中为方便起见常用产品本身或其高沸点副产物做溶剂或稀释剂。

8.4.2　丙烯氢甲酰化合成丁醇、辛醇

1. 丁醇和辛醇的性质、用途及合成途径

丁醇为无色透明的油状液体，有微臭，可与水形成共沸物，沸点为 117.7℃，主要用途是作为树脂、油漆、胶黏剂和增塑剂的原料（如邻苯二甲酸二丁酯），还可用作选矿用消泡剂、洗涤剂、脱水剂和合成香料的原料。

2-乙基己醇简称辛醇，是无色透明的油状液体，有特臭，与水形成共沸物，沸点为 185℃，主要用于制备增塑剂如邻苯二甲酸二辛酯、癸二酸二辛酯、磷酸三辛酯等，也是许多合成树脂和天然树脂的溶剂，还可做油漆颜料分散剂、润滑油的添加剂、消毒剂和杀虫剂的减缓蒸发剂以及在印染等工业中做消泡剂。

丁醇、辛醇可用乙炔、乙烯或丙烯和粮食为原料进行生产。以丙烯为原料的氢甲酰化法，原料价格便宜，合成路线短，是目前生产丁醇和辛醇的主要方法。

以丙烯为原料用氢甲酰化法生产丁醇、辛醇，主要包括下列三个反应过程。

（1）在金属羰基配位化合物催化剂存在下，丙烯氢甲酰化合成丁醛

$$CH_3CH\!=\!\!CH_2+CO+H_2 \longrightarrow CH_3CH_2CH_2CHO \qquad (8-41)$$

（2）丁醛在碱催化剂存在下缩合为辛烯醛

$$2CH_3CH_2CH_2CHO \xrightarrow{OH^-} CH_3CH_2CH_2CH\!=\!\!C(C_2H_5)CHO \qquad (8-42)$$

（3）辛烯醛加氢合成 2-乙基己醇

$$CH_3CH_2CH_2CH\!=\!\!C(C_2H_5)CHO+2H_2 \xrightarrow{\text{镍催化剂}} CH_3CH_2CH_2CH_2CH(C_2H_5)CH_2OH$$
$$(8-43)$$

如用氢酰化法生产丁醇，则只需氢甲酰化和加氢两个过程就可以了。

上述三个过程的关键是丙烯氢甲酰化合成丁醛。当今的羰基合成工业技术已经相当成熟。传统的高压方法仍在一些装置中被采用，主要是以 C_6 以上烯烃为原料生产高碳数醛和醇。

2. 丙烯高压氢甲酰化合成正丁醛

采用羰基钴为催化剂的传统高压羰基合成法曾经是应用最广的方法，至 20 世纪

70 年代中后期,世界上仍有 40 多套采用这种方法的工业装置在运转。高压法被不同公司所采用,一度发展出十余种不同的工艺,但由于都是基于最初的鲁尔技术而发展的,除在催化剂回收方面有较大差别外,这些工艺差别不大。近年高压法生产的装置大部分已被改造为低压铑的方法。中国吉林化学工业公司于 20 世纪 70 年代后期曾引进 BASF 公司高压羰基合成技术建设了 50 kt/a 的丁醇、辛醇装置,目前已改造为采用低压铑生产技术。

3. 丙烯低压氢甲酰化法合成正丁醛

低压羰基合成法以铑膦配位化合物为催化剂。铑比钴原子多一个电子层,原子体积较大,因此价电子易于极化,容易形成高配位数配位化合物,也就是易于发生氧化加成反应。铑作催化剂时的氢甲酰化反应速率比钴高 $10^2 \sim 10^4$ 倍。将有机膦配体引入羰基铑,增加了催化剂的稳定性,使反应在常压下即能进行;又由于有机膦配体的空间位阻效应,在催化剂与丙烯配位时较易与端位的碳原子结合,所以反应产物中正构醛的比例也大大增加。典型的低压羰基合成法的反应压力为 1 MPa $\sim$ 3 MPa,反应温度为 90℃ $\sim$ 120℃,产物的正异比可达 10 以上。

8.5　羰基化反应技术发展趋势

低压氢甲酰化法有许多优越性,但因铑价格昂贵、催化剂制备和回收复杂等因素,目前正从开发新催化体系和改进工艺两个方面加以革新。另外,使用羰基合成方法生产的工业产品已经突破了传统概念中的羰基合成醇类,可乐丽公司的烯丙醇羰基合成制 1,4 -丁二醇以及羰基合成方法在精细化工中的一些应用实例代表着这方面的发展。

8.5.1　均相固相化催化剂的研究

为了克服铑膦催化剂制备和回收复杂的缺点,进一步减少其消耗量,简化产品分离步骤等,进行了均相催化剂固载化的研究,即把均相催化剂固定在有一定表面的固体上,使反应在固定的活性位上进行,催化剂兼有均相和多相催化的优点。

固相化方法主要有两种,一是通过各种化学键合把配位催化剂负载于高分子载体上,称为化学键合法;另一种是物理吸附法,即把催化剂吸附于硅胶、氧化铝、活性炭、分子筛等无机载体上,也可将催化剂溶于高沸点溶剂后,再浸于载体上。

8.5.2　非铑催化剂的研究

铑是稀贵资源,故其利用受到限制。国外除对铑催化剂的回收利用进一步研究外,对非铑催化剂的开发也非常重视。其中铂系催化剂是很好的研究方向。

日本研究了螯形环铂催化剂,于 0.5 MPa $\sim$ 10 MPa、70℃ $\sim$ 100℃ 条件下,反应 3 h,烯烃 100% 转化为醛。另外还报道了钌族离子型配位催化剂 $HRu_3(CO)_{15}$ 丙烯氢甲酰化,正/异醛比例达 21.2。

对钴膦催化剂也在做进一步研究,该催化剂一步可得到醇,若能找到一种合适的配位体,使之有利于醛的生成而不再进一步加氢为醇,就能与铑膦催化剂媲美了。

8.5.3 羰基合成生产1,4-丁二醇

日本可乐丽公司开发的以烯丙醇为原料经羰基合成反应和加氢反应生产1,4-丁二醇的工艺已由美国ARCO公司实现工业化。烯丙醇合成1,4-丁二醇工艺流程示意图见图8-4。

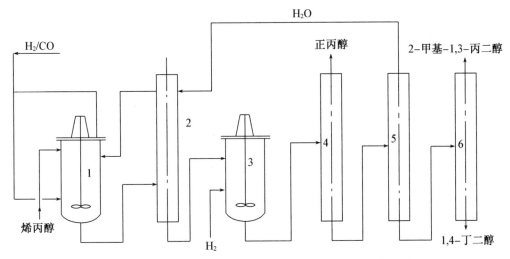

1—羰基化反应器；2—萃取塔；3—加氢反应器；4,5,6—精馏塔。

图8-4 烯丙醇合成1,4-丁二醇工艺流程示意图

该工艺羰基合成采用三苯基膦改性的羰基铑做催化剂，苯做溶剂，反应温度为60℃，反应压力为0.2 MPa～0.3 MPa，反应转化率为98%，主产物的收率约80%。反应后，产物用水进行连续萃取，油相含苯和催化剂，可循环使用；水中产物直接进行液相加氢，然后用精馏法进行产品的分离精制。

8.5.4 羰基化合成在精细化工中的应用

羰基合成在精细化工方面的应用很广。例如，在香料方面，长链醛本身即可做香料，如十一醛、2-甲基十一醛、十九醛、羰基香茅醛等。醛还原为醇或氧化为酸，醇、酸再形成酯，可衍生出许多可作为香料的产品。例如，由丁烯合成的戊醛是制备二氢茉莉酮酸(酯)的原料；双环戊二烯经羰基合成所得产品可作为定香剂及进一步合成香料的中间体。

另外，以天然的萜烯为原料进行羰基合成制备特殊结构的醛和醇，也是重要的香料或香料中间体。

思考题

8-1 羰基合成反应有哪几类？其主要特点是什么？

8-2 羰基合成反应所使用的催化剂有哪几类？各有什么特点？

8-3 比较甲醇羰基合成醋酸的低压法与高压法的工艺特点,并阐述低压法的优缺点。

8-4 采用羰基钴催化剂,为什么使用高压法?

8-5 影响烯烃氢甲酰化反应的因素有哪些?

8-6 简述羰基合成反应催化剂研究方面的最新进展。

8-7 简述配位催化反应的基本原理。

Chapter 8   Carbonylation

8.1   Overview

The addition reaction of olefin and synthesis gas (CO/H_2) or carbon monoxide and hydrogen in a certain ratio occurs under the catalysis of transition metal complex to generate aldehyde with one more carbon atom than the raw olefin. This reaction was first discovered in 1938 by O. Röelen of Ruhrchime, Germany, and was named as oxo synthesis, also known as Röelen reaction.

$$2RCH\!=\!CH_2 + 2CO + 2H_2 \longrightarrow RCH_2CH_2CHO + RCH(CHO)CH_3 \quad (8-1)$$

Since the main industrial use of this reaction is the production of fatty alcohols, it is customary that the synthesis process of alcohols from aldehydes by olefins and synthesis gas followed by hydrogenation (or aldehydes are condensed and then hydrogenated) is also called oxo process. Reaction formula (8 – 1) can be regarded as adding a hydrogen and a formyl group (—HCO) to the carbon atoms at each end of the olefin double bond, so it is also called hydroformylation.

The primary product of oxo synthesis is aldehyde. Among organic compounds, aldehyde group is one of the most active groups, which can undergo a series of reactions such as hydrogenation to alcohol, oxidation to acid, ammoniation to amine, disproportionation, condensation and acetalization. In addition, because of the variety of raw olefins and the subsequent processing of alcohols, acids, amines and other products, they form a very rich product network with oxo synthesis as the core, and the application field involves many aspects of the chemical industry.

With the development of one-carbon chemistry, the types of reactions that carbon monoxide participates in gradually increase. Usually, in the presence of transition metal coordination compound (mainly carbonyl coordination compounds) catalysts, the introduction of carbonyl groups (—CO—) into organic compound molecules is classified into the scope of carbonylation reactions, of which there are mainly two categories, and they are described below.

8.1.1 Carbonylation reaction of unsaturated compounds

1. Hydroformylation of olefins

Preparation of saturated aldehydes or alcohols with one more carbon atom, such as

$$H_2C=CHCH_3+H_2+CO \longrightarrow CH_3CH_2CH_2CHO \xrightarrow{H_2} CH_3CH_2CH_2CH_2OH$$
$$(8-2)$$

2. Hydroformylation of olefin derivatives

Double bonds in unsaturated alcohols, aldehydes, esters, ethers, acetals, halides, nitrogen-containing compounds, can undergo oxo reaction, but functional groups do not participate in the reaction.

Allyl alcohol undergoes oxo reaction under certain conditions to generate hydroxy aldehyde, which is hydrogenated to obtain 1,4 - butanediol. 1,4 - butanediol is a versatile organic raw material.

$$HO-CH_2-CH=CH_2+CO+H_2 \longrightarrow$$
$$HO-CH_2-CH_2-CH_2CHO+HO-CH_2-CH(CHO)-CH_3 \quad (8-3)$$
$$HO-CH_2-CH_2-CH_2CHO+H_2 \longrightarrow HO-CH_2-CH_2-CH_2-CH_2-OH$$
$$(8-4)$$

At present, the technology of producing 1,4 - butanediol by allyl alcohol oxo method with rhodium phosphine complex as catalyst developed by Kuraray Company has been industrialized by American ARCO Company.

Another reaction of industrial interest is the oxo synthesis of acrylonitrile.

$$NC-CH=CH_2+CO+H_2 \longrightarrow NC-CH_2-CH_2-CHO \quad (8-5)$$

The nitrile aldehyde produced by the reaction is further processed to produce dl-glutamic acid.

In addition to the above-mentioned unsaturated compounds as raw materials, some unsaturated compounds with special structures, and even some polymer compounds can also be used for oxo reaction, for example, oxo reaction products of terpenes or steroids can be used as fragrances or pharmaceutical intermediates. Oxo reaction of unsaturated resins is a method for preparing specialty coatings.

3. Hydrocarboxylation of unsaturated compounds

Unsaturated compounds react with CO and H_2O. Such as

$$H_2C=CH_2+CO+H_2O \longrightarrow CH_3CH_2COOH \quad (8-6)$$

$$HC\equiv CH +CO+H_2O \longrightarrow H_2C=CH-COOH \quad (8-7)$$

Since the result of the reaction is that a hydrogen atom and a carboxyl group are added to the atoms at each end of the double bond or triple bond, it is called a hydrocarboxylation reaction. This reaction can be used to produce saturated or unsaturated acids with one more carbon atom.

Acrylic acid can be obtained from acetylene, and the polyacrylate produced from it are widely used as coatings.

4. Hydroesterification of unsaturated compounds

Unsaturated compounds react with CO and ROH. Such as

$$RCH = CH_2 + CO + R'OH \longrightarrow RCH_2CH_2COOR' \qquad (8-8)$$

$$HC \equiv CH + CO + ROH \longrightarrow H_2C = CHCOOR \qquad (8-9)$$

5. Asymmetric synthesis

The oxo reaction of olefins with certain structures can produce aldehydes containing enantiomers. If a special catalyst is used, the two enantiomers produced are not completely equal in content, and ideally only a single enantiomer is produced. Such a reaction is called asymmetric catalytic hydroformylation.

Single enantiomer has a wide range of application prospects in the fields of medicine, spices, pesticides, food additives and so on.

8.1.2   Carbonylation of methanol

1. Methanol carbonylation to synthesize acetic acid

Synthesis of acetic acid by Monsanto acetic acid process:

$$CH_3OH + CO \longrightarrow CH_3COOH \qquad (8-10)$$

2. Synthesis of acetic anhydride by carbonylation of methyl acetate

Synthesis of acetic anhydride by Tennessce eastman method:

$$CH_3COOCH_3 + CO \longrightarrow (CH_3CO)_2O \qquad (8-11)$$

Methyl acetate can be obtained by methanol carbonylation and then esterification:

$$CH_3OH + CO \longrightarrow CH_3COOH \xrightarrow{CH_3OH} CH_3COOCH_3 + H_2O \qquad (8-12)$$

Therefore, this method is actually based on methanol as raw material. The selectivity of methyl acetate to acetic anhydride was 95%, and the conversion of methyl acetate and carbon monoxide was 50%.

The methyl acetate route is superior to the ketene route in terms of investment and material selection.

3. Methanol carbonylation to synthesize formic acid

$$CH_3OH + CO \longrightarrow HCOOCH_3 \qquad (8-13)$$

$$HCOOCH_3 + H_2O \longrightarrow HCOOH + CH_3OH \qquad (8-14)$$

4. Methanol carbonylation and oxidation to synthesize dimethyl carbonate, dimethyl oxalate or ethylene glycol

$$2CH_3OH + CO + \frac{1}{2}O_2 \longrightarrow CO(OCH_3)_2 + H_2O \qquad (8-15)$$

$$2CH_3OH + 2CO + \frac{1}{2}O_2 \longrightarrow (COOCH_3)_2 + H_2O \qquad (8-16)$$

$$(COOCH_3)_2 + 2H_2O \longrightarrow (COOH)_2 + 2CH_3OH \qquad (8-17)$$

$$(COOCH_3)_2 + 4H_2 \longrightarrow (CH_2OH)_2 + 2CH_3OH \qquad (8-18)$$

It can be seen from the above reaction that the reactants of carbonylation are alkenes, alkynes, acids, alcohols, esters and amines. Therefore, carbonylation has become an important means to obtain organic chemicals. Since CO, H_2 and CH_3OH are involved in the carbonylation reaction, and they are the main products of the one-carbon chemical industry, industrial carbonylation is often an important means for the development of downstream products in the one-carbon chemical industry sector, and many of them have been industrialized. For example, synthesis of acetic acid from methanol, synthesis of dimethylformamide from dimethylamine, synthesis of butyraldehyde and butanol from propylene, etc. Among them, the synthesis of acetic acid from methanol through carbonylation has successfully competed with acetaldehyde oxidation and has become an important method for producing acetic acid. There are few bulk organic chemicals produced by coal chemical industry that can compete with petrochemical, so far there is only acetic acid. The next one that can compete with the petrochemical industry is the synthesis of dimethyl oxalate by the reaction of methanol, CO and oxygen through carbonylation, and further hydrogenation to synthesize ethylene glycol. Both acetic acid and ethylene glycol are bulk organic chemicals, and the change of this raw material route is of great significance to the development of the chemical industry in the future.

8.2 Theoretical basis for carbonylation reactions

In the catalytic reaction, the catalyst is activated by coordination with the reaction molecule in the form of a coordination compound, the reaction molecule reacts in the coordination compound to form a product, the product is decoordinated from the ligand, and finally the catalyst is reduced, such a catalyst is called a coordination (complex) catalyst, and such a catalytic process is called a coordination (complex) catalytic process. The oxo reaction is a typical coordination catalyzed reaction.

Catalysts for oxo are often formed "in situ". The so-called "in situ" means that the compound or coordination compound added to the reaction system forms a catalyst right under the reaction conditions, and at the same time produces a catalytic effect. The compound or coordination compound added to the reaction system is called catalyst precursor, and what really works is called catalyst active structure. The typical structure of oxo catalysts is a carbonyl hydride with transition metal (M) as the central atom, which can be modified by a certain ligand (L), and the general form can be expressed as $H_x M_y (CO)_z L_n$. The main object of research on this type of catalyst is the central atom metal (M), ligand (L) and their interaction and the role of the catalytic process.

To evaluate the performance of oxo catalysts, many aspects are included, such as the requirements for reaction conditions and the scope of application, stability and lifespan of catalyst, poisoning resistance and regeneration possibility; and economic aspects such as resources and prices of catalyst raw material, the difficulty of processing and recycling, etc. Of course, the most important thing is the activity and selectivity of the catalyst. The activity of an oxo catalyst is often expressed in terms of the amount of target product catalyzed by a unit metal concentration in a unit time. The selectivity of a catalyst includes chemoselectivity, regioselectivity (position of aldehyde groups), enantioselectivity (asymmetric synthesis).

8.2.1　Central atom

Cobalt metal was first discovered to have catalytic activity for oxo synthesis. It can be added to the reaction system in various forms, such as cobalt oxide, cobalt hydroxide, organic acid cobalt salt or pre-formed cobalt carbonyl [$Co_2(CO)_8$]. A large number of studies have confirmed that the active structure of the catalyst is $HCo(CO)_4$. Later studies found that the oxo catalytic activity of rhodium is $10^2 \sim 10^4$ times than that of cobalt. Based on the recognition that any metal that can form carbonyl hydride may have catalytic activity for oxo synthesis, other metals have also been systematically studied. The oxo catalytic activity sequence of transition metals in group Ⅷ is as follows:

$$Rh \gg Co > Ir \quad Ru > Os > Pt > Pd > Fe > Ni$$

Other metals studied were Mn, Re, Cr, Cu, Mo and even Na and Ca, which showed very low oxo catalyst activity.

8.2.2　Ligands

In coordination compounds, the ligands and the central atom, as well as the ligands themselves, interact with each other. Changing the ligand will inevitably affect the electronic structure and spatial structure of the entirecoordination compound, thereby affecting its catalytic activity.

Classical oxo catalysts are transition metal carbonyl hydrides, in which one or several CO groups can be replaced by other ligands.

$$HM(CO)_m + L \longrightarrow HM(CO)_{m-1}L + CO \tag{8-19}$$

$$HM(CO)_{m-1}L + L \longrightarrow HM(CO)_{m-2}L_2 + CO \tag{8-20}$$

$$HM(CO)_{m-2}L_2 + L \longrightarrow HM(CO)_{m-3}L_3 + CO \tag{8-21}$$

Changing the properties of the catalyst in this way is called catalyst modification, and the new ligands introduced are called modifiers. Obviously, every time a new ligand is introduced, a new catalyst is produced. This approach provides a broad avenue for continuous innovation in catalysts. Therefore, the study of changing ligands constitutes an important aspect of oxo catalyst research.

To date, a considerable number of modifications have been studied, most of which are trivalent compounds of group V elements. This is mainly due to the fact that they can provide lone pair electrons to coordinate with the central atom of the coordination compound. The comparative study shows that the modification effect of trivalent phosphine (PR_3) is the most superior, which has been industrialized.

8.2.3 Phase

Coordination catalysis is classified as homogeneous catalysis because metal coordination compounds often participate in the reaction in a dissolved manner. The disadvantage of this reaction method is that the catalyst and the reaction product are in the same phase and are difficult to separate. To overcome this shortcoming, research on catalyst "applied phases" was initiated. In the research of oxo catalysts, such research on "phase" occupies an important position.

According to the characteristics of this kind of research, it can be roughly divided into two types. One is that the coordination catalyst is supported on a solid support in the form of a solid phase by various physical or chemical methods. The gaseous or liquid raw material undergoes a heterogeneous reaction, and finally achieves the purpose of separating the product from the catalyst. In this kind of method, according to the solid type of the supported catalyst, it can be divided into organic polymer anchoring method, inorganic carrier immobilization method and double immobilization method. If such methods are used in large-scale oxo synthesis industry, a series of technical problems such as loss of catalyst active components should be solved. The other is trying to make the catalyst and the reaction product in two immiscible liquid phases. After the reaction, a simple phase separation can be performed to achieve the purpose of separating the catalyst. This reaction system is also known as a two-phase catalytic system. The most representative one is the water-soluble rhodium phosphine catalyst modified with aqueous phosphine ligands, which has been industrialized. There are also non-aqueous two-phase catalytic systems such as Exxon's fluoride two-phase system in development.

8.3　Methanol carbonylation to acetic acid

Acetic acid is an important organic raw material, mainly used in the production of vinyl acetate, acetic anhydride, terephthalic acid, polyvinyl alcohol, acetate, chloroacetic acid, cellulose acetate, etc. Acetic acid is also used in industries such as pharmaceuticals, pesticides, dyes, coatings, synthetic fibers, plastics and adhesives.

The industrial production methods of acetic acid include acetaldehyde oxidation, butane or light oil oxidation and methanol carbonylation. The process of oxo-synthesizing acetic acid with methanol as raw material not only has cheap and easy-to-obtain raw materials, but also generates acetic acid with a selectivity of more than 99%, which means there's basically no by-products; it saves investment and has low production cost, as well as obvious advantages over acetaldehyde oxidation method. At present, nearly 40% of the acetic acid in the world is produced by this process, and this production process is considered to be used in new production plants. The world's acetic acid production is developing towards large-scale development, and the largest single-set acetic acid production unit has reached 1,000 kt/a.

8.3.1　Rationale of methanol carbonylation to synthesize acetic acid

The BASF high-pressure method and the Monsanto low-pressure method have basically the same chemical principle for the synthesis of acetic acid by methanol carbonylation, and the reaction processes are similar, both having a catalyst cycle and a co-catalyst cycle. They use the Group VIII elements as catalysts and iodine as a co-catalyst. However, due to different specific metal elements, activities and intermediate compositions are different, as well as catalytic effects, reaction kinetics and reaction rate control steps.

1. The basic principle of high-pressure methanol carbonylation to synthesize acetic acid

BASF high-pressure method adopts cobalt-iodine catalytic cycle process, as shown in Figure 8 – 1.

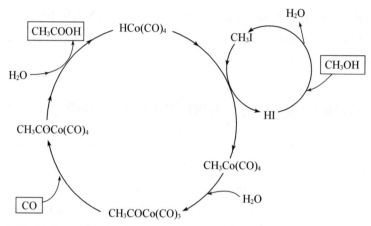

Figure 8 - 1   Schematic diagram of high-pressure carbonyl reaction of methanol catalyzed by cobalt iodine

The entire catalytic reaction equation is as follows.

$$Co_2(CO)_8 + H_2O + CO \longrightarrow 2HCo(CO)_4 + CO_2 \tag{8 - 22}$$

$$CH_3OH + HI \Longleftrightarrow CH_3I + H_2O \tag{8 - 23}$$

$$HCo(CO)_4 \Longleftrightarrow H^+ + [Co(CO)_4]^- \tag{8 - 24}$$

$$[Co(CO)_4]^- + CH_3I \longrightarrow CH_3Co(CO)_4 + I^- \tag{8 - 25}$$

$$CH_3Co(CO)_4 \longrightarrow CH_3CO—Co(CO)_3 \tag{8 - 26}$$

$$CH_3CO—Co(CO)_3 + CO \Longleftrightarrow CH_3CO—Co(CO)_4 \tag{8 - 27}$$

$$CH_3CO—Co(CO)_4 + HI \longrightarrow CH_3COI + H^+ + [Co(CO)_4]^- \tag{8 - 28}$$

$$CH_3COI + H_2O \longrightarrow CH_3COOH + HI \tag{8 - 29}$$

The above series of reaction processes require a higher temperature to maintain a reasonable reaction rate, and in order to stabilize $[Co(CO)_4]^-$ at a higher temperature, the partial pressure of carbon monoxide must be increased, which determines the harsh reaction conditions of the high-pressure production process. The by-products of the reaction are methane, carbon dioxide, ethanol, acetaldehyde, propionic acid, acetate, α - ethyl butanol, etc. In methanol conversion, methane accounts for about 3.55%, liquid by-products are about 4.5%, and waste gas is about 2.0%. About 10% of carbon monoxide is converted to CO_2 through the steam shift reaction.

2. Basic principle of low-pressure methanol carbonylation reaction to synthesize acetic acid

The Monsanto low-pressure method uses rhodium and iodine catalyst system, and the main chemical reactions are as follows.

Main reaction

$$CH_3OH + CO \longrightarrow CH_3COOH \quad \Delta H = -138.6 \text{ kJ/mol} \tag{8 - 30}$$

Side reactions

$$CH_3COOH+CH_3OH \Longrightarrow CH_3COOCH_3+H_2O \qquad (8-31)$$

$$2CH_3OH \Longrightarrow CH_3OCH_3+H_2O \qquad (8-32)$$

$$CO+H_2O \Longrightarrow CO_2+H_2 \qquad (8-33)$$

In addition, there are by-products such as methane and propionic acid (generated from the carbonylation of ethanol contained in the raw material methanol). Since formula (8 - 31) and formula (8 - 32) are reversible reactions, under low-pressure carbonylation conditions, if the generated methyl acetate and dimethyl ether are recycled back to the reactor, they can be carbonylated to form acetic acid. Therefore, the use of rhodium catalyst for low-pressure carbonylation has few side reactions. Based on methanol, the selectivity to acetic acid can be as high as 99%. The side reaction formula (8 - 33) is a CO shift reaction, which can also occur under carbonylation conditions, especially when the temperature and the catalyst concentration are high, and the methanol concentration is decreased. Therefore, based on carbon monoxide, the selectivity to acetic acid is only 90%.

8.3.2   Process flow of methanol carbonylation to acetic acid

1. BASF high-pressure production process

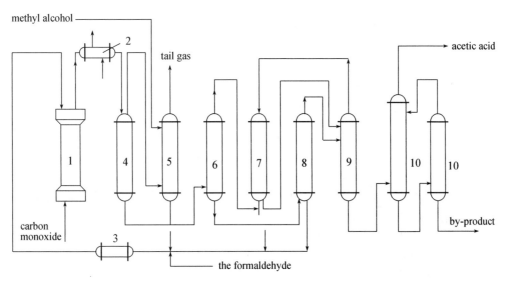

1 - Reactor; 2 - Cooler; 3 - Preheater; 4 - Low pressure separator; 5 - Tail gas scrubber; 6 - Stripper tower; 7 - Separator tower; 8 - Catalyst separator; 9 - Azeotropic distillation column; 10 - Distillation tower.

Figure 8 - 2   Process flow chart of high-pressure oxo synthesis of acetic acid from methanol (BASF method)

The production process flow of BASF high-pressure method is shown in Figure 8 - 2. After methanol passes through the tail gas scrubbing tower, it is continuously added to the high-pressure reactor together with carbon monoxide, dimethyl ether, fresh supplementary catalyst, recycled cobalt catalyst and methyl iodide, and the reaction temperature is maintained at 250℃, with pressure at 70 MPa. The crude acetic acid and unreacted gas drawn from the top of the reactor enter the low-pressure separator after cooling, and the crude acid from the low-pressure separator is sent to the refining section. In the refining section, crude acetic acid is passed through a degassing tower to remove low-boiling substances, and then cobalt iodide is removed in a catalyst separator. Cobalt iodide is removed as a column bottom residue in an aqueous acetic acid solution. The crude acetic acid after removing the catalyst is dehydrated and refined in an azeotropic distillation column, and the acetic acid free of water and formic acid obtained from the column are still processed into pure acetic acid with a purity of more than 99.8% in two rectifying columns. The yield of acetic acid was 90% based on methanol and 59% based on carbon monoxide. By-products are 3.5% methane and 4.5% other liquid by-products.

The high-pressure carbonylation reactor is a tower reactor lined with Hastelloy C alloy steel. A circulating pipe is set in the reactor, the rising gas provides energy to achieve the purpose of stirring and mixing, so as to keep the temperature of the reactor uniform.

2. Monsanto low-pressure production process

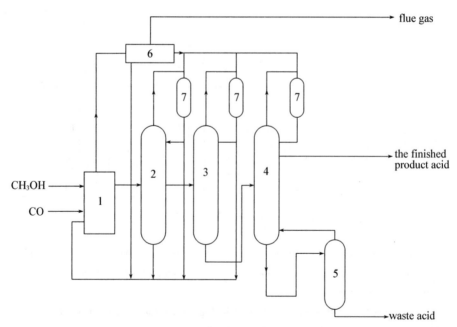

1 – Reaction system; 2 – Dehydrogenation tower; 3 – Dehydration tower; 4 – Weight removal tower; 5 – Waste acid tower; 6 – Scrubber tower; 7 – Distillation condensate tank.

Figure 8 – 3　Process flow diagram of Monsanto low-pressure oxo synthesis of acetic acid

The production flow of Monsanto low-pressure method is shown in Figure 8 - 3. After methanol is preheated, it is added to the bottom of the reactor together with carbon monoxide, the returned catalyst-containing mother liquor, the light fraction returned from the refining system and the water-containing acetic acid. The reaction occurs at 175℃ ~ 200℃, with total pressure at 3 MPa, and the carbon monoxide partial pressure at 1 MPa~1.5 MPa. After reaction, the reaction liquid is drawn out from the upper side line, and the flash pressure is about 200 kPa, so that the reaction product is separated from the mother liquor containing the catalyst. The latter is returned to the reactor, and the reactor is discharged. The gas containing carbon monoxide, methyl iodide, hydrogen and methane is sent to the scrubber. There are four towers in the refining system, and the reaction mixture containing crude acetic acid and light ends is sent to the first tower or light-removing tower in gas phase. At around 80℃, the light ends are removed, and the overhead gas containing methyl iodide, methyl acetate and a small amount of methanol enters the scrubber tower. The still liquid of the light-removing tower is water-containing crude acetic acid, which is sent to the second tower or dehydration tower. The bottom of the tower is anhydrous crude acetic acid, which is sent to the third tower, the weight removal tower, and the product acid is drawn from the side line at the upper part of the tower. The tower bottom liquid contains 40% propionic acid and other higher carboxylic acids. The fourth column is the spent acid distillation column, which is smaller and recovers the acetic acid in the bottom fraction of the weight-removing column. The heavy waste acid discharged from the bottom of the tower is 0.2% of the output, which can be incinerated or recycled. The combined composition of the gases discharged from the four towers and the reactor is CO of 40% ~ 80%, and the remaining 20% ~ 60% is H_2, CO_2, N_2, O_2 and a small amount of acetic acid and methyl iodide, together with cold methanol in the scrubbing tower. After washing and recovering iodine, the gases are incinerated and vented.

8.3.3   Advantages and disadvantages of low-pressure synthesis of acetic acid from methanol carbonylation

The low-pressure carbonylation of methanol to produce acetic acid has great technical and economic advantages, and its advantages are as follows.

(1) Using coal, natural gas, heavy oil as raw materials, the raw material routes are diversified, and are not affected by crude oil supply and price fluctuations.

(2) Conversion, selectivity and process energy efficiency are high.

(3) The catalytic system is stable, with less dosage and long service life.

(4) The reaction system and the refining system are integrated, and the engineering and control are ingenious with compact structure.

(5) Although acetic acid and iodide corrode equipment seriously, Hastelloy C, a Ni-Mo alloy and a corrosion-resistant material with excellent performance has been found that is able to solve equipment material problem.

(6) The reaction system is controlled by a computer so that the operating conditions are always kept optimal.

(7) There are very few by-products with little three wastes, and the production environment is clean.

(8) Safe and reliable operation.

The main disadvantage of low-pressure carbonylation is that the resources of catalyst rhodium are limited, and the corrosion-resistant materials used in the equipment are expensive.

8.4　Butanol and octanol synthesis by propylene carbonylation

8.4.1　Basic principles of hydroformylation of olefins

1. Reaction process

The main reaction of olefin hydroformylation is to generate normal aldehyde. Since both the raw material olefin and the product aldehyde have high reactivity, there are a series of side reactions and parallel side reactions. The parallel side reactions are mainly the formation of isomerized aldehydes and the hydrogenation of raw olefins. These two reactions are important indicators to measure the selectivity of catalysts. The main chains of side reactions are the hydrogenation of aldehydes to alcohols and the formation of acetals. Take propylene hydroformylation as an example.

Main reaction

$$H_2C\!=\!CHCH_3 + CO + H_2 \longrightarrow CH_3CH_2CH_2CHO \qquad (8-34)$$

Side reactions

$$H_2C\!=\!CHCH_2 + CO + H_2 \longrightarrow (CH_3)_2CHCHO \qquad (8-35)$$

$$H_2C\!=\!CHCH_2 + H_2 \longrightarrow CH_3CH_2CH_3 \qquad (8-36)$$

$$CH_3CH_2CH_2CHO + H_2 \longrightarrow CH_3CH_2CH_2CH_2OH \qquad (8-37)$$

$$2CH_3CH_2CH_2CHO \longrightarrow CH_3CH_2CH_2CH(OH)CH(CHO)CH_2CH_3 \qquad (8-38)$$

$$CH_3CH_2CH_2CHO + (CH_3)_2CHCHO \longrightarrow (CH_3)_2CHCH(OH)CH(CHO)CH_2CH_3$$
$$(8-39)$$

In the presence of excess butyraldehyde, under the reaction conditions, butyral can further combine with butyraldehyde to form cyclic acetal and chain trimer, and acetal is easily dehydrated to form another by-product, aldehyde.

$$CH_3CH_2CH_2CH(OH)CH(CHO)CH_2CH_3 \xrightarrow{-H_2O} CH_3CH_2CH_2CH = C(C_2H_5)CHO$$

$$(8-40)$$

2. Catalyst

Various transition metal carbonyl complex catalysts can catalyze the hydroformylation reaction. Cobalt carbonyl and rhodium carbonyl catalysts are often used in industry, which are discussed below respectively.

(1) Cobalt carbonyl and phosphine carbonyl cobalt catalysts

Various forms of cobalt such as powdered metal cobalt, Raney cobalt, cobalt oxide, cobalt hydroxide and cobalt salt can be used, and oil-soluble cobalt salt and water-soluble cobalt salt are mostly used, such as cobalt naphthenate, oleic acid cobalt, cobalt stearate and cobalt acetate. These cobalt salts are relatively easy to dissolve in the olefin and raw material solvent, so that the reaction is carried out in a homogeneous system.

(2) Phosphine carbonyl rhodium catalyst

In 1952, Schiller reported for the first time that carbonyl hydrogen rhodium catalysts $[HRh(CO)_4]$ could be used for hydroformylation. Its main advantage is good selectivity, and the products are mainly aldehydes. It has few side reactions. A series of side reactions such as aldehyde-aldehyde condensation and aldol condensation rarely occur or not at all, and the activity is also $10^2 \sim 10^4$ times higher than cobalt carbonyl hydrogen. It also has high n-aldehyde/isoaldehyde ratio. In the early days, $Rh_4(CO)_{12}$ was used as a catalyst, which was formed by Rh_2O_3 or $RhCl_3$ in the presence of syngas. The main disadvantage of rhodium carbonyl catalysts is the high isomerization activity with a n-aldehyde/isoaldehyde ratio of only 50/50. Later, some carbonyl groups such as $HRh(CO)(PPh_3)_3$ are replaced with organic phosphine ligands (similar to rhodium arsine carbonyl complexes), the isomerization reaction can be greatly suppressed, and the ratio of n-aldehyde to isoaldehyde reaches 15 : 1. The catalyst has stable performance, which can operate at low CO pressure, and can withstand high temperature $(150℃)$ and (1.87×10^3) Pa vacuum distillation, and it can be used repeatedly.

3. Reaction thermodynamics, kinetics and mechanisms

Carbonylation is an exothermic reaction, the heat varies with the structure of the raw materials, and the equilibrium constant of the reaction is large.

There are many factors that affect the reaction rate of hydroformylation, including reaction temperature, catalyst concentration, species and concentration of raw olefins, H_2 and CO pressure, ligand concentration, solvent and product concentration.

(1) The position of the double bond is closely related to the reaction rate, and the reaction rate of straight-chain $α$ - olefins is the fastest. The reaction rate slows

down slightly when the chain grows. The reaction rate of straight chain non - α - olefins is slow.

(2) The branched chain of the olefin will reduce the reaction rate, the more branched chains are, the slower the reaction rate is, and the closer the branched chain is to the double bond, the more the reaction rate is slowed down. It may be that the branched chain creates a steric barrier, making it difficult for the olefin to interact with the catalyst and reducing the reaction rate.

In addition, the olefin structure also affects the n-aldehyde/isoaldehyde ratio. In general, all olefins can undergo hydroformylation. Except for olefins that are not isomerized due to double bond migration, such as cyclopentene and cyclohexene, which do not produce isoforms, other olefins give two or more isomers. Under normal hydroformylation conditions, the position of the double bond does not have a significant effect on the n-aldehyde/isoaldehyde ratio. During the reaction process, isomerization may occur at the same time. Regardless of a terminal alkene or an internal alkene, almost the same product composition is obtained. For branched alkenes, the double bond carbon atoms are hindered by the space of the branched chain, and the aldehyde group is mainly added to the α - atom, such as the hydroformylation of isobutylene, 95% of the products is 3 - methylbutyraldehyde.

The kinetic equations measured with carbonyl rhodium and triphenylphosphine modified carbonyl rhodium as catalysts show that the reaction rate is still proportional to the first power of the metal concentration and olefin concentration in the catalyst, while the exponents of $p(H_2)$ and $p(CO)$ are different; some mechanism-based kinetic equations also have different equation forms.

4. Factors affecting hydroformylation reaction

(1) Influence of temperature

The reaction temperature has an effect on the reaction rate, the n/iso ratio of the product aldehyde, and the amount of by-products formed. When the temperature rate increases, the reaction rate increases, but the ratio of n-aldehyde to isoaldehyde decreases, and the production of heavy components and alcohols increases.

The temperature of hydroformylation reaction should not be too high. When using cobalt carbonyl catalyst, the temperature is generally controlled at 140℃ ~ 180℃. When using phosphine carbonyl rhodium catalyst, it is more suitable to be controlled at 100℃ ~ 110℃, and requires the reactor to have good heat transfer conditions.

(2) Influence of CO partial pressure, H_2 partial pressure and total pressure

It can be seen from the kinetic equation and reaction mechanism of olefin hydroformylation that increasing the partial pressure of carbon monoxide will slow down the reaction rate, but too low partial pressure of carbon monoxide is also unfavorable for the reaction, because the metal carbonyl complex catalyst will decompose, precipitate metals, and lose catalytic activity when the partial pressure is depressed at a certain value. The required partial pressure of carbon monoxide is related to the stability of the metal carbonyl complex, as well as the reaction temperature and the concentration of the catalyst.

(3) Influence of solvent

The hydroformylation reaction often uses a solvent, and the main function of the solvent is as follows. ① Dissolving catalyst. ② When the raw material is a gaseous hydrocarbon, the use of a solvent enables the reaction to proceed in the liquid phase, which is favorable for gas-liquid mass transfer. ③ As a diluent, it can take away the heat of reaction. Aliphatic hydrocarbons, naphthenic hydrocarbons, aromatic hydrocarbons, various ethers, esters, ketones and aliphatic alcohols can be used as solvents. In industrial production, the product itself or its high-boiling by-products are often used as solvents or diluents for convenience.

8.4.2　Synthesis of butanol and octanol by propylene hydroformylation

1. Properties, uses and synthetic routes of butanol and octanol

Butanol is a colorless and transparent oily liquid with a slight odor, which can form an azeotrope with water. It has a boiling point of 117.7℃, and the main use is as a raw material for resins, paints, adhesives and plasticizers (such as dibutyl phthalate). Besides it can be used as a raw material for mineral processing defoaming agents, detergents, dehydrating agents and synthetic fragrances.

2-Ethylhexanol is referred to as octanol. It is a colorless and transparent oily liquid with a special odor. It forms an azeotrope with water, and the boiling point is 185℃. It is mainly used for the preparation of plasticizers such as dioctyl phthalate, dioctyl sebacate, and trioctyl phosphate. It is also a solvent for many synthetic resins and natural resins. Besides it can be used as paint and pigment dispersants, lubricant additives, anti-evaporative agents for disinfectants and insecticides, and as defoaming agents in printing and dyeing industries.

Butanol and octanol can be produced from acetylene, ethylene or propylene and grains. The hydroformylation method using propylene as a raw material has cheap raw materials and short synthetic route, and is currently the main method for producing butanol and octanol.

Using propylene as raw material to produce butanol and octanol by hydroformylation mainly includes the following three reaction processes.

(1) Synthesis of butyraldehyde by hydroformylation of propylene in the presence of metal carbonyl complex catalysts

$$CH_3CH=CH_2+CO+H_2 \longrightarrow CH_3CH_2CH_2CHO \qquad (8-41)$$

(2) Butyraldehyde is condensed to octenal in the presence of a base catalyst

$$2CH_3CH_2CH_2CHO \xrightarrow{\quad OH^- \quad} CH_3CH_2CH_2CH = C(C_2H_5)CHO \qquad (8-42)$$

(3) Synthesis of 2 - ethylhexanol by hydrogenation of octenal

$$CH_3CH_2CH_2CH = C(C_2H_5)CHO + 2H_2 \xrightarrow{\quad Nickel\ catalyst \quad}$$
$$CH_3CH_2CH_2CH_2CH(C_2H_5)CH_2OH \qquad (8-43)$$

If the butanol is produced by the hydroacylation method, only hydroformylation and hydrogenation are required.

In the above processes, the key is to synthesize butyraldehyde by hydroformylation of propylene. Today oxo industrial technology is quite mature. Traditional high-pressure methods are still used in some devices, mainly producing high-carbon-number aldehydes and alcohols from olefins above C_6 as raw materials.

2. Synthesis of n-butyraldehyde by high-pressure hydroformylation of propylene

The traditional high-pressure oxo synthesis method using cobalt carbonyl as a catalyst was once the most widely used method. By the mid-to-late 1970s, there were still more than forty sets of industrial plants using this method in operation in the world. The high-pressure method has been adopted by different companies, and once developed into more than ten different processes, but since they are all developed based on the original Ruhr technology, these processes are not very different except for the large differences in catalyst recovery. Most of the devices produced by the high-pressure method in recent years have been converted to the low pressure rhodium method. In the late 1970s, China Jilin Chemical Industry Company introduced BASF's high-pressure oxo synthesis technology to build a 50 kt/a butanol and octanol plant, which has been transformed into a low-pressure rhodium production technology.

3. Low-pressure chloroformylation of propylene to synthesize n-butyraldehyde

The low-pressure oxo method uses a rhodium phosphine complex as a catalyst. Rhodium has one more electron layer than cobalt, and the atomic volume is larger, so the valence electrons are easily polarized, and it is easy to form a high coordination number compound, that is, an oxidative addition reaction is easy to occur. When rhodium is used as a catalyst, the hydroformylation reaction rate is $10^2 \sim 10^4$ times higher than that of cobalt. The organophosphine ligand was introduced into the rhodium carbonyl. The stability of the catalyst is increased, so that the reaction can be carried out under normal pressure. In addition, due to the steric hindrance effect of the organophosphine ligand, it is easier to combine with the terminal carbon atom when the catalyst is coordinated with propylene, so the proportion of normal aldehyde in the reaction product is also greatly increased. The reaction pressure of a typical low-pressure oxo synthesis method is 1 MPa~3 MPa, and the reaction temperature is

90℃～120℃. The n/iso ratio of the product can reach more than 10.

8.5　Development trend of carbonylation reaction technology

The low-pressure hydroformylation method has many advantages, but due to the high price of rhodium and the complexity of catalyst preparation and recovery, it is currently innovating from two aspects: developing a new catalytic system and optimizing the process. In addition, the industrial products produced by the oxo formation is more than the traditional concept of oxo alcohols. Kuraray's allyl alcohol oxo synthesis of 1,4 - butanediol and some applications examples of the oxo method in fine chemicals represent developments in this regard.

8.5.1　Research on homogeneous solidification catalyst

In order to overcome the shortcomings of complex preparation and recovery of rhodium phosphine catalysts, to further reduce its consumption, and to simplify product separation steps, research on the immobilization of homogeneous catalysts was carried out, that is, the homogeneous catalysts were fixed on a solid with a certain surface, so that the reaction is carried out on fixed active sites, and the catalyst has the advantages of both homogeneous and heterogeneous catalysis.

There are two main solidification methods. One is to load the coordination catalyst on the polymer carrier through various chemical bonds, which is called chemical bonding method. The other is the physical adsorption method, that is, the catalyst is adsorbed on inorganic carriers such as silica gel, alumina, activated carbon, molecular sieve, or the catalyst can be dissolved in a high boiling point solvent and then immersed on the carrier.

8.5.2　Research on non-rhodium catalysts

Rhodium is a rare resource, so its utilization is limited. In addition to further research on the recycling and utilization of rhodium catalysts, the development of non-rhodium catalysts is also very important. Among them, platinum-based catalysts are a good research direction.

Japan has studied the chelated cyclic platinum catalyst. Under 70℃～100℃ and 0.5 MPa～10 MPa, the reaction takes place for three hours, and all the olefins are converted into aldehydes. In addition, a ruthenium group ionic complex catalyst $HRu_3(CO)_{15}$ was reported for the hydroformylation of propylene, and the ratio of n-aldehyde to isoaldehyde reached 21.2.

Further research is also carried out on the cobalt phosphine catalyst. This catalyst can obtain alcohol in one step. If a suitable ligand can be found to facilitate the formation of aldehyde without further hydrogenation to alcohol, it can be comparable with rhodium phosphine catalyst.

8.5.3 Production of 1,4 - butanediol by carbonylation

The process of producing 1,4 - butanediol by carbonylation and hydrogenation reaction developed by Japan Kuraray Company has been industrialized by American ARCO Company. The schematic diagram of the process flow is shown in Figure 8 - 4.

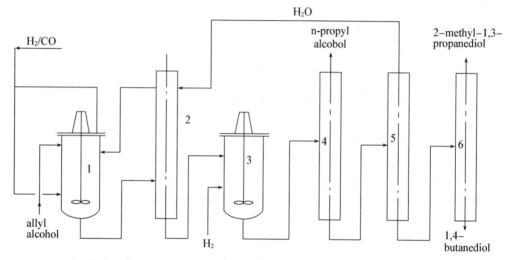

1 - Carbonylation reactor; 2 - Extraction tower; 3 - Hydrogenation reactor;
4,5,6 - Distillation column.

**Figure 8 - 4   Schematic diagram of the process flow for the synthesis of
1,4 - butanediol from allyl alcohol**

The carbonyl synthesis of this process adopts a carbonyl rhodium catalyst modified by triphenylphosphine. It uses benzene as a solvent, and the reaction temperature is 60℃ with pressure of 0.2 MPa～0.3 MPa. The reaction conversion rate is 98%, and the yield of the main product is about 80%. After the reaction, the product is continuously extracted with water. The oil phase containing benzene and the catalyst is recycled. The water product is directly subjected to liquid phase hydrogenation, and then the product is separated and refined by rectification.

8.5.4 Applications of carbonylation in fine chemicals

Oxo synthesis has a wide range of applications in fine chemicals. For example, in terms of fragrances, long-chain aldehydes can be used as fragrances themselves, such as undecanal, 2 - methylundecanal, nonadecanal, and carbonyl citronellal. Aldehydes are reduced to alcohols or oxidized to acids, then alcohols and acids form esters, they can lead to many products that can be used as fragrances. For example, valeraldehyde synthesized from butene is the raw material for the preparation of dihydrojasmonic acid (ester). The products obtained from dicyclopentadiene through oxo synthesis can be used as fixatives and intermediates for further synthesis of perfumes.

In addition, aldehydes and alcohols with special structures are prepared by oxo synthesis with natural terpenes as raw materials, which are also important fragrances or fragrance intermediates.

Questions

8 - 1   What are the types of oxo reactions? What are their main characteristics?

8 - 2   What types of catalysts are used in oxo reaction? What are the characteristics of each?

8 - 3   Compare the process characteristics of the low-pressure method and the high-pressure method for acetic acid by methanol oxo synthesis, and describe the advantages and disadvantages of the low-pressure method.

8 - 4   Why is high-pressure method used when using cobalt carbonyl catalyst?

8 - 5   What are the factors affecting olefin hydroformylation reactions?

8 - 6   Briefly describe the latest progress in the research of oxo synthesis catalysts.

8 - 7   Briefly describe the basic principles of coordination catalysis reactions.

第9章　聚合物生产方法与工艺

9.1　聚合物合成工业生产过程概述

聚合物合成工业的任务是将简单的小分子单体，经聚合反应，合成为聚合物。能够发生聚合反应的单体分子应当含有两个或两个以上能够发生聚合反应的活性原子或官能团。一个单体分子中含有的活性原子或官能团数称为官能度。根据单体分子化学结构和官能度的不同，合成的产品结构、分子量和用途也有所不同。仅含有两个聚合活性官能度（包括双键）的单体可以合成高分子量的线形结构聚合物。分子中含有两个以上聚合活性官能度的单体生产上则要求先合成分子量较低的具有反应活性的低聚物，在后期加工过程中进一步反应形成交联的聚合物，前者主要用来进一步加工为热塑性塑料和合成纤维，后者则主要用来加工为热固性塑料制品。

聚合物合成工业技术的发展和研究都是以线形高分子量合成树脂与合成橡胶的生产为主要对象的，合成聚合物的反应主要包括不饱和单烯烃和二烯烃类单体的连锁加成聚合反应和具有活性官能团单体的逐步聚合反应两大类。

当前由于线形高分子量合成树脂和合成橡胶的需求量日益扩大，生产这些合成树脂与橡胶的每套生产装置，规模小者年产数千吨，规模大者年产量则达数万吨乃至十万吨以上。这些生产装置不仅规模大，而且自动化程度高。

大型化的合成聚合物生产，主要包括以下生产过程和完成这些生产过程的相应设备与装置。

（1）原料准备与精制过程，包括单体、溶剂、去离子水等原料的贮存、洗涤、精制、干燥、调整浓度等过程和设备。

（2）催化剂或引发剂的配制过程，包括聚合用催化剂或引发剂和助剂的制造、溶解、贮存、调整浓度等过程和设备。

（3）聚合过程，包括聚合和以聚合釜为中心的有关热交换设备及反应物料输送过程与设备。

（4）分离过程，包括未反应单体和溶剂、催化剂残渣、低聚物等物质脱除的过程与设备。

（5）聚合物后处理过程，包括聚合物的输送、干燥、造粒、均匀化、贮存、包装等过程与设备。

（6）回收过程，主要是指未反应单体和溶剂的回收和精制过程及其设备。

9.2 聚合方法

9.2.1 本体聚合

本体聚合的定义为单体在有少量引发剂(甚至不加引发剂,而是在光、热、辐射能下)的作用下聚合为聚合物的过程。本体聚合依据生成的聚合物是否溶于单体分为均相本体聚合与非均相本体聚合。均相本体聚合指生成的聚合物溶于单体,如苯乙烯、甲基丙烯酸甲酯的本体聚合;非均相本体聚合指生成的聚合物不溶于单体,沉淀出来成为新的相,如氯乙烯的本体聚合。根据单体的相态,本体聚合还可分为气相本体聚合和液相本体聚合。气相本体聚合是指单体状态为气相的聚合,如乙烯本体聚合制备高压聚乙烯;液相本体聚合是指单体状态为液相的聚合,如甲基丙烯酸甲酯、苯乙烯的本体聚合等。

工业上采用本体聚合生产的聚合物品种有高压法聚乙烯、聚苯乙烯、聚甲基丙烯酸甲酯及一部分聚氯乙烯等。在甲基丙烯酸甲酯和苯乙烯的本体聚合中,单体可以完全转化,无须未反应单体回收工序,而乙烯及氯乙烯的本体聚合均需要未反应单体的回收利用工序。

本体聚合反应器大致分为形状一定的聚合反应器、釜式聚合反应器、管式聚合反应器和塔式聚合反应器等。

9.2.2 溶液聚合

溶液聚合是指单体、引发剂溶于适当溶剂中并聚合为聚合物的过程。依据生成的聚合物是否溶于溶剂分为均相溶液聚合与非均相溶液聚合。均相溶液聚合指生成的聚合物溶于溶剂,如丙烯腈的浓硫氰化钠水溶液聚合、丙烯腈的二甲基甲酰胺溶液聚合、丙烯酰胺的水溶液聚合、醋酸乙烯酯的甲醇溶液聚合、苯乙烯的甲苯溶液聚合等;非均相溶液聚合指生成的聚合物不能溶解在溶剂中,聚合至一定转化率时聚合物从溶剂中析出成为新的相,也可形象地称之为沉淀聚合或淤浆聚合,如丙烯腈的水溶液聚合、丙烯酰胺的丙酮溶液聚合、苯乙烯和马来酸酐的甲苯溶液共聚合等。

工业上,溶液聚合产品如涂料、胶黏剂、浸渍液、合成纤维纺丝液等,可在使用场合直接使用。溶液聚合的工艺流程如图9-1所示。

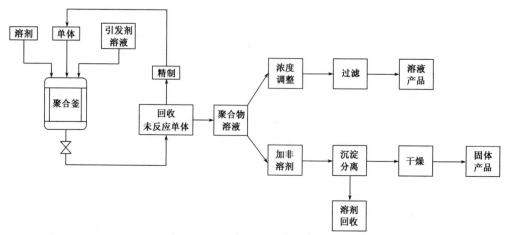

图 9-1　溶液聚合的工艺流程

9.2.3　悬浮聚合

　　悬浮聚合是指溶有引发剂的单体,借助悬浮剂的悬浮作用和机械搅拌,使单体分散成小液滴的形式在介质中的聚合过程。一个单体小液滴相当于一个本体聚合单元,因此悬浮聚合也称小本体聚合。一般悬浮聚合体系以大量水为介质,因此不适合阴离子聚合、阳离子聚合、配位聚合机理的高分子合成反应,因为这些反应的引发剂遇水会剧烈分解。

　　依据聚合物是否溶于单体,悬浮聚合分为均相悬浮聚合和非均相悬浮聚合。均相悬浮聚合是指聚合物溶于单体的聚合,产物呈透明小珠,也称珠状聚合,如苯乙烯和甲基丙烯酸甲酯的悬浮聚合。非均相悬浮聚合是指聚合物不溶于单体的聚合,以不透明小颗粒沉淀出来,呈粉状,也称沉淀聚合或粉状聚合,如氯乙烯、偏二氯乙烯、三氟氯乙烯、四氟乙烯等的悬浮聚合。

　　悬浮聚合工艺过程简单,聚合热易于排除,操作控制方便,聚合物易于分离、洗涤、干燥,产品也较纯净,且可直接用于成型加工,特别适于大规模的工业生产。悬浮聚合工艺流程如图 9-2 所示。

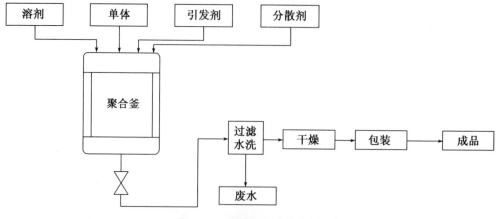

图 9-2　悬浮聚合工艺流程

9.2.4 乳液聚合

乳液聚合是指在单体、反应介质、乳化剂形成的乳状液中进行的自由基聚合反应过程。

经典乳液聚合体系主要由单体、水、乳化剂、引发剂和其他助剂所组成。自由基乳液聚合时，液态的乙烯基单体或二烯烃单体在乳化剂存在下分散于水中成为乳状液，此时是液-液乳化体系，然后在引发剂分解产生的自由基作用下，液态单体逐渐发生聚合反应，最后生成固态的聚合物分散在水中的乳状液，此时转变为固-液乳化体系。这种固体微粒的粒径一般在 1 μm 以下，静置时不会沉降析出。

9.2.5 熔融缩聚

熔融缩聚是指单体和缩聚物均处于熔融状态时进行的缩聚反应过程。熔融聚合体系中仅有单体、产物及少量催化剂，就这方面而言，其与本体聚合有着相似之处，但是两者适用的聚合反应机理不同。熔融缩聚的特点是聚合热不大，聚合过程的热效应没有本体聚合显著，因此聚合温度的控制相对容易；聚合体系简单，产物纯净，提高单体转化率时可以免去后续分离工序；聚合设备的利用率高、产能高，缩聚物可连续直接纺丝，生产成本低。但是，熔融缩聚需要很高的聚合温度，一般在 200℃～300℃，比生成的聚合物的熔点高 10℃～20℃。

熔融缩聚方法不适合高熔点的缩聚物、易挥发单体，以及热稳定性不良的单体和缩聚物；制备高分子量的缩聚物需要严格的等当量单体配比，计量操作难度大。反应物料黏度高，反应后期生成的小分子不容易脱除；局部过热易导致物料受热不匀、甚至焦化。长时间的高温缩聚过程易发生副反应使分子链结构和聚合物组成复杂化，长时间高温缩聚物易氧化降解、变色，为避免高温时缩聚产物的氧化降解，常需在惰性气体中进行。

9.2.6 溶液缩聚

溶液缩聚是指缩聚单体溶解在适当溶剂中进行的缩聚反应过程。聚合体系中有单体、溶剂、产物及少量催化剂，就这方面而言与溶液聚合有着相似之处，但是两者适用的聚合反应机理不同。溶液缩聚包括均相溶液缩聚和非均相溶液缩聚两种。均相溶液缩聚单体及生成的聚合物均能溶解在溶剂中，聚合过程体系为均匀一相，而非均相溶液缩聚单体能溶解在溶剂中，生成的聚合物不能溶解，聚合至一定程度时，聚合物从体系析出，形成聚合物相。

工业上一些缩聚物的生产过程常常采用前期溶液缩聚、后期熔融缩聚的方法，如尼龙-66 的合成，先是己二酸己二胺盐的水溶液缩聚，后是溶剂水不断排除，通过熔融缩聚得到产物。溶液缩聚在普通的聚合釜中进行，物料黏度不大，采用框式或釜式搅拌器即可。

9.2.7 界面缩聚

界面缩聚是指将两种单体分别溶解在两种互不相溶的溶剂中，在两相界面处进行的缩聚反应。界面缩聚体系中虽然也有溶剂,但是与溶液聚合及溶液缩聚相比,聚合场所不同。溶液聚合及溶液缩聚的聚合场所是在整个溶剂中,而界面缩聚的聚合场所在两种溶剂的界面区域,并且生成的聚合物必须及时地排除才能使缩聚反应进一步进行下去。

由于界面缩聚需要活性高的单体,因此界面缩聚只适用于少数缩聚物的合成,工业生产实例较少,目前工业上较为成熟的界面缩聚合成工艺是聚碳酸酯的合成。

9.2.8　固相缩聚

固相缩聚(SSP)是指单体及聚合物处于固相状态下进行的缩聚反应。固相缩聚的温度一般在聚合物的玻璃化温度以上、熔点以下。此阶段聚合物的大分子链段能自由活动,活性端基能进行有效碰撞发生化学反应。固相缩聚与熔融缩聚、本体聚合的相同点是没有溶剂或反应介质的参与,但是熔融缩聚是在单体及聚合物熔点之上反应的,本体聚合没有明确的反应温度范围。固相缩聚工艺的优点是反应温度较低,温度低于熔融缩聚温度,反应条件相对熔融缩聚而言较温和。固相缩聚工艺的缺点是反应原料需要充分混合,固体粒子粒径要达到一定细度;反应速率低;生成的小分子副产物不易脱除。经固相缩聚获得的聚合物可以是单晶或多晶聚集态。

9.2.9　逐步加成聚合

单体官能团之间通过相互加成而逐步形成高聚物的过程称为逐步加成聚合(step-growth-addition-polymerization),相应的产物为逐步加成聚合物。逐步加成聚合反应具有的特征是不生成小分子副产物,产物分子量随聚合时间逐步增加,聚合物结构类似缩聚物。

Chapter 9　Polymer production method and process

9.1　Overview of production processes in the polymer synthesis industry

The task of the polymer synthesis industry is to synthesize simple small molecular monomers into polymers through polymerization reactions. A monomer molecule capable of polymerization shall contain two or more active atoms or functional groups capable of polymerization. The number of active atoms or functional groups in a monomer molecule is called the functional degree. According to the different chemical structure and functional degree of monomer, the structure, molecular weight and the use of synthetic products are also different. Linear high molecular weight polymers can be synthesized from monomers containing only two polymeric active functions (including double bonds). The production of monomers with more than two polymeric active functions in the molecule requires the synthesis of reactive oligomer with lower molecular weight and further reaction to form cross-linked polymers in the post-processing process. The former is mainly used for further processing into thermoplastics and synthetic fibers, while the latter is mainly used for processing into thermosetting plastic products.

The development and research of industrial technology of polymer synthesis are mainly focused on the production of linear high molecular weight synthetic resins and synthetic rubber. The reactions of synthetic polymers mainly include the linkage addition polymerization of unsaturated monomers and diolefin monomers and the stepwise polymerization of monomers with active functional groups.

At present, due to the increasing demand for linear high molecular weight synthetic resin and synthetic rubber, in each set of production equipment for the production of these synthetic resin and rubber varieties, the small annual output of thousands of tons, the large annual output of tens of thousands of tons and even more than 100,000 tons. These production units are not only large in scale, but also highly automated.

Large-scale synthetic polymer production mainly includes the following production processes and the corresponding equipment and devices.

（1）Raw material preparation and refining process, including the storage, washing, refining, drying, concentration adjustment process of monomer, solvent, deionized water and other raw materials and equipment.

（2）Preparation process of catalyst or initiator, including manufacture, dissolution, storage, concentration adjustment process of polymerization catalyst or initiator and auxiliary and the corresponding equipment.

（3）Polymerization process, including polymerization and the related heat exchange equipment centered around the polymerization kettle, as well as transport process and equipment of reaction material .

（4）Separation process, including the removal process of unreacted monomer and solvent, catalyst residue, oligomer and other substances and equipment.

（5）Polymer post-treatment process, including polymer transport, drying, granulation, homogenization, storage, packaging and other processes and equipment.

（6）Recovery process, mainly refers to unreacted monomer and solvent recovery and refining process and equipment.

9.2 Polymerization process

9.2.1 Bulk polymerization

Bulk polymerization is defined as the polymerization of a monomer into a polymer in the presence of a small amount of initiator （or even without initiator, but with light, heat or radiant energy）. According to whether the resulting polymer is dissolved in monomer, it can be divided into homogeneous bulk polymerization and heterogeneous bulk polymerization. Homogeneous bulk polymerization refers to the bulk polymerization of the resulting polymer dissolved in monomers, such as the bulk polymerization of styrene and methyl methacrylate. Heterogeneous bulk polymerization means that the resulting polymer is not dissolved in the monomer and precipitates out to become a new phase, such as bulk polymerization of vinyl chloride. According to the phase state of monomer, bulk polymerization can also be divided into gas phase bulk polymerization and liquid phase bulk polymerization. Gas phase bulk polymerization refers to the polymerization of the monomer in the gas phase, such as ethylene bulk polymerization to produce high pressure polyethylene. Liquid phase bulk polymerization refers to the polymerization of liquid monomer state, such as bulk polymerization of methyl methacrylate and styrene.

Industrial polymer varieties produced by bulk polymerization are high-pressure polyethylene, polystyrene, polymethyl methacrylate and a part of polyvinyl chloride. For the bulk polymerization of methyl methacrylate and styrene, the monomer can be completely transformed without the recovery process of unreacted monomer, while the bulk polymerization of ethylene and vinyl chloride has the recovery process of unreacted monomer.

Bulk polymerization reactors are generally divided into polymerization reactors with a certain shape model, tank type polymerization reactor, tube type polymerization reactor and tower type polymerization reactor.

9.2.2 Solution polymerization

Solution polymerization refers to the polymerization process of monomer and initiator into polymer by dissolving in appropriate solvent. According to whether the polymer is dissolved in solvent, it can be divided into homogeneous solution polymerization and heterogeneous solution polymerization. Homogeneous solution polymerization refers to the generated polymer dissolved in solvent, such as concentrated sodium thiocyanide aqueous polymerization of acrylonitrile, dimethyl formamide solution polymerization of acrylonitrile, aqueous polymerization of acrylamide, methanol solution polymerization of vinyl acetate, toluene solution polymerization of styrene, etc. Heterogeneous bulk polymerization generated cannot be dissolved in a solvent, to a certain conversion, the polymer precipitated from the solvent into a new phase, which is vividly called precipitation polymerization or slurry polymerization, such as aqueous solution polymerization of acrylonitrile, acrylamide acetone solution polymerization, toluene solution copolymerization of styrene and maleic anhydride, etc.

In industry, solution·polymerization products such as coatings, adhesives, impregnation solution, synthetic fiber spinning solution and so on, can be used directly in the application. The solution polymerization process is shown in Figure 9 − 1.

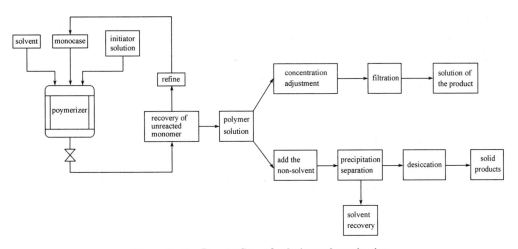

Figure 9 − 1   Process flow of solution polymerization

9.2.3 Suspension polymerization

Suspension polymerization refers to the polymerization process in which the monomer is dissolved with initiator, and the monomer is dispersed into small droplets in medium by suspension of suspension agent and mechanical stirring. A monomer droplet is equivalent to a bulk polymerization unit, so suspension polymerization is also called small bulk polymerization. General suspension polymerization system takes a large amount of water as the medium, so it is not suitable for anionic polymerization, cationic polymerization, coordination polymerization mechanism of polymer synthesis reaction, because the initiator of these reactions will be violently decomposed in water.

According to whether the monomer dissolves the polymer, it can be divided into homogeneous suspension polymerization and heterogeneous suspension polymerization. Homogeneous suspension polymerization refers to the polymerization that the polymer dissolved in monomer, also known as bead polymerization, because the products are transparent beads, such as styrene and methyl methacrylate suspension polymerization. Heterogeneous suspension polymerization refers to the polymerization that the polymer is insoluble in monomer, with opaque small particles precipitated out, powdery, so it is also known as precipitation polymerization or powdery polymerization, such as vinyl chloride, vinylidene chloride, chlorotrifluoroethylene, tetrafluoroethylene and other suspension polymerization.

The suspension polymerization process is simple, the polymerization heat is easy to remove, the operation and control is convenient, the polymer is easy to separate, wash and dry, the product is pure and can be directly used for molding processing, so it is especially suitable for large-scale industrial production. Figure 9 - 2 shows the suspension polymerization process.

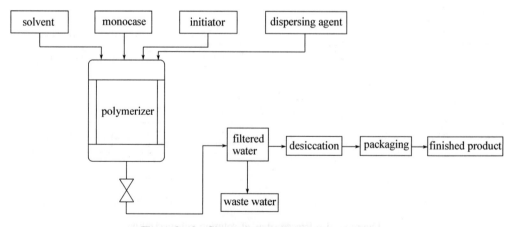

Figure 9 - 2 Suspension polymerization process

9.2.4 Emulsion polymerization

Emulsion polymerization refers to the free radical polymerization process in the emulsion formed by monomer, reaction medium and emulsifier.

The classical emulsion polymerization system consists of monomer, water, emulsifier, initiator and other additives. In free radical emulsion polymerization, the liquid vinyl monomer or alkadiene monomer dispersed in water in the presence of emulsifier, at this point is liquid-liquid emulsification system, and then under the effects of the free radical initiator's decomposition, liquid monomers polymerize gradually, finally form the solid polymer dispersed in water, which is solid-liquid emulsification system at this time. The particle size of this solid particle is generally below 1 μm, and it will not precipitate when standing.

9.2.5 Melting polycondensation

Melting polycondensation refers to the polycondensation process in which both monomer and polycondensate are in the molten state. The polymerization system contains only monomer, product and a small amount of catalyst, which is similar to bulk polymerization in this respect, but the applicable polymerization mechanism is different. The characteristic of melting polycondensation is that the polymerization heat is not large, and the thermal effect of polymerization process is not as significant as that of bulk polymerization, so the control of temperature is relatively easy. The polymerization system is simple, the product is pure, and the subsequent separation process can be avoided when the monomer conversion is increased. The polymerization equipment has high utilization rate, high productivity, and the polycondensation polymer can be spun continuously and directly, and the production cost is low. However, melting polycondensation requires a very high polymerization temperature, generally between 200℃ and 300℃, which is 10℃ to 20℃ higher than the melting point of the resulting polymer.

Melting polycondensation method is not suitable for high melting point polycondensate, not suitable for volatile monomers, not suitable for monomers and polycondensation with poor thermal stability. The preparation of high molecular weight polycondensate requires strict ratio of equivalent monomers, which is difficult to measure. The viscosity of the reaction material is high, and the small molecules generated in the later stage of the reaction are not easy to remove. Local overheating leads to uneven heating of materials and even coking. Long-term high temperature polycondensation process is prone to side reactions, which complicate the molecular chain structure and polymer composition. Long-term high temperature polycondensation is prone to oxidative degradation and discoloration. In order to avoid oxidative degradation of polycondensation products at high temperature, it is often necessary to carry out in inert gas.

9.2.6　Solution polycondensation

Solution polycondensation refers to the polycondensation process in which the monomer is dissolved in an appropriate solvent. The polymerization system consists of monomer, solvent, product and a small amount of catalyst, which is similar to solution polymerization in this respect, but the applicable polymerization mechanism is different. Solution polycondensation includes homogeneous solution polycondensation and heterogeneous solution polycondensation. Both the monomer of homogeneous solution polycondensation and the generated polymer can be dissolved in the solvent, and the polymerization process system is a homogeneous phase, while the monomer of heterogeneous solution polycondensation can be dissolved in the solvent, but the generated polymer cannot be dissolved. When the polymerization reaches a certain reaction degree, the polymer is precipitated from the system to form the polymer phase.

In industry, the production process of some polycondensation often adopts the method of early solution polycondensation and later melting polycondensation, such as the synthesis of nylon - 66, first is the aqueous solution polycondensation of hexandiamine adipate salt, and finally the solvent water is continuously excluded, and the product is obtained by melting polycondensation. The solution polycondensation is carried out in the ordinary polymerization kettle, the material viscosity is not big, and the frame or kettle type agitator can be used.

9.2.7　Interfacial polycondensation

Interfacial polycondensation refers to the polycondensation reaction carried out at the interface of two phases by dissolving two monomers in two insoluble solvents. Although there are solvents in the interfacial polycondensation system, the polymerization sites are different compared with solution polymerization and solution polycondensation. The polymerization site of solution polymerization and solution polycondensation is the whole solvent, while the polymerization site of interface polycondensation is the interface region of the two solvents, and the generated polymer must be removed in time to make the polycondensation further.

Because the polycondensation reaction can only occur when there is active monomer, so the polycondensation is only suitable for the synthesis of a few polycondensates, and there are few industrial production examples. At present, the most mature industrial polycondensation synthesis process is the synthesis of polycarbonate.

9.2.8　Solid phase polycondensation

Solid phase polycondensation (SSP) refers to the polycondensation reaction of monomers and polymers in the solid-phase state. The temperature of solid phase polycondensation is generally above the glass transition temperature and below the melting point of the polymer. At this stage, the macromolecular chain segments of the polymer can move freely, and the active end groups can carry out effective collisions and chemical reactions. The similarity between solid phase polycondensation, melting polycondensation and bulk polymerization is no solvent or reaction medium involved, but the temperature of melting polycondensation is above the monomer and polymer melting point, bulk polymerization does not have a clear reaction temperature range. The advantages of the solid phase polycondensation process are that the reaction temperature is lower than the melting polycondensation, and the reaction conditions are relatively mild. The disadvantages of solid phase polycondensation process are that the reaction material needs to be fully mixed and the particle size of solid particles needs to reach a certain fineness; low reaction rate; the resulting small molecule by-products are not easy to remove. The polymer obtained by solid phase polycondensation can be single crystal or polycrystalline aggregate state.

9.2.9　Step growth addition polymerization

The process of polymerization by mutual addition of monomer functional groups is referred to as step growth addition polymerization. The corresponding products are stepwise addition polymers. The characteristics of stepwise addition polymerization are that no small molecular by-products are formed, the molecular weight of the products gradually increases with polymerization time, and the structure of the polymer is similar to a polycondensation polymer.

第10章 绿色化工概论

10.1 绿色化工概念

"绿色"化工是指在化工生产中要实现生态"绿色"化,采用化工产品为相关行业服务时,也要追求使相关行业的生产实现生态"绿色"化,也就是要模拟动植物、微生物生态系统的功能,建立相当于"生态者、消费者和还原者"的化工生态链,以低消耗(物耗和水、电、汽、冷等能耗及工耗)、无污染(至少低污染)、资源再生、废物综合利用、分离降解等方式实现无毒化工产品生产的"生态"循环和"环境友好"及清洁和安全生产的"绿色"结果。

绿色化学与化工是21世纪化学工业可持续发展的科学基础,其目的是将现有化工生产的技术路线从"先污染,后治理"改变为"从源头上根治污染"。我国"十一五"中长期科技规划中就已提出建设资源节约型、环境友好型社会。资源节约型、环境友好型社会具有丰富的内涵,包括有利于环境的生产和消费方式,无污染或低污染的技术和产品,少污染与低损耗的产业结构,符合生态条件的生产力布局,倡导人人关爱环境的社会风尚和文化氛围。按照我国的发展目标,中国化学科学与工程的发展也必须走绿色化道路,实现由分子水平去研究、设计、创造新的有用物质,直至完成其工业制造与转化过程的全程目标,最终实现资源的生态化利用,建立生态工业园区,实现循环经济,促进并保证经济发展与资源、能和环境相协调。

10.2 绿色化工特点

化学可以粗略地看作研究一种物质向另一种物质转化的科学。传统的化学虽然可以得到人类需要的新物质,但在许多场合中却未能有效地利用资源,并产生大量排放物造成严重的环境污染。

绿色化学早期称为环境无害化学、环境友好化学、洁净化学。目前又称绿色化学为可持续发展化学或绿色可持续发展化学,以强调绿色化学化工与可持续发展的关系。从可持续发展科学的观点出发,绿色化学化工不仅要考虑是否产生对人类健康和生态环境有害的污染物,还要考虑原料资源是否有效利用,是否可以再生,是否可以促进经济、社会的可持续发展等,简单地讲即是否符合可持续发展的原理和规律。目前这种观点已为大多数人所接受,是绿色化学化工的真正内涵。

经验表明,环境的污染可能较快地形成,但要消除其危害则需较长时间,况且有的危害是潜在的,要在几年甚至几十年后才能显现出来。因此,实现化工生产与生态环境协调发展的绿色化学化工是化学工业今后的发展方向。

　　绿色化学的主要特点是原子经济性,即在获取新物质的转化过程中充分利用每个原子,实现"零排放",既充分利用资源,又不产生污染。传统化学向绿色化学的转变可以看作化学从粗放型向集约型的转变。绿色化学可以变废为宝,可使经济效益大幅度提高,是环境友好技术或清洁技术的基础,但它更注重化学的基础研究。

10.3　绿色化工原则

　　2000 年,Paul T Anastas 概括了绿色化学的 12 条原则,得到了国际化学界的公认。绿色化学的十二条原则是:

　　(1) 防止废物产生,而不是待废物产生后再处理。

　　(2) 合理地设计化学反应和过程,尽可能提高反应的原子经济性。

　　(3) 尽可能少使用、不生成对人类健康和环境有毒有害的物质。

　　(4) 设计高功效、低毒害的化学品。

　　(5) 尽可能不使用溶剂和助剂,必须使用时则采用安全的溶剂和助剂。

　　(6) 采用低能耗的合成路线。

　　(7) 采用可再生的物质为原材料。

　　(8) 尽可能避免不必要的衍生反应(如屏蔽基,保护/脱保护)。

　　(9) 采用性能优良的催化剂。

　　(10) 设计可降解为无害物质的化学品。

　　(11) 开发在线分析监测和控制有毒有害物质的方法。

　　(12) 采用性能安全的化学物质以尽可能减少化学事故的发生。

　　以上 12 条原则从化学反应角度出发,涵盖了产品设计、原料和路线选择、反应条件等方面,既反映了绿色化学领域所开展的多方面研究工作内容,也为绿色化学未来的发展指明了方向。

　　一个理想的化工过程,应该是用简单、安全、环境友好和资源有效的操作,快速、定量地把廉价易得的原料转化为目的产物。绿色化学工艺的任务就是在原料、过程和产品的各个环节渗透绿色化学的思想,运用绿色化学原则研究、指导和组织化工生产,以创立技术上先进、经济上合理、生产上安全、环境上友好的化工生产工艺。这实际上也指出了实现绿色化工的原则和主要途径(参见图 10-1)。

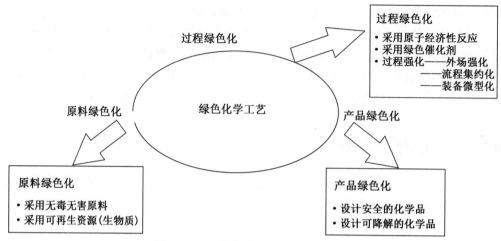

图 10-1　绿色化学工艺的原则和方法

10.4　绿色化工途径

如何实现绿色化学的目标,是当前化学、化工界研究的热点问题之一。绿色化工技术的研究与开发主要围绕"原子经济"反应,提高化学反应的选择性,无毒无害原料、催化剂和溶剂,可再生资源为原料和环境友好产品几个方面开展,如图 10-2 所示。

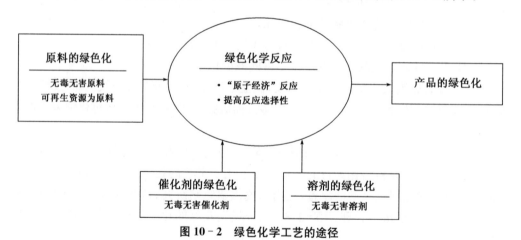

图 10-2　绿色化学工艺的途径

具体包括:

(1) 开发"原子经济"反应。

(2) 提高烃类氧化反应的选择性。

(3) 采用无毒、无害的原料。

(4) 采用无毒、无害的催化剂。

(5) 采用无毒、无害的溶剂。

(6) 采用生物技术从可再生资源合成化学品。

(7) 有机电化学合成方法。

Chapter 10   Introduction to green chemical industry

10.1   Concept of green chemical industry

"Green" chemical industry means to realize ecological "green" in chemical production. When chemical products are used to serve related industries, they should also pursue to realize ecological "green" in production of related industries. In other words, they should simulate the functions of animal, plant and microbial ecosystem and establish chemical ecological chain equivalent to "ecologist, consumer and reducer". With low consumption (material consumption, water, electricity, steam, cold and other energy consumption and industrial consumption), no pollution (at least low pollution), resource regeneration, comprehensive utilization of waste, separation and degradation and other ways to achieve the production of non-toxic chemical products of the chemical industry "ecological" cycle, "environmentally friendly" and clean and safe production of the "green" results.

Green chemistry and chemical industry are the scientific basis for the sustainable development of chemical industry in the 21st century. The aim is to change the existing technical route of chemical production from "pollution first, treatment later" to "root out pollution at the source". Construction of a resource-saving and environment-friendly society has been proposed in the medium-term and long-term science and technology planning of our country's "Eleventh Five-year Plan". Resource-saving and environmentally friendly society has rich connotations, including environmentally friendly production and consumption modes, pollution-free or low-pollution technologies and products, industrial structure with less pollution and low loss, layout of productive forces in line with ecological conditions, and social fashion and cultural atmosphere in which everyone cares about the environment. According to the development goal of China, the development of chemical science and engineering in China must also take the road of green; realize the whole goal of researching, designing and creating new useful substances from the molecular level until the completion of their industrial manufacturing and transformation process; finally realize the ecological utilization of resources, establish ecological industrial parks, realize circular economy, promote and ensure the coordination of economic development with resources, energy and environment.

10.2 Characteristics of green chemical industry

Chemistry can roughly be seen as the study of the transformation from one substance to another. Although traditional chemistry can obtain new substances needed by human beings, in many cases it fails to use resources effectively, resulting in a large number of emissions that cause serious environmental pollution.

Green chemistry was also called environmentally sound chemistry, environmentally friendly chemistry and clean chemistry in the early stage. At present, green chemistry is also called sustainable development chemistry or green sustainable development chemistry to emphasize the relationship between green chemistry and sustainable development. From the viewpoint of sustainable development of science, green chemistry and chemical engineering should not only consider whether harmful to human health and ecological environment of the pollutants, but also consider whether the raw material resources are effectively used, whether they can be regenerated, whether can promote the sustainable development of economy, society and so on, simply conformed to the sustainable development principle and law of science. At present, this view has been accepted by most people, which is the real connotation of green chemistry.

Experience has shown that environmental pollution may be formed quickly, but it will take a long time to eliminate the damage. Some of the damage is potential, and it will take years or even decades to become apparent. Therefore, it is the future development direction of chemical industry to realize the coordinated development of chemical production and ecological environment.

The main characteristic of green chemistry is atomic economy, which means that every atom is fully utilized in the conversion process to obtain new substances, achieving "zero emission", so that resources can be fully utilized without pollution. The change from traditional chemistry to green chemistry can be seen as the change from extensive chemistry to intensive chemistry. Green chemistry can turn waste into treasure and greatly improve economic benefits, it is the basis of environmentally friendly technology or clean technology, and it is more emphasis on the basic research of chemistry.

10.3 Principles of green chemical industry

In 2000, Paul T Anastas outlined twelve principles of green chemistry, which have been recognized internationally. The twelve principles of green chemistry are:

(1) Preventing waste generation, rather than treating waste after generation.

(2) Designing chemical reactions and processes rationally to improve atomic economy of reactions as much as possible.

(3) Using as little as possible, do not generate toxic and harmful substances to human health and the environment.

(4) Designing high efficacy, low toxicity of chemicals.

(5) Do not use solvents and auxiliaries as far as possible, and use safe solvents and auxiliaries when they must be used.

(6) Using low energy consumption synthesis route.

(7) Using renewable materials as raw materials.

(8) Avoiding unnecessary derivative reactions as far as possible (such as shielding group, protection/removal of protection).

(9) Adopting catalyst with good performance.

(10) Designing chemicals that can be degraded into harmless substances.

(11) Developing methods for on-line analysis, monitoring and control of toxic and hazardous substances.

(12) Using safe chemicals to minimize the occurrence of chemical accidents.

From the perspective of chemical reactions, the above twelve principles cover product design, selection of raw materials and routes, reaction conditions and other aspects, which not only reflect the various research content carried out in the field of green chemistry, but also point out the direction for the future development of green chemistry.

An ideal chemical process should be a simple, safe, environmentally friendly and resource efficient operation to quickly and quantitatively convert cheap and readily available raw materials into target products. The task of green chemical technology is to infiltrate the idea of green chemistry into each link of raw materials, processes and products, and use the principles of green chemistry to study, guide and organize chemical production, so as to create a technologically advanced, economically reasonable, safe and environmentally friendly chemical production technology. This actually points out the principles and main ways to achieve green chemistry (see Figure 10 - 1).

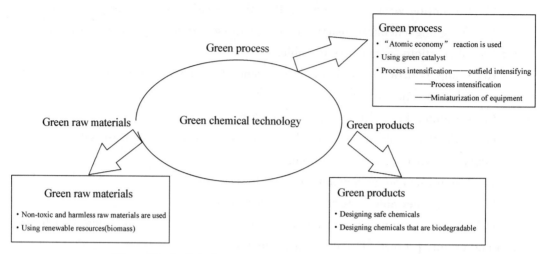

Figure 10 - 1　Principles and methods of green chemical processes

10.4　Green chemical approach

How to realize the goal of green chemistry is one of the hot issues in the field of chemistry and chemical industry. The research and development of green chemical technology mainly focuses on "atomic economy" reactions, improving the selectivity of chemical reactions, selecting non-toxic and harmless raw materials, catalysts and solvents, and developing environmentally friendly products with renewable resources as raw materials, as shown in Figure 10 - 2.

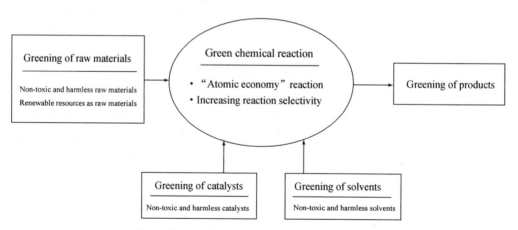

Figure 10 - 2　The ways of green chemical processes

Specific contents include:

(1) Developing "atomic economic" reactions.

(2) Improving the selectivity of hydrocarbon oxidation reactions.

(3) Using non-toxic, harmless raw materials.

（4）Using non-toxic, harmless catalysts.

（5）Using non-toxic, harmless solvents.

（6）Using biotechnology to synthesize chemicals from renewable resources.

（7）Using organic electrochemical synthesis method.

参考文献

[1] 米镇涛. 化学工艺学[M]. 2版. 北京:化学工业出版社,2006.

[2] 刘晓勤. 化学工艺学[M]. 2版. 北京:化学工业出版社,2016.

[3] 张秀玲,邱玉娥. 化学工艺学[M]. 北京:化学工业出版社,2012.

[4] 傅承碧,沈国良. 化工工艺学[M]. 北京:中国石化出版社,2014.

[5] 朱志庆. 化工工艺学[M]. 2版. 北京:化学工业出版社,2017.

[6] 曾之平,王扶明. 化工工艺学[M]. 北京:化学工业出版社,2007.

[7] 徐绍平,殷德宏,仲剑初. 化工工艺学[M]. 2版. 大连:大连理工大学出版社, 2012.

[8] JACOB A M, MICHIEL M, ANNELIES V D. Chemical Process Technology [M]. 2nd ed. Hoboken:A John Wiley & Sons, Ltd. , Publication, 2013.

[9] 姚颂东,余江龙. 化工专业英语(英汉双语版)——化学工程与能源化学工程方向 [M]. 北京:化学工业出版社,2015.

[10] 王保国,田建华. 化工英语教程[M]. 北京:化学工业出版社,2004.

[11] 张媛媛,杨昌炎. 化工英语阅读教程[M]. 北京:化学工业出版社,2011.

[12] 丁丽,王志萍. 化工专业英语[M]. 2版. 北京:化学工业出版社,2019.

[13] 李争. 化学工程英语[M]. 北京:外语教学与研究出版社,2017.

[14] 胡鸣,刘霞. 化学工程与工艺专业英语[M]. 北京:化学工业出版社,1998.

[15] 李文玲. 化学工程与工艺专业英语[M]. 兰州:兰州大学出版社,2012.

[16] 王沛. 化学工程专业英语[M]. 哈尔滨:哈尔滨工业大学出版社,2008.

[17] 张裕平,姚树文,龚文君. 化学化工专业英语[M]. 2版. 北京:化学工业出版社, 2014.

[18] 孟祥海,段爱军. 石油化学化工专业英语[M]. 北京:中国石化出版社,2014.

[19] 邵荣,许伟,吕慧华. 新编化学化工专业英语[M]. 2版. 上海:华东理工大学出版 社,2017.

[20] 宁春花,左明明,左晓兵. 聚合物合成工艺学[M]. 2版. 北京:化学工业出版社, 2019.

[21] 赵德仁,张慰盛. 高聚物合成工艺学[M]. 3版. 北京:化学工业出版社,2013.